Integrated Research in Science and Technology

Integrated Research in Science and Technology

Edited by **Amber Hall**

Willford Press
New York

Published by Willford Press,
118-35 Queens Blvd., Suite 400,
Forest Hills, NY 11375, USA
www.willfordpress.com

Integrated Research in Science and Technology
Edited by Amber Hall

International Standard Book Number: 978-1-68285-207-1 (Hardback)

Printed in China.

Contents

Preface

The rapid advancement in the field of science and technology has wide ramifications for future innovations and development in different fields. This book aims to integrate concepts and applications from different fields of science and engineering. The significant topics discussed herein are related to different disciplines like computer simulation and modeling, manufacturing engineering, materials science, industrial management, computer engineering, etc. The case studies and researches compiled in this book are bound to provide innovative insights into the significant applications of these fields. Students and academicians from various interdisciplinary fields will find it an invaluable source of reference.

After months of intensive research and writing, this book is the end result of all who devoted their time and efforts in the initiation and progress of this book. It will surely be a source of reference in enhancing the required knowledge of the new developments in the area. During the course of developing this book, certain measures such as accuracy, authenticity and research focused analytical studies were given preference in order to produce a comprehensive book in the area of study.

This book would not have been possible without the efforts of the authors and the publisher. I extend my sincere thanks to them. Secondly, I express my gratitude to my family and well-wishers. And most importantly, I thank my students for constantly expressing their willingness and curiosity in enhancing their knowledge in the field, which encourages me to take up further research projects for the advancement of the area.

Editor

1

ACTIVE DIMENSIONAL CONTROL OF LARGE-SCALED STEEL STRUCTURES

Radosław Rutkowski[1]

[1] Faculty of Maritime Technology and Transport, West Pomeranian University of Technology, 41 Piastów Str., 71-065 Szczecin, Poland, e-mail: radoslaw.rutkowski@zut.edu.pl

ABSTRACT
The article discusses the issues of dimensional control in the construction process of large-scaled steel structures. The main focus is on the analysis of manufacturing tolerances. The article presents the procedure of tolerance analysis usage in process of design and manufacturing of large-scaled steel structures. The proposed solution could significantly improve the manufacturing process.

Keywords: large-scaled steel structures, dimensional quality, tolerances, statistical dimensional control, measurements, shipbuilding.

INTRODUCTION

The ever-increasing requirements for dimensional quality of prefabricated components force measurement processes to resign from passive dimensional control (which finds the state with essentially no possibility of changing) in favour of active control, having influence on the formation of the desired features of prefabricated components.

The use of active control allows, among others, carrying out the analysis of the impact of measurement error (estimated by standard deviation) on an ongoing basis on the dimensional quality of prefabricated structure components. The concept of such control is shown schematically in Figure 1.

The basic idea of the solution presented in the figure above is the systematic complementation of tolerance equation of a given chain with actual dimensions (obtained through the measurement process) and the analysis of the impact of the results obtained on the final (dependent) dimension of a given dimensional chain.

In order to present the concept further in this document its basic components will be discussed: tolerance equations with dimensional chains issues and the process of analysis.

TOLERANCE EQUATIONS

In assembly process of each type of structure, also large-sized, some dimensions depend on others. These are called dependent dimensions. The literature also specifies the terms: result dimen-

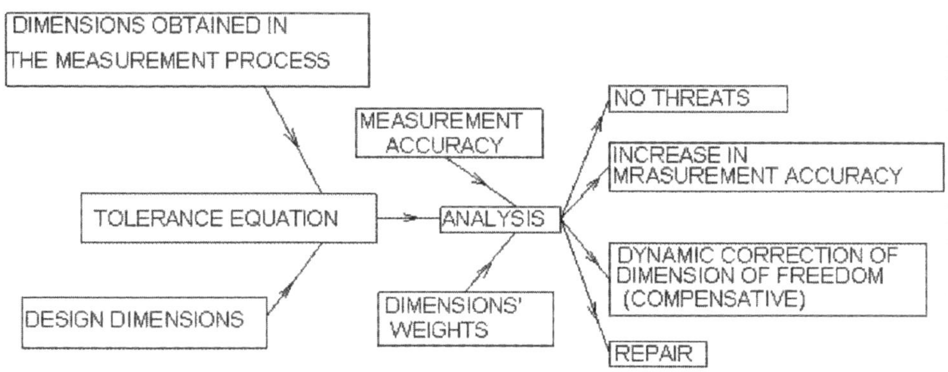

Fig. 1. The concept of active dimensional control

sions, closed dimesions and total dimensions. The interrelation between dimensions can be described analytically using the following equation:

$$Z = F(X_1, X_2, ..., X_n) \quad (1)$$

where: Z – dependent tolerated dimension,
X_i – independent i-th tolerated dimension, $i = 1, ..., n$.

Function (1) can be interpreted as:

- the relationship between actual dimensions,
- the relationship between random dimensions of known probability distributions.

Thus, when calculating tolerated dimensions the following methods may be used:

- deterministic,
- stochastic.

Tolerance equation based on deterministic model

Among deterministic methods of calculating tolerated dimensions there are three different methods:

- differential calculus method,
- arithmetic method (also called min-max method, border case method, worst case method) ,
- logarithmic method.

Differential calculus method is based on the study of dimensional function (1) increase. It is a general method, used in all arithmetic operations. It is particularly effective for cases of non-linear form of the function (1). The equation for dependent dimension tolerance developed on the basis of differential calculus method takes the following form:

$$T_Z = \sum_{i=1}^{n} |Q_i| \cdot T_{X_i} \quad (2)$$

where: T_Z – dependent dimension tolerance,
$Q_i = \frac{\partial F}{\partial X_i}$ – influence coefficient for the i-th independent dimension,
F – dimensional function,
T_{X_i} – i-th independent dimension tolerance.

Coefficient Q_i determines the impact of independent dimension X_i changes on dependent dimension Z changes. Values of Q_i coefficients are determined based on nominal dimensions.

In order to determine the upper and lower deviation of dimension Z it is necessary to group all independent dimensions, separating them into two groups:

- increasing dimensions marked with index from 1 to k,
- decreasing dimensions marked with index from $k + 1$ to n.

Using the above division the upper and lower deviation can be calculated from the following equations:

$$z2 = \sum_{i=1}^{k} Q_i \cdot x2_i + \sum_{i=k+1}^{n} Q_i \cdot x1_i$$
$$z1 = \sum_{i=1}^{k} Q_i \cdot x1_i + \sum_{i=k+1}^{n} Q_i \cdot x2_i \quad (3)$$

where: $z2$ – upper deviation of dependent dimension Z,
$z1$ – lower deviation of dependent dimension Z,
$x2_i$ – upper deviation of i-th independent dimension X_i,
$x1_i$ – lower deviation of i-th independent dimension X_i.

The arithmetic method is based on the analysis of extreme values achieved by simple arithmetic operations (addition, subtraction) on the tolerated dimensions. However, in case of multiplication, division and roots, logarithmic method is used. Formulas for selected operations on tolerated dimensions are given in Table 1.

A detailed description of the above methods of tolerance calculations along with calculation examples can be found among others in [2, 3, 4].

Tolerance equations based on stochastic model

Deterministic rules for determining tolerance equations are based on the following statement: *"threshold values of independent component dimensions cause threshold values of dependent dimensions"* [2]. It is obvious that actual independent dimensions generated in manufacturing process may be considered as random variables with specified probability distributions. Given the nature of the most common statistical distributions, it can be easily seen that the probability of occurrence of dimension value near their threshold value is rather low. It can be concluded that the tolerance calculations using deterministic methods do not meet the practice and significantly reduce required tolerances of components.

Table 1. Operations on tolerated dimensions

Operation	Tolerance function	No
$A_{a1}^{a2} + B_{b1}^{b2} = (A+B)_{a1+b1}^{a2+b2}$	$T_{A+B} = T_A + T_B$	(4)
$A_{a1}^{a2} - B_{b1}^{b2} = (A-B)_{a1-b2}^{a2-b1}$	$T_{A-B} = T_A - T_B$	(5)
$A_{a1}^{a2} \times B_{b1}^{b2} = (A\times B)_{A\times b1+B\times a1}^{A\times b2+B\times a2}$	$T_{A\times B} = A\times T_B + B\times T_A$	(6)
$k\times A_{a1}^{a2} = (k\times A)_{k\times a1}^{k\times a2}$	$T_{k\times A} = k\times T_A$	(7)
$\left(A_{a1}^{a2}\right)^2 = (A^2)_{2\times A\times a1}^{2\times A\times a2}$	$T_{A^2} = 2\times A\times T_A$	(8)
$\sqrt{A_{a1}^{a2}} = (\sqrt{A})_{\frac{a1}{2\times\sqrt{A}}}^{\frac{a2}{2\times\sqrt{A}}}$	$T_{\sqrt{A}} = \frac{1}{2\times\sqrt{A}}\times T_A$	(9)

Much more realistic results can be obtained by using statistical methods for the tolerated dimension calculations. In addition, a major advantage of these methods is achievement of much larger fields of independent dimensions tolerances. The paper [4] derived the formula for the coefficient showing the increase value of tolerance of components determined using stochastic methods in relation to deterministic methods:

$$G = \frac{\sqrt{m-1}}{k_{śr}} \tag{10}$$

where: m – number of chain components,
$k_{śr}$ – average value of variation coefficients of dimension chain components.

Variation coefficient is expressed by the following formula:

$$k_j = \frac{6\cdot\sigma_j}{T_J} \tag{11}$$

where: σ_j – standard deviation of the j-th dimension distribution,
T_J – tolerance of j-th dimension.

Tolerance increase coefficient G should be greater than one ($G > 1$) for the calculation of tolerance based on stochastic model was justified.

In statistical method of determining the tolerance, independent tolerated dimensions are considered to be independent random variables of any probability distributions, which are characterized by corresponding standard deviations:

$$\sigma_Z = \sqrt{\sum_{i=1}^{n} Q_i^2\cdot\sigma_i^2} \tag{12}$$

where: Q_i – influence coefficient for i-th independent dimention,
σ_i – standard deviation characterizing i-th independent dimension.

Assuming that the variation of all dimensions can be characterized by a normal symmetrical distribution, tolerance function developed on basis of (12) takes the following form:

$$T_Z = \lambda\cdot\sqrt{\sum_{i=1}^{n} Q_i^2\cdot T_{X_i}^2} \tag{13}$$

where: λ – correction coefficient, $\lambda = 1$ for normal distribution.

In the case where distributions of certain considered dimensions differ from the symmetric normal distribution, the value of correction coefficient should be greater than unity. In the literature, the proposed values range from 1.4 to 1.8.

In situation where independent dimensions have different known probability distributions, for determining tolerance equations the relation taking into account variation coefficients of particular dimensions should be used:

$$T_Z = \frac{1}{k_Z}\sqrt{\sum_{i=1}^{n}\left(\frac{\partial Z}{\partial X_i}\right)^2\cdot k_{X_i}^2\cdot T_{X_i}^2} \tag{14}$$

where: k_Z – variation coefficient of dependent dimension Z,
k_{X_i} – variation coefficient of i-th independent dimension X_i.

The variation coefficient, for example, has a value of 1.0 for a normal distribution, 1.225–1.414 for the triangular distribution to c.a. 1.732 for the uniform distribution.

Analysis of dependent dimension distribution is often difficult, therefore, in the calculations in most cases of approximate methods are used. They are based on the so-called. threshold theorem, which states that the dependent dimension distribution of tends asymptotically to a normal distribution, regardless of independent dimensions distributions, when the number of the independent component dimensions of dimensional function dimensions is large enough (at least four components).

COMPENSANTION CELLS IN DIMENSIONAL CHAINS

It is known that the dimension chain is a closed set of subsequent, connected to each other (common databases) dimensions that describe the position and orientation of the elements of the geometrical structure of large-scale structures and form a closed circuit.

For example, in shipbuilding (large-scale structures) the dimensional chain consists mostly of compensation cells, also known as dimensions of freedom. The values of these dimensions of freedom can be varied within a set range during an ongoing production process. They are supposed to compensate for the deviation of particular cells of a given dimension chain. The dimensions of freedom mainly include technological reserves and the size of welding joints. Dimensions of freedom can be seen as an independent chain cells, dimension of which shall be equal to zero.

Rules for the use of technological reserves depend on the technology of hull production and measurement techniques used in a shipyard There are mainly two types of reserves:

- prefabrication ("cutting off" in section prefabrication process),
- assembly („cutting off" in process of assembly of sections and blocks into a hull on a slipway or in a dry dock).

In rare cases, one can find also treatment reserves on bent elements. Location of those reserves mainly depends on:

- production technology,
- assembly sequence,
- structural and operational properties of a given element and the entire hull.

Currently, most shipyards, including Polish shipyards, use, technological reserves to a greater or lesser extent. This is an unfavourable situation, since this considerably extends the construction cycle of the hull, and increases energy and material expenditures. Particularly troublesome is significant the increase in engagement of lifting equipment - section, which holds the assembly reserve must be initially set to cut the reserve off and then re-positioned for assembly.

CALCULATIONS OF TOLERANCES OF COMPONENT DIMENSIONS

In the process of determining tolerances of component dimensions based on the deterministic and statistical model, the following calculation methods are used:

- The method of equal tolerance – consists in calculating the tolerances of chain component cells on the assumption that tolerances of all component cells are equal.

$$T_{X_i} = T_{X_2} = \ldots = T_{X_n} = T$$

Thus the calculation uses the following relationships:

– in deterministic model:

$$T_Z = \sum_{i=1}^{n} \left(|Q_i| \cdot T \right) \quad (15)$$

where:

$$Q_i = \frac{\partial F}{\partial X_i}$$

F – dimensional function,

X_i – independent i-th tolerated dimension.

– in statistical model:

$$k_Z^2 \cdot T_Z^2 = \sum_{i=1}^{n} \left(Q_i^2 \cdot k_{X_i}^2 \cdot T^2 \right) \quad (16)$$

- The method of equal accuracy class – consists of calculating the tolerance of component cells depending on the size of nominal dimension of the component cell. In this method it is assumed, that accuracy class coefficient "h" is approximately constant for all chain cells. The value of dimension tolerance in this method can be presented as:

$$T_{X_i} = h \cdot \sqrt[3]{X_i} \quad (17)$$

where: h – accuracy class coefficient

The calculation uses the following relationships:

– in deterministic model:

$$T_Z = h \cdot \sum_{i=1}^{n} \left(|Q_i| \cdot \sqrt[3]{X_i} \right) \quad (18)$$

– in statistical model:

$$k_Z^2 \cdot T_Z^2 = h^2 \cdot \sum_{i=1}^{n} \left(Q_i^2 \cdot k_{X_i}^2 \cdot \left(\sqrt[3]{X_i} \right)^2 \right) \quad (19)$$

The unknown in equations (18) and (19) is a coefficient "*h*". After it is determined the tolerance of particular cells can be calculated according to formula (17).

The method of equal accuracy is used when analyzing dimensional chain there are dimensions with significantly differing nominal values.

- Method of equal effect – consists in identifying the component cell tolerance with the assumption that change in tolerance of resultant cell is equally dependant on the value of the product of the partial derivative of a given cell multiplied by tolerance of this component cell. For a parallel chain, this method is no different from the method of equal accuracy class.
- The method of least costs – is to calculate the component cells tolerance at minimal costs of manufacturing a component or at minimal costs of its assembly. This method requires knowledge on the function of manufacturing costs of a given component according to changes in tolerance of a given dimension.

Detailed description of these methods can be found, among others, in positions [2, 3, 4].

A simple example of determining tolerances of component dimensions for dimensional chain presented in Figure 2. The values of particular dimensions are presented in Table 2.

The chain equation (dimensional function) in this case takes the following form:

$$b = \sum_i a_i \quad i = 1, 2, \dots, 8$$

In the calculations method of equal tolerance (formulas (16) (16)) and method of equal accuracy class were used (equation (17) (18) (19)). Partial derivatives (impact coefficient) for particular dimensions are equal to one. The calculations were made for the deterministic and the stochastic models. It was assumed that the distributions of chain component cells are identical according to the normal distribution. The results obtained are summarized in Table 3.

The presented example illustrates benefits of using statistical method of determining tolerance – significant increase in tolerance of complement cells of dimensional chain.

ANALYSIS OF TOLERANCE EQUATIONS IN ACTIVE DIMENSIONAL CONTROL

This analysis should be carried out in two stages. It begins by identifying an independent dimensional chain of danger coming from the position of measurement result in uncertainty field for a given cell (Fig. 3).

This is done based on weight of determined dimension. The weight is determined in the design process of production technology for the specific structure. If the risk is high, re-measurement of higher precision should be done, or the item is to be sent for corrections.

The second stage is to determine the impact of the resulting independent dimension on the dependent dimension. This is achieved by checking the following condition:

$$T_Z \geq T_{Z_R} \tag{20}$$

where: T_Z – tolerance of dependent dimension determined on design stage,
T_{Z_R} – tolerance of dependent dimension determined on actual (real) data.

Value T_{Z_R} is determined on the basis of design tolerance equation through exchange of design data for actual real data acquired during an ongoing production process.

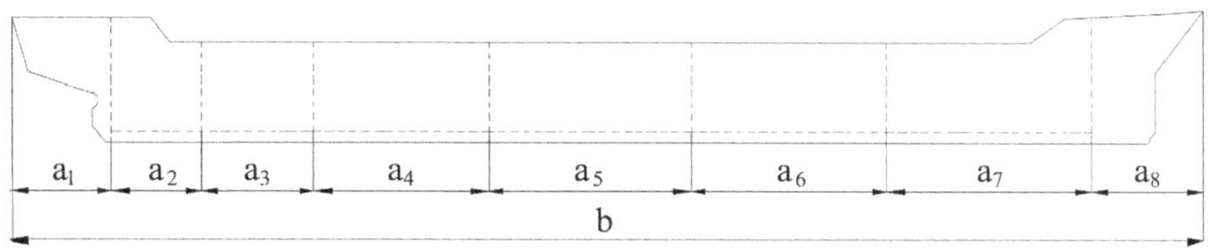

Fig. 2. The basic dimensional chain of a ship's hull

Table 2. Dimensions of the basic dimensional chain of a ship hull

Component	b	a_1	a_2	a_3	a_4	a_5	a_6	a_7	a_8
Dimension [m]	$150.0^{+0,200}_{-0,200}$	12.5	11.2	14.4	24.1	24.1	24.1	25.6	14.0

Table 3. Results of the determination of component dimensions tolerances of dimensional chain of ship hull

Method	Model			
	Deterministic		Stochastic	
Equal tolerance	$a_1 = 12.5^{+0,025}_{-0,025}$	$a_2 = 11.2^{+0,025}_{-0,025}$	$a_1 = 12.5^{+0,070}_{-0,070}$	$a_2 = 11.2^{+0,070}_{-0,070}$
	$a_3 = 14.4^{+0,025}_{-0,025}$	$a_4 = 24.1^{+0,025}_{-0,025}$	$a_3 = 14.4^{+0,070}_{-0,070}$	$a_4 = 24.1^{+0,070}_{-0,070}$
	$a_5 = 24.1^{+0,025}_{-0,025}$	$a_6 = 24.1^{+0,025}_{-0,025}$	$a_5 = 24.1^{+0,070}_{-0,070}$	$a_6 = 24.1^{+0,070}_{-0,070}$
	$a_7 = 25.6^{+0,025}_{-0,025}$	$a_8 = 14.0^{+0,025}_{-0,025}$	$a_7 = 25.6^{+0,070}_{-0,070}$	$a_8 = 14.0^{+0,070}_{-0,070}$
Equal accuracy class	$a_1 = 12.5^{+0,022}_{-0,022}$	$a_2 = 11.2^{+0,021}_{-0,021}$	$a_1 = 12.5^{+0,062}_{-0,062}$	$a_2 = 11.2^{+0,060}_{-0,060}$
	$a_3 = 14.4^{+0,023}_{-0,023}$	$a_4 = 24.1^{+0,027}_{-0,027}$	$a_3 = 14.4^{+0,065}_{-0,065}$	$a_4 = 24.1^{+0,077}_{-0,077}$
	$a_5 = 24.1^{+0,027}_{-0,027}$	$a_6 = 24.1^{+0,027}_{-0,027}$	$a_5 = 24.1^{+0,077}_{-0,077}$	$a_6 = 24.1^{+0,077}_{-0,077}$
	$a_7 = 25.6^{+0,028}_{-0,028}$	$a_8 = 14.0^{+0,023}_{-0,023}$	$a_7 = 25.6^{+0,079}_{-0,079}$	$a_8 = 14.0^{+0,064}_{-0,064}$

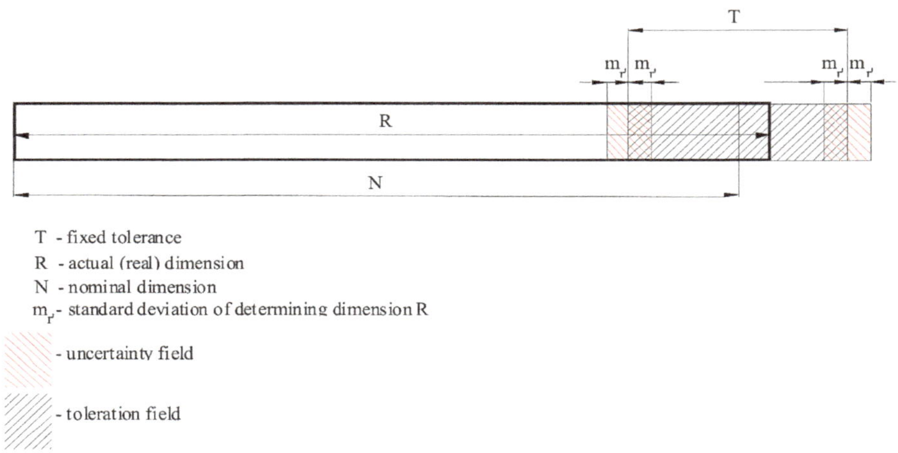

Fig. 3. Position of dimension in relative to nominal dimension

In case the condition (20) is met, a decision should be made to direct the item to the next stage of production. Otherwise actions should be taken resulting from Figure 1, which are:

- increase measurement accuracy,
- perform correction of dimensions of freedom (compensative),
- send the item to repairs.

Various tasks should be performed in the order presented. When one does not produce the intended effect, or they are unsatisfactory, one should move to the next activity. It should be noted that the adjustment of dimensions of freedom should be carefully analyzed technological, strength and economic terms. Economic issues concern mainly labour-intensive welding process involving a change in the size of welding gaps. For example, the relationship between man-hours and the distance between elements when welding T-joints is presented in Figure 4 [2].

As it can be seen from the above chart, excessive increase in the size of welding gaps significantly extends the welding process.

Dimentional weights

Active dimensional control requires determining the weights of component cell of all dimensional chains making up the ship hull. Its goal is to set limits for “diverging” a given chain cell from the fixed tolerance.

Weights of particular cell result mainly from these types of:

- contract,
- strength,
- proper functioning of the machinery and equipment fitted into hull structure,
- associated with block equipment of the hull,
- technological, minimizing fit mounting of the structure.

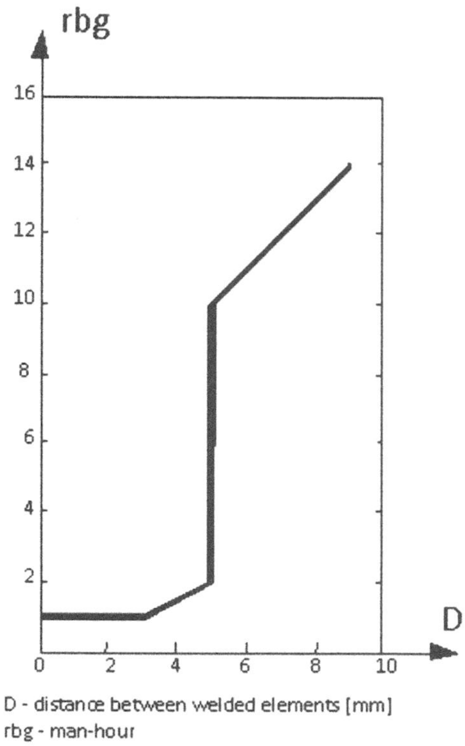

Fig. 4. Relationship between the workload of welding and distance between welded elements [2]

Particular groups of demands determining weights of dimensional chains cells constitute a separate complex research topics. Issues concerning determining tolerance limits are discussed, among others, by works [9].

MATHEMATICAL DESCRIPTION OF APPLIED PRODUCTION TOLERANCES

In practice of production plants manufacturing large-scale steel structures, it is common to encounter a situation, when tolerances are set on the basis of years of experience and only in small degree do they result from calculation analysis. In such cases application of active dimensional control requires performing mathematical description of applied tolerances.

The description should be performed starting from the ultimate dimension of the structure, such as a ship, through dimensions of smaller and smaller groups of elements, to the individual components. Proceeding according to this policy is recommended by most researchers and engineers dealing with these issues. This way of tolerance preparation is called "top down" method, but it can also be found in the literature under the term "from the whole to the detail method" or "modern synthesis of dimensional chains". The advantages of establishing tolerances according to this rule are discussed in detail, inter alia, in [9].

The development of mathematical description of applied tolerances and synthesis of results that are received when the description is complemented with real values may bring simultaneously a lot of information concerning, among other things, manufacturing process control.

Stages of proceedings

The mathematical description process should start with identification of dimensional chains of the whole hull strength and determining the relationships between them. This task can be accomplished through the graphs of dimensional chains hierarchy (Fig. 5). By creating various graphs one should come from "the top" independent dimension (e.g. the length of the cable) and then divide it into dimensions of sections, subsections and down to individual elements. Detailed fragmentation of the dimensions of the hull structure is the only way to allow the analysis of dimensional changes in different phases of the technological process.

Based on graphs equations of toleration of all dimensional chains must be developed. This can be divided into two stages involving determining tolerance of dependent dimension on the basis of applied tolerances of independent dimension. In the first stage the calculation should be performed based on a deterministic model. In case when the condition:

$$T_Z \geq T_D \tag{21}$$

where: T_Z – currently applied tolerance of dependent dimension,

T_D – tolerance of dependent dimension determined based on deterministic model,

is not met, go to step two – make calculation using a statistical model, and check the condition:

$$T_Z \geq T_S \tag{22}$$

where: T_S – tolerance of dependent dimension determined on basis of statistical model.

When the condition (22) is not met it is necessary to immediately analyze the controlled technological process to identify hazards arising from this fact. Process discussed above is presented schematically in Figure 6.

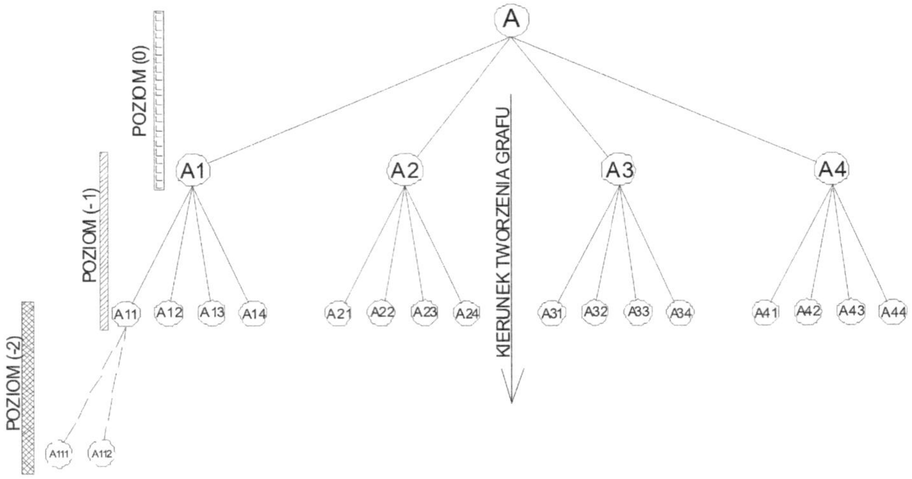

Fig. 5. The graph of dimensional chains hierarchy

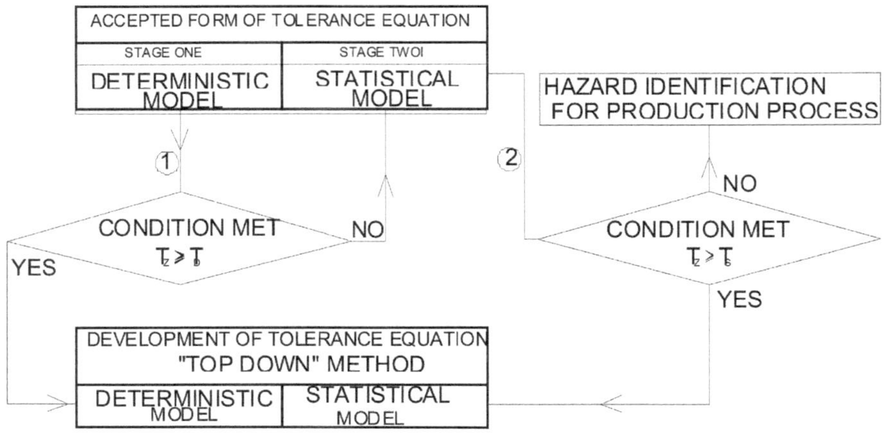

Fig. 6. Procedure of mathematical description of manufacturing tolerance

The mathematical description of tolerance demonstrates the ability of the manufacturing process due to the dimensional quality. If it is greater than the applied tolerances, it is possible to reduce or even, in some cases, eliminate the technological reserve.

REFERENCES

1. Fujita Y., Done H.: General Philosophy Behind Japanese Shipbuilding Quality Standard (JSQS). The Naval Architect, March 1980.
2. Humienny Z. (ed.), Osanna P. H., Tamre M., Weckenmann A., Blunt L., Jakubiec W.: Specyfikacje geometrii wyrobu (GPS) – wykład dla uczelni technicznych. Oficyna Wydawnicza Politechniki Warszawskiej, Warszawa 2001.
3. Janusz W.: Obsługa geodezyjna budowli i konstrukcji. PPWK, Warszawa 1975.
4. Jezierski J.: Analiza tolerancji i niedokładności pomiarów. WNT, Warszawa 1983.
5. Rutkowski R.: Analiza dokładności pomiarów wysokościowych w procesie budowy wielkogabarytowych konstrukcji stalowych. Postępy Nauki i Techniki, nr 5, 2010.
6. Rutkowski R.: Bazy pomiarowe w procesie kontroli jakości wymiarowej wielkogabarytowych konstrukcji stalowych. Postępy Nauki i Techniki, nr 9, 2011.
7. Rutkowski R.: Dynamical control of dimensional quality of large steel structures in production enterprises of low – level technological support. Polish Maritime Research, 17, 1(63), 2010.
8. Rutkowski R.: Modelowanie systemów kontroli geometrycznej przestrzennych konstrukcji stalowych w aspekcie wymaganych standardów dokładnościowych rozprawa doktorska, Politechnika Szczecińska, Wydział Techniki Morskiej, Szczecin 2005.
9. Rosochowicz K. (ed.): Zasady określania granic tolerancji (Projekt celowy nr 9 T12C 060 97 c/3480 pn.). Politechnika Gdańska, Wydział Oceanotechniki i Okrętownictwa, Gdańsk 1999.

2

ANALYSIS OF TROLLEYBUS ENERGY CONSUMPTION

Ewa Siemionek[1]

[1] Department of Automotive Vehicles, Faculty of Mechanical Engineering, Lublin University of Technology, 36 Nadbystrzycka Str., 20-618 Lublin, Poland, e-mail: e.siemionek@pollub.pl

ABSTRACT
Developing energy-saving solutions in vehicle construction and operation has become a priority in the automotive sector. For this reason, zero-emission and low-noise trolleybuses have started to play a significant role in urban public transportation. The paper discusses the methodology and results of energy consumption analysis for trolleybus travel on a determined route. For a given section of the route, energy recovery under braking was calculated.

Keywords: energy consumption, trolleybuses.

INTRODUCTION

Public transportation is globally undergoing a series of reforms aimed at the reduction of environmental pollution and noise caused by various means of transport. From an ecological standpoint, it is favorable to launch successively the vehicles with electric drive in the market, and presently, renewed interest in such vehicles can be observed due to the technological progress in the production of electric energy from renewable sources. In recent years, power engineering problems have become the most important aspect of transportation development policies and the selection of transport means. Therefore, zero-emission vehicles, such as trolleybuses, have acquired particular significance in this respect.

There have been actions undertaken to reduce exhaust emissions and energy consumption used to meet the transport needs of the Lublin region. In order to calculate the energy balance for the public transportation in Lublin, trolleybus energy consumption in real operating conditions was investigated [2, 3, 4].

TEST OBJECT CHARACTERISTICS

The first trolleybuses were built in Germany at the end of the 19th century. The name "trolleybus" is a combination of two words: "trolley" (a pulley or truck travelling on an overhead track) and "bus" (from the Latin word "omnibus" meaning "for all"). A trolleybus is a road vehicle that is driven by electric energy supplied from an external source. Its motion is, to a large extent, dependent on electric traction system. In Poland, trolleybuses are usually powered by a direct current of approximately 600 V. The main advantages of these vehicles include: zero emissions, silent-running, high performance and overload capacity, shock elimination as well as a long operating time. The disadvantage of a trolleybus is the dependence of its routing on the electric traction supplying power to this vehicle.

Trolleybuses can be powered by different types of electric motors and control systems, from direct-current motors (series and compound ones) to alternating-current motors. Nowadays, all newly produced trolleybuses are equipped with a kinetic energy recovery system for recovering kinetic energy under braking. In the areas with high-density housing and low driving speed, trolleybuses with battery-powered traction motors are used, while on long straight sections of routes these vehicles draw power from overhead traction wiring.

EXPERIMENTAL DETAILS AND RESULTS

Experimental tests were conducted to determine energy consumption of trolleybuses. The test object was a Solaris Trollino 12 AC (Figure 1). Table 1 lists basic technical data of the vehicle.

The following values were measured:

- supply voltage,
- current strength drawn from the traction by the motor and control system,
- vehicle speed,
- acceleration and time delay,
- energy consumption for the designated route.

During the measurements, geographic coordinates were recorded.

Fig. 1. Tested vehicle – Solaris Trollino 12

Fig. 2. Route traveled by the trolleybus

Table 1. Technical date of Trolleybus

Solaris Trollino 12 – Technical Data	
Length/Width/Height	12000 / 2550 / 3135 – 3490* [mm]
Interior Height	2370 [mm]
Front Overhang /Rear Overhang	2700/3400 [mm]
Angle of Approach/Departure	7/7 degrees
Curb Weight / Gross Vehicle Weight	11 600 – 14 100 / 18 000 [kg]
Wheelbase	5900 [mm]
Turning circle	~21,4 [m]
Motor	4-pole asynchronous motor (Škoda 4ML3444 K/4 160 kW)
Control System	TV Europulse (IGBT) CEGELEC ANT 175-600 (IGBT) MEDCOM Škoda BlueDrive (IGBT)

The route (Figure 2) was determined by the AB-BA section with a length of 7.2 km. The tested vehicle was in good working order. At vehicle running, all roof vents and windows were closed. The devices necessary for safe vehicle operation were switched on, while others were off. The working temperature of particular units had a steady value owing to the preliminary test run. During the tests, changes in voltage, current strength, vehicle speed (Figure 3) and acceleration as a function of time were measured [9].

As the time history for traction voltage demonstrates, a sudden voltage drop occurred in the 520th section, caused by sudden vehicle braking. As a result, voltage supply lines were disconnected. Higher voltage variations occurred when the vehicle was at a longer distance from the substation supplying power to the contact system. The energy was recovered under brak-

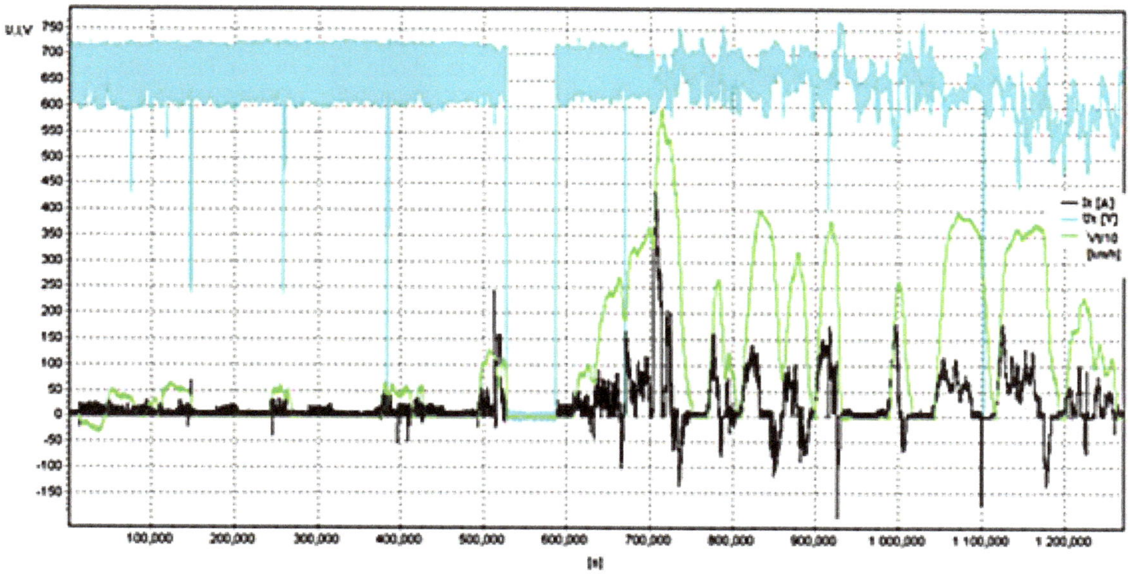

Fig. 3. Voltage, current strength and vehicle speed as a function of time

Table 2. Comparison of data, energy consumption results and kinematic parameters for section AB

Parameter	1	2	3	4	5	6	7	8
L_c [m]	410.76	708.453	581.45	209.44	132.97	580.26	731.37	270.48
T_c [s]	66	70	89	36	52	80	69	80
V_c [m²/s²]	16.28	13.4	20.069	10.85	12.67	19.75	22.82	8.38
T_a [s]	23	30	36	13	15	28	18	26
T_u [s]	5	E	3	1	0	12	16	5
T_z [s]	14	21	17	8	4	13	20	13
L_z [m]	143,3	280.83	204.44	87.5	36.38	151.11	237.77	50.5
L_n [m]	338.13	679.37	500.82	176.5	98	504.71	237.77	50.55
L_h [m]	72.63	29.083	80.63	32.94	34.97	75.55	48.61	75.3
T_h [s]	18	8	19	10	10	18	13	28
L^*	0.823181	0.958949	0.861329	0.842723	0.737008	0.8698	0.325102	0.18689
V^* [m²/s²]	13.40139	12.84991	17.28602	9.143549	9.337896	17.17854	7.418832	1.566138
a^* [m/s²]	0.198169	0.094572	0.172577	0.259024	0.476423	0.170182	0.156009	0.15491
E_t [J]	437851.3	879729.8	648521.8	228553.4	126902.2	653559.1	307893.1	65458.21
E_p [J]	17064.74	28221.03	31157.97	5936.578	3849.146	30900.87	16820.33	1313.188
E_k [J]	976800	804000	1204140	651000	760200	1185000	1369200	502800
E [J]	1431716	1711951	1883820	885490	890951.3	1869460	1693913	569571.4
E_t / E	0.305823	0.513876	0.344259	0.25811	0.142434	0.349598	0.181764	0.114925
E_p / E	0.011919	0.016485	0.01654	0.006704	0.00432	0.016529	0.00993	0.002306
E_k / E	0.682258	0.46964	0.639201	0.735185	0.853245	0.633873	0.806306	0.882769
ψ	3435.529	2416.464	3239.866	4227.893	6700.393	3221.763	2316.083	2105.78
Φ	0.290461	0.201372	0.269989	0.352324	0.558366	0.26848	0.193007	0.175482

ing, which is demonstrated by the negative values of current strength as a function of time. The highest energy recovery occurred at frequent changes in speed profiles.

Based on the technical data and recorded changes in voltage, current strength, vehicle speed and acceleration values, selected traction parameters were calculated.

The following data were determined: total distance covered, total time, mean speed, drive time, time of steady motion, slow-down time, as well as energy balance elements between particular stops on the route. Table 2 shows an example of the comparison made for the data and calculation results.

CONCLUSIONS

Analyzing the available studies, it can be observed that one of the basic criteria for the evaluation of public transportation involves operating costs, which depend on energy consumption in real operating conditions[1, 5, 6, 7, 8].

Based on the results obtained, it has been found that energy consumption level is the same for a particular route if the operating conditions are maintained the same.

The energy balance for the tested object demonstrates that energy consumption of the trolleybus on the traveled route section was 5 kWh. The energy recovered on the designated section of the route was approximately 20% of the total energy.

REFERENCES

1. Ciastoń A., Sapań G.: Prędkość komunikacyjna pojazdów w miejskim transporcie zbiorowym. Transport Miejski i Regionalny, 10, 2006.
2. Dębicki M.: Teoria samochodu. Teoria napędu. WNT, Warszawa 1974.
3. Grochowski L., Żuchowski A.: Samochody ciężarowe i autobusy. WKŁ, Warszawa 2004.
4. Kasprzak J. Kelles-Krauz M.: Podstawy trakcji elektrycznej w komunikacji miejskiej. WSI, Opole 1995.
5. Siłka W.: Analiza energetyczna procesu rozpędzania samochodu. Zeszyty Naukowe WSI, Opole 1993.
6. Siłka W.: Analiza wpływu parametrów cyklu jezdnego na energochłonność ruchu samochodu. Monografia, zeszyt 14, 2. PAN, Oddział Kraków 1998.
7. Siłka W.: Wpływ intensywności procesu rozpędzania samochodu na jego energochłonność.
8. Ubysz A.: Energochłonność samochodu a zużycie paliwa w złożonych warunkach ruchu. WPŚ Gliwice 2003.
9. Wołyńczuk P.: Energochłonność, porównanie autobusu i trolejbusu, Praca dyplomowa PL, 2011.

List of websites:

- http://pl.wikipedia.org/wiki/Silnik_pr%C4%85du_sta%C5%82ego
- http://www.jp555.republika.pl/troll/uklnap.html
- http://www.mpk.lublin.pl

3

DATA PROCESSING FROM THE MEASURING DEVICE BALLBAR QC20

Matúš Košinár[1], **Ivan Kuric**[1]

[1] Department of Automation and Production Systems, Faculty of Mechanical Engineering, University of Žilina, SK-010 26 Žilina, Slovak Republic, e-mail: matus.kosinar@gmail.com; ivan.kuric@fstroj.uniza.sk

ABSTRACT

The paper presents an innovative method of data processing from the measurement device – Ballbar QC20W. It was created with a program for data transformation (Visual Basic.NET) and it used Fourier transformation. The paper deals with the measuring method of CNC machine tools using Ballbar QC20W. There is an influence between qualitative parameters of machine tools and qualitative parameters of products (tolerances, roughness, etc.). It is very important to hold the stability of qualitative parameters of products as a key factor of production quality. Therefore, is also important to evaluate the accuracy of machine tools and make prediction of possible accuracy.

Keywords: data processing, measurement device, CNC machine tools.

INTRODUCTION

To improve the accuracy of machined parts, it is necessary to increase the accuracy of the machine tool. There are various techniques available for error and calibration of the CNC machine but every technique has its own limitations. The total ballbar device is useful for measurement in all axes [5].

Leading manufacturers of machine tools try to ensure that the machine will have the same properties (positioning accuracy, quality, etc.) in and out of cut, under certain conditions (tool wear, cutting speed, feed and etc.) [4]. However, it is questionable whether the machine tool maintains this property for machining parts in different places worktable and at different technological parameters.

The machining accuracy of the part is not only influenced by technological system (machine – tool – workpiece), as well as external environment (environment temperature, pressure, vibration and etc.) [1, 4–6]. In identifying geometric accuracy is usually measured with an unloaded (machine unload machining) machine tool.

When we measure the precision of a CNC machine tool not only one diagnostic method can be used, but the multiparametrical approach as well. It is difficult to select the suitable measurement methods by multiparametrical diagnostics to achieve the rating of the machine in the shortest time, as well as at the lowest cost. These methods are independent and their evaluation has a synthesized character. One more expensive method can be replaced with a less cost-intensive method.

Geometric accuracy of CNC machine will depend on various techno-logical conditions as well as the location of the machined part on a worktable. Geometrical accuracy of a produced part and its course time should be correlated with the precision of the machine tool.

The state of a machine tool has an enormous impact on the quality of the piece, on which the machining process takes place. Therefore, it is important to keep the machine tool in such conditions, that it will be able to produce parts that meet the demanded accuracy. Very low tolerances or very high quality surface can cause unnecessary production costs, hence a high cost product. high reliability and long-time use is positive. In contrast, products with low prices have a positive impact on enterprise competitiveness [3–7].

DIRECT MEASUREMENT METHOD

Direct measurement method is suitable for unloaded machine, which is not affected by impacts of the machining process. This measurement method obtains information about the current machine state.

Two devices were used for measuring unloaded machine (not during machining). Both devices have almost the same construction; one of these is a newer version. Full identification of device is Renishaw Ballbar QC10 a QC20W (see Figure 3). Ballbar QC20W has a higher reading speed and a bluetooth port. Another advantage of the newer version is the possibility of concluding a working space of the machine.

Fig. 1. Measuring device Ballbar QC20: 1 – Magnetic center holder, 2 – Magnetic centre cup attached to rack, 3 – Magnetic centre cup clamped in collet, 4 – Measuring device

BALLBAR QC20W

The Renishaw Ballbar QC20 and software is used to measure geometric errors present in a CNC machine tool and detect inaccuracies induced by its controller and servo drive systems. If the machine had no errors, the plotted data would show a perfect circle. The presence of any errors will distort this circle, for example, by adding peaks along its circumference and possibly making it more elliptical. These deviations from a perfect circle reveal problems and inaccuracies in the numerical control, drive servos and the machine's axes. During the data capture session, the Ballbar moves in a clockwise and counter-clockwise direction through 360° data capture arcs with 180° overshoot arcs. The items of hardware that you will use during your test with a QC20-W ballbar are shown in the Figure 1.

The QC20-W allows machine calibrations to be conducted in the XY, ZX and YZ planes without having to setup and centre the machine between each test. Therefore, with one setup, machine volumetric performance can be evaluated using the volumetric analysis software. Renishaw diagnostics can be performed on the data captured from three planes, allowing machine errors to be diagnosed.

The Ballbar 20 software can automatically analyse 360° and 220° ballbar plots and diagnose machine errors (Figure 2).

Plot errors can be caused by machine errors and test errors. Machine errors are errors or faults in the machine under test. Test errors or are errors or faults in either the ballbar or the way the test was carried out.

INNOVATIVE METHOD OF DATA PROCESSING

In the evaluation of ballbar QC20 results data from the software Ballbar 20 were used. Final processing was done in Microsoft Excel software,

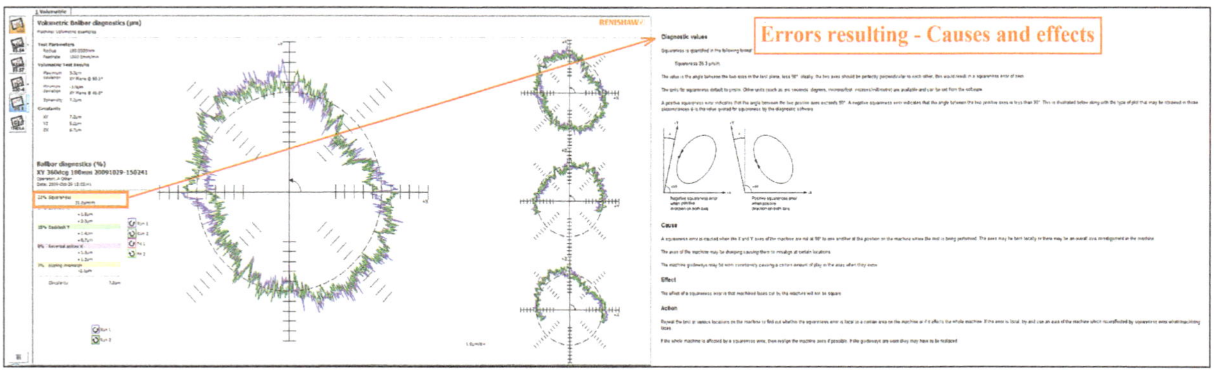

Fig. 2. Background software Ballbar – Error resulting

where the data is decomposed to harmonic components of the profile (Figure 4). In the program Ballbar 20 the data was evaluated according to ISO 230-4:2005 and the results from the program were used to verify the results processed in Excel. Because of incompatibility of measurement records from software Ballbar 20 and Microsoft Excel was created convertor in the programming language Visual Basic.NET. The program is able to convert *.B5R files to *.Sig (Figure 3). During

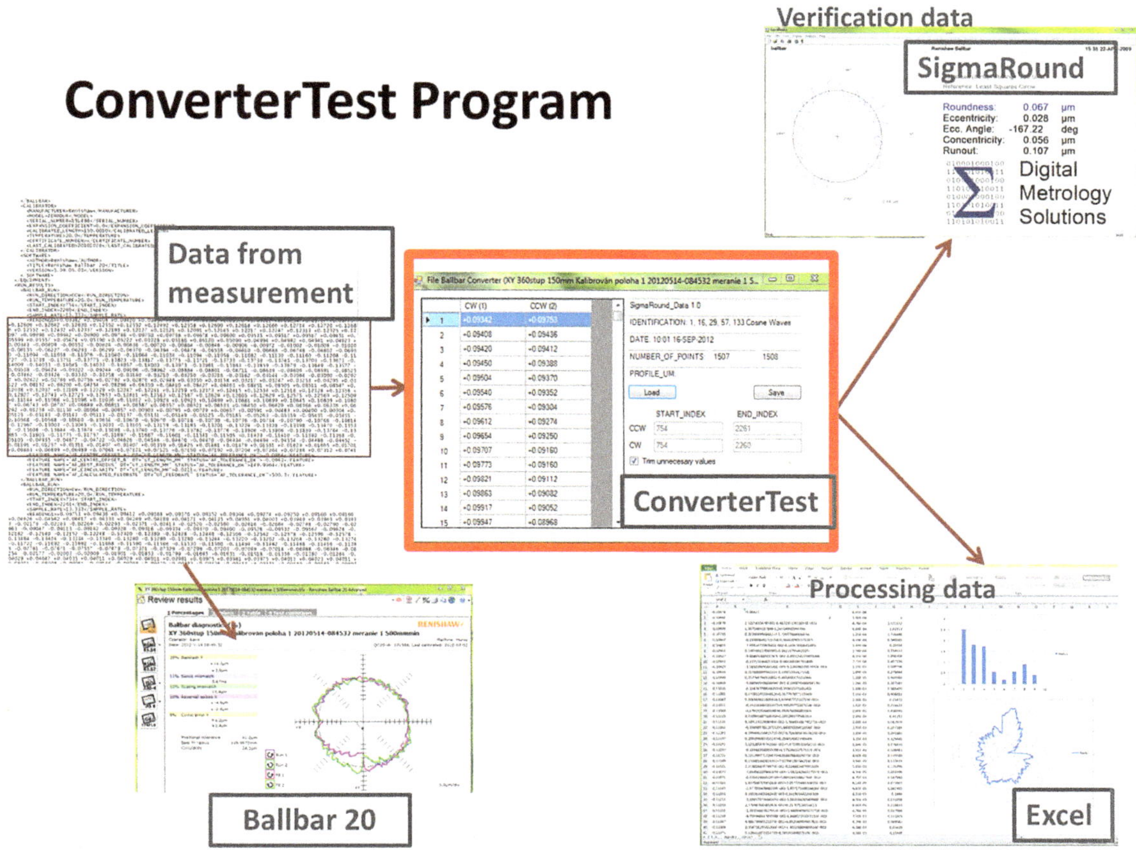

Fig. 3. ConverterTest program

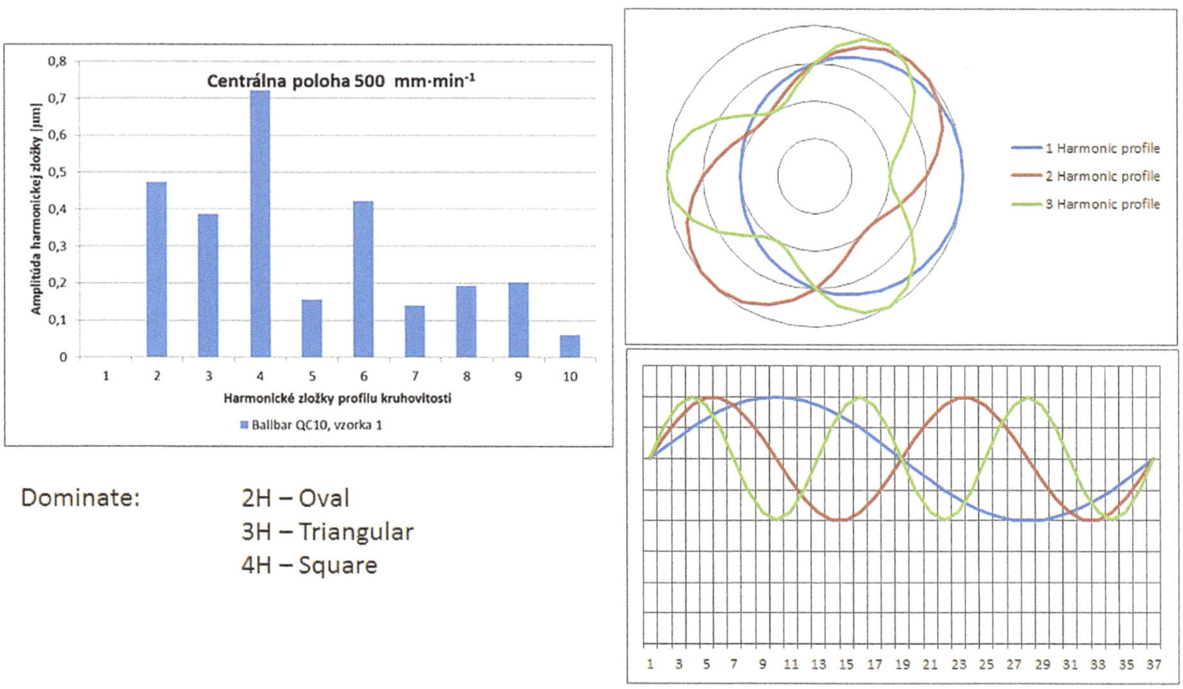

Fig. 4. Processing of harmonic analysis

the conversion the program filters input data file because the file contains data for multiple measurements. For evaluation only data gained during one turn without starting and ending part of the machine tool movement were needed.

Machine can be affected by an angular error that causes a run-out of X or Y axis to the plane of the test during movement. That can be caused by lack of stiffness or obliqueness of feed guiding elements. This error results to dimensional errors of the machined parts. That can be solved by verification of used compensations, checking of the feed guiding elements and leadscrews, etc. Accuracy trends can be predicted by using error values in the percent obtained with software Ballbar 20. FFT analysis was used to show the real profile. Using these data, it is possible to get a more accurate idea of the future development of machine tool errors and it's accuracy.

CONCLUSION

The quality of every component produced on a CNC machine is highly dependent on the machine's performance. Many inspection procedures take place after the component is produced. This is too late. To avoid scrap it is better to check the machine before cutting any metal. Determining a machine tool's capabilities before machining, and subsequent post-process part inspection, can greatly reduce the potential for scrap, machine downtime and as a result, lower manufacturing costs. The age of machine is irrelevant as every machine makes errors. The processes of control and improvement are the key to raising quality and productivity.

The development of various methods for measuring of machine tools is still a hot topic. There are a number various methods deployed in practice, where individual devices are constantly improved. A significant development is the measuring of geometric parameters machine tool where multiple measuring devices are replaced with a universal one alone. The progress in the development does not necessarily ensure wide use in the practice in Slovak manufacturing companies. The experiment consisted of measuring precision of the machine tool with Ballbar QC20W, which is used to measure the circular error. Machine tool during measuring circular error was not machining. The output data from measuring device files are written in a different format, so it was necessary to create a program that allows data processing and export in a readable form. For this purpose the program ConverterTest for Ballbar devices has been developed. This program is also applicable in other experiments performed by the aforementioned mentioned measuring devices. The data obtained were processed by measurement in Microsoft Excel, which were compared using the graphical display. Roundness profile was using Fast Fourier decomposed into harmonic components, next correlations indexes between each measurement were calculated. It was necessary to verify the procedures for the processing data measured and evaluation. Graphical results were compared through the SigmaRound for Ballbar devices.

REFERENCES

1. Aronson R.B. Monitoring the machine. Manufacturing Engineering, 1995.
2. Demeč P. Presnosť obrábacích strojov a jej matematické modelovanie. Košice 2001.
3. Košinár M. Innovative method of data processing from measurement on Talyrond 73. [In:] 14th International Conference "Automation in Production Planning and Manufacturing" Turčianske Teplice, 2013.
4. Svoboda O. Volumetric positioning accuracy of a vertical machining center equipped with linear motor drives. Research center of Manufacturing Technology, CVUT.
5. Sata A. Error Measurement and Calibration of Five axis CNC machine using ball Bar device. International Conference and Workshop on Emerging Trends in Technology (ICWET 2010) – TCET, Mumbai, India, 2010, p. 660.
6. Józwik J., Pieśko P., Krajewski G. Ocena testu QC10 do kontroli off-line obrabiarek sterowanych numerycznie CNC. Eksploatacja i Niezawodność, 3(47), 2010: 10.
7. Józwik J. Ocena odchyłki prostopadłości osi obrabiarki sterownej numerycznie z wykorzystaniem systemu QC10 Ballbar. Postępy Nauki i Techniki, 4, 2010: 91-102.
8. Beňo M., Zvončan M., Kováč M. Dynamic characteristics of machining of flexible parts. Journal CA Systems in Production Planning, 12, 2012.

4

ANALYSIS OF FREE TORSIONAL VIBRATION IN CARBON NANOTUBES EMBEDDED IN A VISCOELASTIC MEDIUM

Mustafa Arda[1], Metin Aydogdu[1]

[1] Trakya University, 22030 Balkan Yerleşkesi, Edirne, Turkey, e-mail: mustafaarda@trakya.edu.tr

ABSTRACT

Carbon Nanotubes (CNTs) have a great potential in many areas like electromechanical systems, medical application, pharmaceutical industry etc. The surrounding physical environment of CNT is very important on torsional vibration behavior of CNT. Damping and elastic effect of medium to the torsional vibration of CNTs are investigated in the present study. Governing equation of motion of nanotube is obtained using Eringen's Nonlocal Elasticty Theory. The effects of some parameters like nonlocal parameter, stiffness parameter and nanotube length are studied in detail.

Keywords: torsional vibration, carbon nanotubes, nonlocal elasticity, viscoelastic medium.

INTRODUCTION

Carbon Nanotubes (CNTs) have been used in many areas because of their superior physical properties. In recent years, medical and pharmaceutical industries have great interest in CNTs. Drug transport problem to a specific tissue or cell, especially in cancer, may be solved using CNTs. Torsional dynamic behavior of CNTs in different physical environments is gaining importance in such application.

Physical environments like fluid flow or biological tissue can be modelled as a viscoelastic medium, which has damping and elastic characteristics, and the effect to the torsional vibration of CNT is a very interesting topic for researchers.

In general, CNTs can be modelled using continuum theories. Unlike from classical theory, Eringen's Nonlocal Elasticity Theory includes size effects. Eringen [1, 2] stated that: "The stress at a point is a functional of the strain field at every point of the continuum". Especially in nano dimensions, size dependency gains much importance because of atomic interactions.

Scientists have shown great interest in the dynamic analysis of CNTs in viscoelastic medium for years. Lots of works have been carried out. Ghavanloo et al. investigated the vibration and instability analysis of a CNT resting on a linear viscoelastic Winkler foundation based on the classical Euler-Bernoulli beam model [3]. Shen presented buckling and postbuckling analysis for axially compressed microtubules embedded in an elastic matrix of cytoplasm. He modelled the microtubule as a nonlocal shear deformable cylindrical shell which contains small scale effects and the surrounding elastic medium as a Pasternak foundation [4]. Soltani et al. developed the transverse vibrational model of a viscous-fluid-conveying SWCNT embedded in biological soft tissue. They used the nonlocal Euler–Bernoulli beam theory to investigate fluid-induced vibration of the SWCNT, while viscoelastic behavior of the surrounding tissue which is simulated by the Kelvin–Voigt model [5]. Ghavanloo et al. investigated in-plane vibration analysis of curved CNTs conveying fluid embedded in viscoelastic medium [6]. Zhen et al. studied transverse vibration of fluid-conveying DWCNTs embedded in biological soft tissue. Daneshmand studied coupled cytosol-microtubule mechanical vibrations of microtubules [7]. Kazemi-Lari et al. investigated the influence of viscoelastic foundation on the non-conservative instability of cantilever CNTs under the action of concentrated follower force [8]. Rafiei et al. investigated vibration characteristics of non-uniform SWCNTs conveying fluid embedded in viscoelastic medium using nonlocal Euler-Bernoulli beam theory [9]. Ghorbanpour and Arani studied nonlocal vibration of double of CNTs system conveying fluid coupled by visco-Pasternak medium based on nonlocal elasticity theory where CNTs are placed in uniform temperature change and magnetic field [10]. Kazemi-Lari et al. used the nonlocal Euler-Bernoulli beam model to predict the static and dynamic structural instability of CNTs subjected to a distributed tangential compressive load. They considered the CNT to be embedded in a Kelvin-Voigt viscoelastic medium [11]. Lei et al. investigated the dynamic behavior of nonlocal viscoelastic damped nanobeams. They employed the Kelvin-Voigt viscoelastic model, velocity dependent external damping and Timoshenko beam theory to establish the governing equations and boundary conditions for the bending vibration of nanotubes [12]. Rezaee and Maleki considered a more realistic model for the CNT conveying viscous fluid which is embedded in a viscoelastic medium [13]. Wang and Li studied the nonlinear free vibration of the nanotube with damping effects. They obtained the governing equation of the nonlinear free vibration for the nanotube based on the nonlocal elastic theory and Hamilton principle [14]. Pang et al. formulated the general governing equation of transverse wave motion in a viscoelastic SWCNT adhered by surface material on the basis of the nonlocal elasticity theory and the Kelvin model [15]. Daneshmand presented a gradient elasticity shell formulation for free vibration analysis of single-walled carbon nanotube placed on Winkler/Pasternak foundation. The proposed formulation is based on the combined strain-inertia gradient elasticity [16].

Nanocomposites are also very popular and innovative topic and some new studies are performed. Arani et al. investigated the nonlinear vibration and instability analysis of a bonded double-smart composite microplate system (DSCMPS) conveying microflow based on nonlocal piezoelasticity theory [17]. Karličić et al. studied the free longitudinal vibration of a nonlocal viscoelastic double-nanorod system (VDNRS). They assumed that a light viscoelastic layer continuously couples two parallel nonlocal viscoelastic nanorods. Their model is aimed at representing dynamic interactions in nanocomposite materials [18].

Also boron nitride nanotubes can be used instead of carbon nanotubes. Arani et al. studied nonlinear free vibration and instability of fluid-conveying double-walled boron nitride nanotubes (DWBNNTs) embedded in viscoelastic medium. They considered the effects of the transverse shear deformation and rotary inertia by utilizing the Timoshenko beam theory. Arani and Roudbari developed the nonlocal longitudinal and transverse vibrations of coupled boron nitride nanotube (BNNT) system under a moving nanoparticle using piezoelastic theory and surface stress based on Euler-Bernoulli beam [19].

Torsional dynamic analysis of CNTs can be useful for modelling in nano-products. According to author's limited literature knowledge, present topic has not been investigated yet. In next section, governing equation of motion of the CNT will be obtained using nonlocal elasticity theory.

GOVERNING EQUATION OF CNT WITH VISCOELASTIC MEDIUM

A nanotube of length L and diameter d is considered. If the Newton's second law is applied to nanotube for torsional vibration, the governing equation is obtained [20]:

$$GI_P \frac{\partial^2 \theta}{\partial x^2} = \rho I_P \frac{\partial^2 \theta}{\partial t^2} - T_e(x,t) \qquad (1)$$

where: G is the shear modulus,
ρ is the density,
I_p is the polar moment of inertia,
θ is the angular displacement of CNT.

The I_p is defined as:

$$I_P = \pi \frac{(R_2{}^4 - R_1{}^4)}{2} \qquad (2)$$

where: R_1 and R_2 is the inner and outer radius of CNT.

The nonlocal constitute relation in a general form can be given as [1, 2]:

$$(1 - \mu\nabla^2)\tau_{kl} = \lambda\varepsilon_{rr}\delta_{kl} + 2G\varepsilon_{kl} \qquad (3)$$

where: τ_{kl} is the nonlocal stress tensor,
ε_{kl} is the strain tensor,
λ and G are the Lame constants,
$\mu = (e_0 a)^2$ is called the nonlocal parameter,
a is an internal characteristic length.
e_0 is a constant – value is very important for the validity of nonlocal models.

Eringen [1, 2] determined this parameter with matching the dispersion curves based on the atomic models.

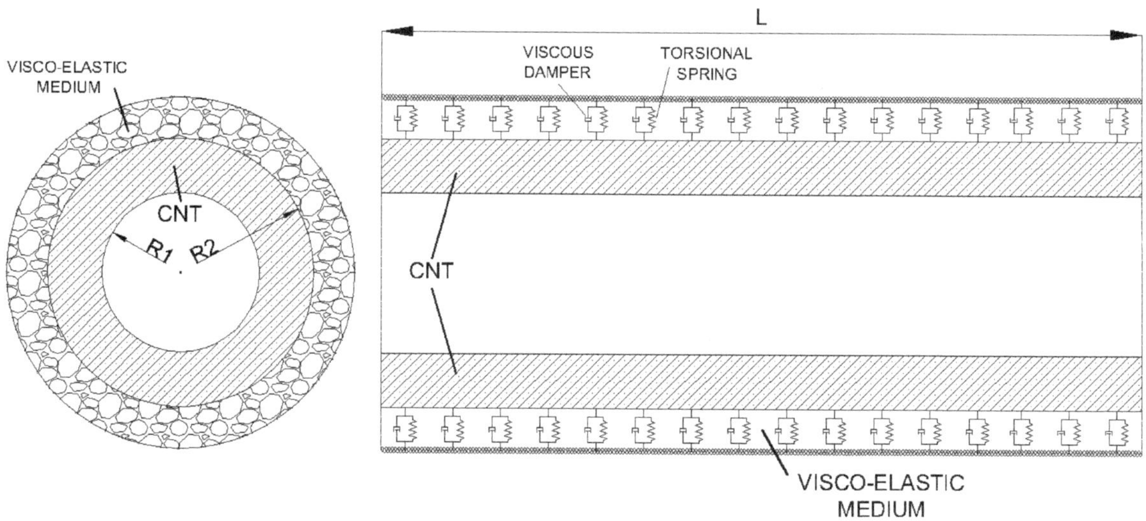

Fig. 1. CNT model with viscoelastic medium

For the torsional deformation of uniform CNT, Eq. (3) can be written in a one dimensional form:

$$\left(1 - \mu\frac{\partial^2}{\partial x^2}\right)\tau = G\gamma \tag{4}$$

where: γ is the shear strain,
τ is the shear stress of CNT.

The stress resultant due to the shear stress is expressed as:

$$S = \int_A \tau \, dA \tag{5}$$

where: A is the cross-section of the CNT.

The torque relation is given as:

$$T = \int_A \tau z \, dA \tag{6}$$

By using the Eq. (4–6), we get the constitute relation as:

$$S - (e_0 a)^2 \frac{\partial^2 S}{\partial x^2} = GA\gamma \tag{7}$$

$$T - (e_0 a)^2 \frac{\partial^2 T}{\partial x^2} = GI_P \frac{\partial \theta}{\partial x} \tag{8}$$

If Eq. (8) is inserted into Eq. (1) one obtains:

$$GI_P \frac{\partial^2 \theta}{\partial x^2} = \rho I_P \frac{\partial^2 \theta}{\partial t^2}\left(1 - \mu\frac{\partial^2}{\partial x^2}\right) - T_e\left(1 - \mu\frac{\partial^2}{\partial x^2}\right) \tag{9}$$

If the nonlocal parameter is chosen as m=0 in Eq. (9), the classical elasticity equation is obtained respectively. T_e defines the external torque load and means elastic and damping effect of viscoelastic medium in this study.

$$T = k_t\theta + c_t\frac{\partial \theta}{\partial t} \tag{10}$$

where: k_t is the stiffness,
c_t is the damping coefficient of viscoelastic medium.

If Eq. (10) is inserted into Eq. (9), Eq. (11) is obtained:

$$\begin{aligned} GI_P \frac{\partial^2 \theta}{\partial x^2} - \rho I_P \frac{\partial^2 \theta}{\partial t^2} + \mu\rho I_P \frac{\partial^4 \theta}{\partial x^2 \partial t^2} + k_t\theta + \\ + c_t\frac{\partial \theta}{\partial t} - \mu k_t \frac{\partial^2 \theta}{\partial x^2} - \mu c_t \frac{\partial^3 \theta}{\partial x^2 \partial t} = 0 \end{aligned} \tag{11}$$

Eq. (11) is the governing equation of motion for CNT embedded in a viscoelastic medium. With harmonic vibration assumption, angular displacement of CNT can be defined as in Eq. (12):

$$\theta(x,t) = A(x)\, e^{\lambda t} \tag{12}$$

where: $A(x)$ is the amplitude of torsional displacement,
λ is the characteristic parameter for CNT.

If Eq. (12) is inserted into Eq. (11) with dimensionless parameter assumption, Eq. (13) is obtained:

$$\begin{aligned} \frac{\partial^2 \theta}{\partial \bar{x}^2}\left[1 + \frac{\mu}{L^2}\frac{\rho I_P \lambda^2 L^2}{GI_P} - \frac{\mu}{L^2}\frac{c_t \lambda L^2}{GI_P} - \frac{\mu}{L^2}\frac{k_t L^2}{GI_P}\right] + \\ + \theta\left[-\frac{\rho I_P \lambda^2 L^2}{GI_P} + \frac{c_t \lambda L^2}{GI_P} + \frac{k_t L^2}{GI_P}\right] = 0 \end{aligned} \tag{13}$$

If Eq. (13) is reorganized, Eq. (14) is obtained:

$$\frac{\partial^2 \theta}{\partial \bar{x}^2} + \beta^2\theta = 0 \tag{14}$$

where:

$$\beta^2 = \frac{-\lambda^2\Omega + \lambda C + K}{1 - \frac{\mu}{L^2}[-\lambda^2\Omega^2 + \lambda C + K]} \tag{15}$$

where: Ω is the frequency parameter coefficient,
C is the damping parameter coefficient,
K is stiffness parameter of CNT as defined in Eq. (16):

$$\Omega = \frac{\rho I_P L^2}{GI_P}, C = \frac{c_t L^2}{GI_P}, K = \frac{k_t L^2}{GI_P} \tag{16}$$

for Clamped-Clamped (C-C) boundary condition, can be found. If Eq. (16) is rearranged, Eq. (17) can be obtained:

$$\left[\left(1+\beta^2\frac{\mu}{L^2}\right)(\lambda^2\Omega-\lambda C-K)\right]+\beta^2=0 \quad (17)$$

Roots of the characteristic equation in Eq. (17), are given the characteristic value for CNT embedded in a viscoelastic medium. can be defined as non-dimensional characteristic parameter. Its imaginary part can be defined as the non-dimensional damping (NDD) and real part can be defined as the non-dimensional frequency (NDF), respectively.

NUMERICAL RESULTS AND DISCUSSION

In this section, torsional vibration analysis of CNT embedded in a viscoelastic medium is carried out for different stiffness parameters, nanotube length and nonlocal parameters. Material properties accepted as shear modulus G = 0.46 TPa, density ρ = 1300 kg/m^3, poisson ratio v = 0.19, nanotube inner radius R_1 = 0.68 nm and thickness of nanotube h = 0.066 nm.

Validation of the present nonlocal Euler-Bernoulli model is investigated by authors in Ref. [20]. According to longitudinal wave propagation, similar results are found. In Figure 2, changing of torsional propagation frequency with wave number can be seen. End of the first Brillion zone, good agreement is observed with Lattice Dynamic results.

In Figures 3–5, changing of NDF with nonlocal parameter (μ) for different stiffness parameter (K) and nanotube length (L) is seen. NDF is decreased with increasing nonlocal parameter, stiffness parameter (K) and nanotube length (L).

In Figure 6, changing of NDD with nonlocal parameter (μ) for different stiffness parameters

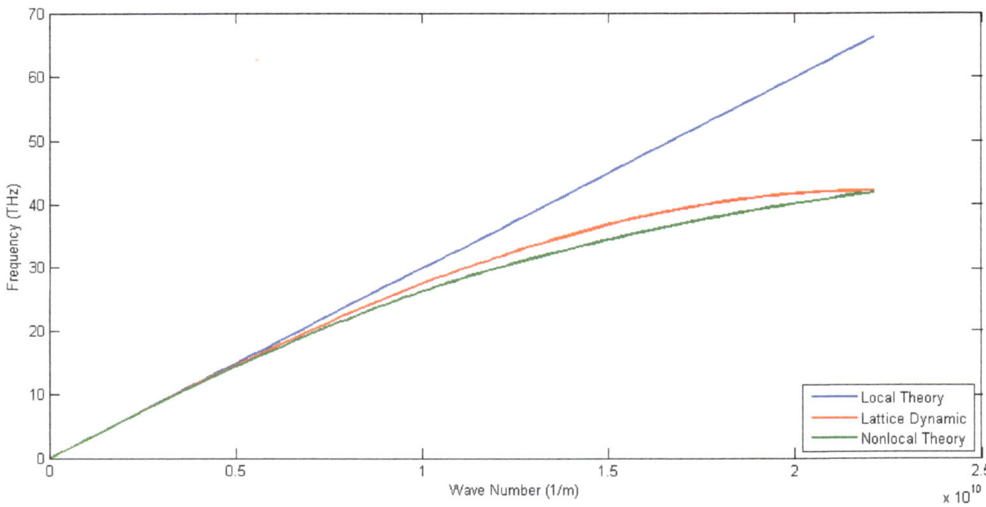

Fig. 2. Torsional wave propagation results according to different theories

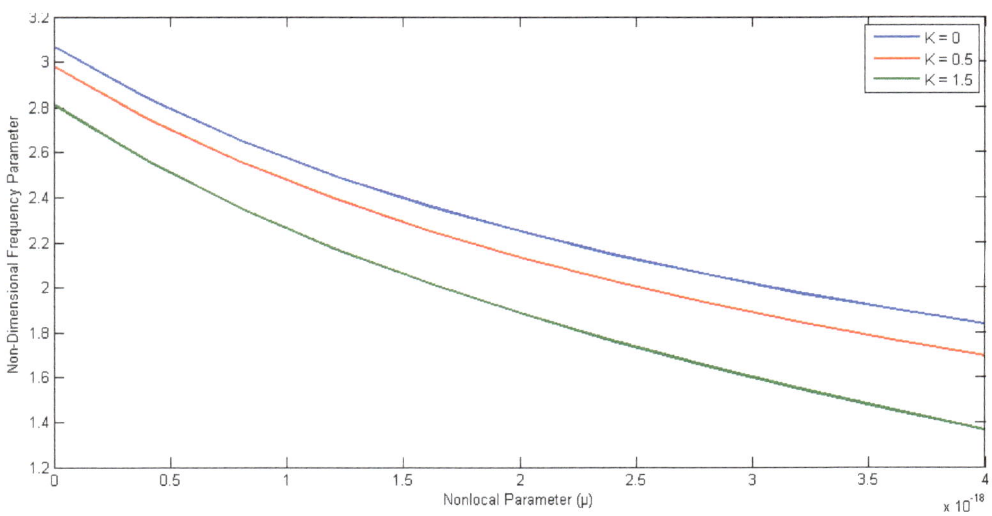

Fig. 3. Variation of NDF with nonlocal parameter (L = 5 nm)

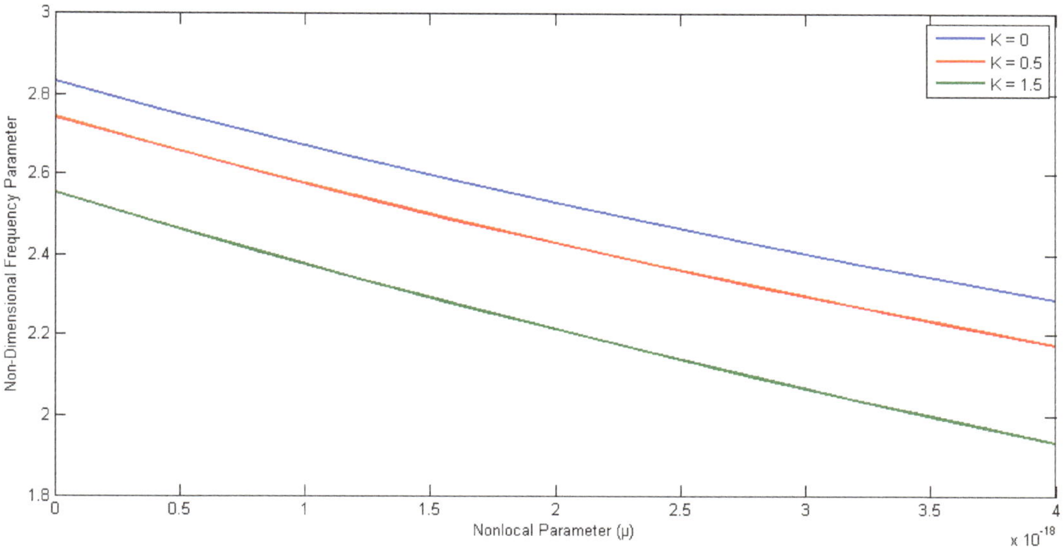

Fig. 4. Variation of NDF with nonlocal parameter (L = 10 nm)

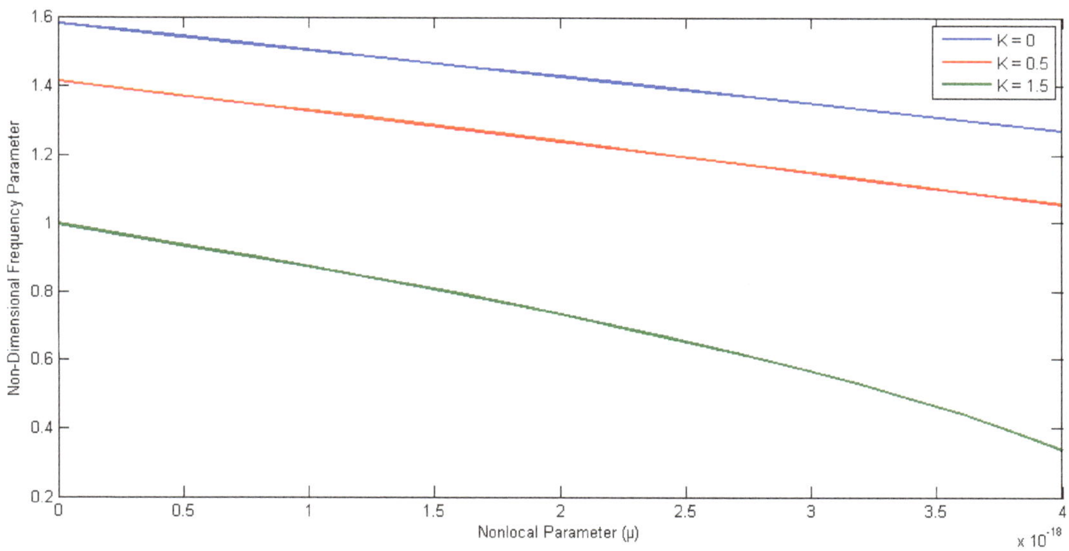

Fig. 5. Variation of NDF with nonlocal parameter (L = 20 nm)

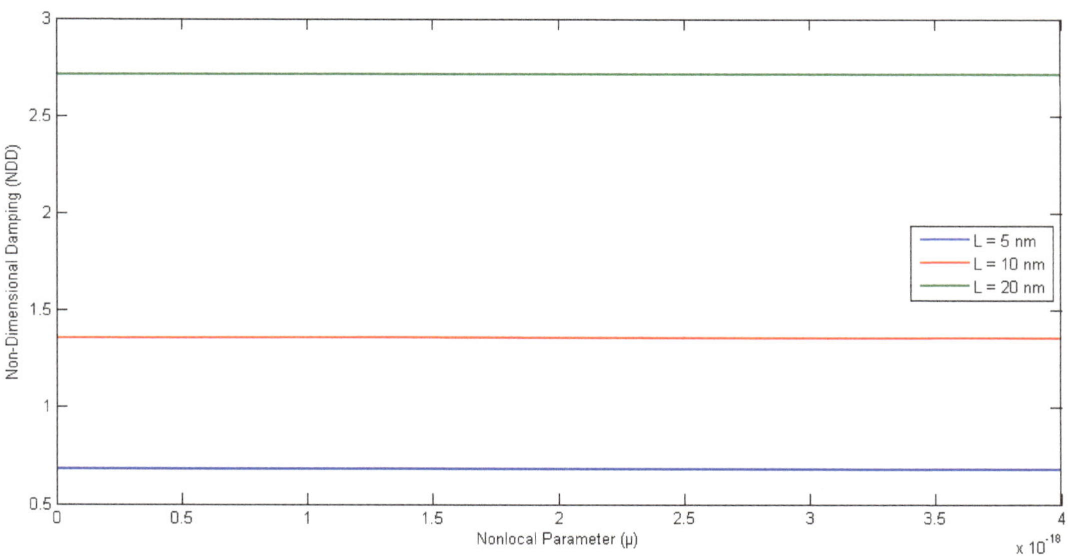

Fig. 6. Variation of NDD with nonlocal parameter

(K) and nanotube length (L) is seen. NDD is not changed with increasing nonlocal parameter (μ) and stiffness parameter (K); but with increased nanotube length, NDD is increased, respectively.

CONCLUSION

Free torsional vibration analysis of CNT embedded in a viscoelastic medium is investigated in detail. Governing equation of motion is obtained using Eringen's Nonlocal Elasticity Theory. Nonlocal parameter and stiffness parameter can be effected only NDF in decreasing way. Nanotube length can be effected NDF decreasingly and NDD increasingly in contrast. Present results can be useful in design of nano-electromechanical products, nano-bearings, medical applications and pharmaceutical industry, possibly.

REFERENCES

1. Eringen A.C. Nonlocal polar elastic continua. Int. J. Eng. Sci. 10, 1972, 1–16.
2. Eringen A.C. On differential equations of nonlocal elasticity and solutions of screw dislocation and surface waves. J. Appl. Phys. 54, 1983, 4703–4710.
3. Ghavanloo E., Daneshmand F., Rafiei M. Vibration and instability analysis of carbon nanotubes conveying fluid and resting on a linear viscoelastic Winkler foundation. Phys. E Low-Dimensional Syst. Nanostructures 42, 2010, 2218–2224.
4. Shen H.S. Nonlocal shear deformable shell model for postbuckling of axially compressed microtubules embedded in an elastic medium. Biomech. Model Mechanobiol. 9, 2010, 345–357.
5. Soltani P., Taherian M.M., Farshidianfar A. Vibration and instability of a viscous-fluid-conveying single-walled carbon nanotube embedded in a visco-elastic medium. J. Appl. Phys. 43, 2010, 401–425.
6. Ghavanloo E., Rafiei M., Daneshmand F. In-plane vibration analysis of curved carbon nanotubes conveying fluid embedded in viscoelastic medium. Phys. Lett. A, 375, 2011, 1994–1999.
7. Daneshmand F. Microtubule circumferential vibrations in cytosol. Proc. Inst. Mech. Eng. H, 226, 2012, 589–599.
8. Kazemi-Lari M.A., Fazelzadeh S.A., Ghavanloo E. Non-conservative instability of cantilever carbon nanotubes resting on viscoelastic foundation. Phys. E Low-Dimensional Syst. Nanostructures 44, 2012, 1623–1630.
9. Rafiei M., Mohebpour S.R., Daneshmand F. Small-scale effect on the vibration of non-uniform carbon nanotubes conveying fluid and embedded in viscoelastic medium. Phys. E Low-Dimensional Syst. Nanostructures 44, 2012, 1372–1379.
10. Ghorbanpour Arani A. Nonlocal vibration of embedded coupled cnts conveying fluid under thermo-magnetic fields via ritz method. J. Solid Mech. 5, 2013, 206–215.
11. Kazemi-Lari M.A., Ghavanloo E., Ahmad Fazelzadeh S. Structural instability of carbon nanotubes embedded in viscoelastic medium and subjected to distributed tangential load. J Mech Sci Technol 27, 2013, 2085–2091.
12. Lei Y, Adhikari S, Friswell M.I. Vibration of nonlocal Kelvin-Voigt viscoelastic damped Timoshenko beams. Int. J. Eng. Sci. 66-67, 2013, 1–13.
13. Rezaee M., Maleki V.A. An analytical solution for vibration analysis of carbon nanotube conveying viscose fluid embedded in visco-elastic medium. Proc. Inst. Mech. Eng. Part C, J. Mech. Eng. Sci. 229, 2014, 644–650.
14. Wang Y-Z., Li F-M. 2014, Nonlinear free vibration of nanotube with small scale effects embedded in viscous matrix. Mech Res Commun 60, 45–51.
15. Pang M, Zhang Y.Q, Chen W.Q. Transverse wave propagation in viscoelastic single-walled carbon nanotubes with small scale and surface effects. J. Appl. Phys. 117, 2015, 305–324.
16. Daneshmand F. 2014, Combined strain-inertia gradient elasticity in free vibration shell analysis of single walled carbon nanotubes using shell theory. Appl Math Comput 243, 856–869.
17. Ghorbanpour Arani A., Vossough H., Kolahchi R. Nonlinear vibration and instability of a visco-Pasternak coupled double-DWBNNTs-reinforced microplate system conveying microflow. Proc. Inst. Mech. Eng. Part C, J. Mech. Eng. Sci. 0, 2015, 1–17.
18. Karličić D., Cajić M., Murmu T., Adhikari S. Nonlocal longitudinal vibration of viscoelastic coupled double-nanorod systems. Eur J Mech - A/ Solids 49, 2015, 183–196.
19. Ghorbanpour Arani A., Roudbari M.A. Nonlocal piezoelastic surface effect on the vibration of visco-Pasternak coupled boron nitride nanotube system under a moving nanoparticle. Thin Solid Films 542, 2013, 232–241.
20. Arda M., Aydogdu M. Torsional statics and dynamics of nanotubes embedded in an elastic medium. Compos Struct 114, 2014, 80–91.

5

THE USE OF EVENT DATA RECORDER (EDR) – BLACK BOX

Gabriel Nowacki[1], Anna Niedzicka[2], Cezary Krysiuk[3]

[1] Military University of Technology, Cybernetics Faculty, Sylwester Kaliski Street 2, 00-908 Warsaw 49, Poland, e-mail: gnowacki@wat.edu.pl

[2] Motor Transport Institute, Management and Telematics Center, Jagiellońska 80, 03-301 Warsaw, Poland, e-mail: anna.niedzicka@its.waw.pl

[3] Motor Transport Institute, Section of Planning and Budget, Jagiellońska 80, 03-301 Warsaw, Poland, e-mail: cezary.krysiuk@its.waw.pl

ABSTRACT

The paper refers to the registration of road events by a modern device called EDR – black box for all types of the motor vehicles. The device records data concerning vehicle's technical condition, the way it was driven and RTS. The recorder may be used in private and commercial cars, taxies, buses and trucks. The recorder may serve the purpose of a neutral witness for the police, courts and insurance firms, for which it will facilitate making the reconstruction of the road accidents events and will provide a proof for those who caused them. The device will bring efficient driving, which will significantly contribute to decreasing the number of road accidents and limiting the environmental pollution. In the end in the last year German parliament backed a proposal to the European Commission to put black boxes, which gather information from vehicles involved in accidents, in all the new cars from 2015 on.

Keywords: Event Data Recorder, EDR – black box.

INTRODUCTION

The number of fatal road accidents' victims in European Union was reduced (Figure 1) from 54 000 in 2001 to nearly 31 000 in 2010 and 28 000 in 2012 [14]. The European Commission (EC) adopted an ambitious Road Safety Program which aims at cutting road deaths by half by 2020. The program sets out a mix of initiatives, at European and national level, focusing on improving vehicle safety, the safety of infrastructure and road users' behaviour.

The statistics are invariably devastating for Poland, which with respect to the number of victims and mortality in the road accidents is always at the end of the list. The threat of death in the road accident in Poland is three times higher than in the rest of EU countries, and in the statistics of the fatalities per 100 accidents Poland is even worse than Lithuania. For Poland this number is 10.3, in Lithuania – 9.9, Greece – 8.8, while the average in the EU is 3. In this context, a shockingly low mortality coefficient is noted in such countries as Germany and Great Britain (1.4), as well as in Austria (1.7) or Italy and Sweden (1.9) [19]. The probability of death of the accident victim in Poland is, on average, four times higher.

Moreover, in the statistics concerning the number of people killed per 1 million inhabitants we also hold last place, since the EU average in 2011 was 61, while in Poland – 109 (Table 1).

The event data recorder (EDR) can meet the above mentioned requirements and improve the level of transport safety with the reduction of death number and accident reconstruction. EDR records events which, in case of an accident or any event, registers and saves data describing the movement of the vehicle (speed, acceleration, using the brake etc.) before, during and after the event occurred [3, 9].

One of primary tasks of accident reconstruction is to determine the values of the event participants motion parameters prior to its occurrence.

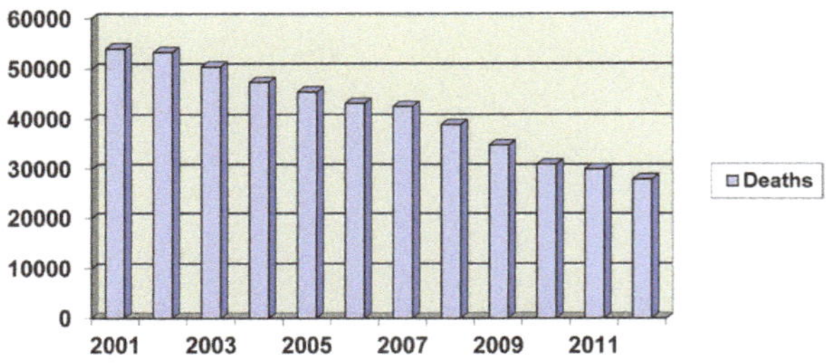

Fig. 1. Road fatalities in the EU since 2001 to 2011 [Based on CARE, EU road accidents database, 06.03.2013, http://ec.europa.eu/roadsafety]

The correctness of their behaviours is assessed on their basis, and then the court makes a decision whether the event participants are guilty or innocent. One of frequently encountered tools, which are useful to determine the values of the aforementioned parameters, is a device recording some selected parameters of vehicle motion. They are so called EDR or just the car 'black boxes' (devices used for accident reconstruction) [5].

The goal is to integrate the functions of monitoring vehicle's behaviour prior to the accident, during the crash and after the crash to the current or developed motor vehicles systems, for the purpose [10]:

- create an instrument for support to make clear specific road traffic accident (chain accident, etc.),
- make easer the guilty and innocence clarification,
- make easer the process of insurance event liquidation,
- increase the active safety (psychological subconscious of driver about the possibility to documentation behaviour of vehicle),
- increase the process of legislation to embed system in vehicles (e.g. in police, fire brigade vehicles, driver's school).

Table 1. Fatalities by population in Member State of EU from 2010 to 2012 [Based on CARE, EU road accidents database, 06.03.2013, http://ec.europa.eu/roadsafety and EC report from 19.03.2013, http://www.eurofundsnews.pl/content/view/20442]

No	Country	2010	2011	2012	The average from 2010 to 2012
1.	United Kingdom	31	31	28	30
3.	Sweden	28	33	31	31
2.	Nederland	32	33	32	32
4.	Denmark	46	40	32	38
5.	Malta	36	41	26	37
6.	Ireland	46	41	36	41
7.	Deutschland	46	49	32	42
8.	Spain	54	45	41	47
9.	Finland	51	54	48	51
11.	France	63	61	56	60
12.	Austria	66	62	64	61
10.	Slovak Republic	68	60	55	61
16.	Luxemburg	64	64	65	64
14.	Italy	68	64	62	65
15.	Slovenia	67	69	59	65
13.	Hungary	74	64	60	66
19.	Estonia	58	75	65	66
22.	Cyprus	73	85	59	72
17.	Czech Republic	76	74	71	74
18.	Portugal	79	74	71	75
20.	Belgium	77	78	73	76
21.	Latvia	97	86	86	90
23.	Bulgaria	103	89	82	91
24.	Lithuania	90	97	100	96
25.	Romania	111	94	96	100
26.	Greece	111	101	92	101
27.	Poland	102	109	93	101
	The average of EU	62	60	55	59

CHARACTERIZATION OF EDR

The first EDRs or *black boxes* were used in the aviation industry in late 1950s. In 1958, the Federal Aviation Act and corresponding regulations issued by the Civil Aeronautics Administration (the predecessor of the Federal Aviation Administration) made the use of black boxes or *flight data recorders* mandatory for commercial aircraft. In 1976, National Transportation Safety Board (NTSB) issued regulations requiring the use of EDRs in commercial marine vehicles. In May 1995, the Federal Railroad Administration issued regulations requiring EDRs on heavy rail transportation. While the use of EDRs in automobiles and light trucks is currently voluntary, vehicle manufacturers such as General Motors and Ford have installed EDRs in many of their newer models [6].

The recorder proposed by the Motor Transport Institute, to a large extent, may help reduce the number of accidents, a significantly shortening the travel time and energy consumption, thereby improving the quality of the environment, and will be useful as evidence in the disputable matters.

The proposed recorder – car black box, can be used to record data concerning technical conditions of the vehicle, driving technique, and drivers' compliance with traffic rules and road traffic safety in the following cars:

- passenger, service and privileged cars – will allow the registration of the earlier indicated data and will provide evidence in case of accident,
- buses and taxis – apart from the recorded data, it will help ensure the safety of the driver and passengers, will enable the location of vehicles in the event of theft,
- Trucks – will ensure registration of the data on technical condition will enable the localization, will contribute to the reduction in the number of accidents and ensuring safety of the driver.

A significant disadvantage is the fact that currently there are few standards referring to such devices. SAE J1698 Standards Committee was established to develop common data output formats and definitions for those data elements that could be used for analyzing vehicle events, including accidents. The standard also specifies common connectors and network protocols to improve data extraction activities [21].

The IEEE P1616 Standards Committee was established to define a protocol for motor vehicle event data recorder output data capability and export protocols for data elements. The committee established these protocols for both light- and heavy-duty vehicles [11]. SAMOVAR project [16] was developed within DRIVE II program by, among others, participants:

- Queen Mary & Westfield College (QMW) drom UK,
- IMPETUS Consultants Ltd from Greece,
- ICS Black Box (UK) Ltd.,
- Man Technologie AG from Germany,
- The Motor Industry Research Association (MIRA) from UK,
- Transport & Road Research Labaratory (TRRL)from UK,
- SWOV Institute for Road Safety Research from Netherlands,
- Royal Mail (RM) from UK,
- IMRS University of Bremen from Germany,

The standard focuses on low cost in-vehicle electronic systems for recording data related to vehicle and its communications to other systems and Databases. A SAMOVAR system will comprise a central interface that integrates several sub-systems, as needed by any vehicle or fleet operator. The complete system will include any of the following functions:

- monitor and record vehicle system parameters,
- monitor and guide a driver's performance or condition,
- warn a driver of unsafe vehicle, driving or environment,
- advise a driver of location, route, other information,
- detect and record details of an accident.

European Commission Directorate-General for Energy and Transport developed a project of vehicle event recording based on intelligent crash assessment – Veronica I (2005–2006) and Veornica II (2007-2009) [17].

VERONICA II is to specify the technical and legal requirements for a possible implementation of Event or Accident Data Recorders in vehicles in Europe. The definition of trigger sensitivity is of major importance in order to capture not only hard crash data but also data from collisions with 'soft objects', i.e. vulnerable road users who represent a relevant part of road users and victims in accidents.

EDR data will be used not only to improve accident investigation and accelerate court procedures but also for enhanced research with in-depth data bases as these data provide real-life information on the vehicles' and drivers' actions, immediately before the crash.

Research based on enhanced real-life data will allow for better evaluation of road safety measures in all fields: active and passive vehicle safety, infrastructure, training, regulation and enforcement. With regard to research with EDR data there are two options for European action in the field of in-depth data bases:

- addition of a new in-depth-chapter to CARE by means of a new or an extended mandate,
- coordination of research with dispersed national data bases, possibly under the ERSO umbrella.

The requirements referring to frequency/ range, accuracy, resolution and phases are fulfilled by the NHTSA standard. A large number of signals to be fed into EDR are already standardised according to SAE J1939-71 standard.

The EDR should be able to download data from digital tachographs. This allows a complete record of activities to be obtained for drivers and vehicles. In addition, the digital data offers many opportunities for monitoring driver activities and evaluation to support improved fleet management. PC NET Service with Motor Transport Institute developed in 2008 TachoScan solution for analyzing data from digital tachograph can generate driver's general report, which shows distance, working hours, waiting hours and rest hours, start and stop work hours per day for individual driver.

Furthermore Motor Transport Institute with Signal Institute and Automex from 2009 to 2011 developed the project NR10 0016 06 refers eCall system. The system is based on 112 emergency call. Yhe call is made automatically by the car as soon as on-board sensors (e.g. the airbag sensors) register a serious accident in the car; or it is done manually. Voice plus minimal set data (MSD) is send to relevant PSAP (Public Safety Answering Point). During the project, the simulator of eCall unit was developed consisting of accident detection module and transmission module to PSAP.

The International Congress on European Association for Accident Research and Analysis (21 EVU, Brasov, Romania 2012) developed the main topics of accidents of two wheel vehicles (reconstruction and safety), human factors in road accidents (also in relationship with active safety systems) and the use of data from EDRs in reconstruction. The EVU is a pan-European association of experts in the field of accident research and reconstruction.

During the conference, two authors: Spek and Bot presented data of the total of nine modern cars crashed in five high speed crash tests. Up to and into the crash, the engines were running. Data communication on the drive train CAN bus was monitored during the crash, and evaluated in order to assess the integrity of speed messages within the crash. Both freeze frame data and EDR data, if applicable, were captured after the crash. The data was compared against the measured pre-crash speed [18].

LEGISLATION REQUIREMENTS OF EDR

The US National Highway Traffic Safety Administration – NHTSA [12, 23] required from the manufacturers who install EDRs to include a minimum standard set of data to be recorded: at least 15 types of crash data including pre-crash speed, engine throttle, brake use, measured changes in forward velocity, driver safety belt use, airbag warning lamp status and airbag deployment times.

NTHSA wants to reconstruct what most vehicles do:

- Use pre-crash data to obtain travelling speed before braking, prior to impact – especially when ABS braking does not leave clear road evidence of braking.
- Use driver throttle and brake inputs to gain insight into driver's intent & causation.
- Use Delta V data as a check on momentum analysis or crush analysis for what happened during impact.
- Challenges to establish probable causes of accidents warrant EDRs becoming more common.

The data collected and registered by EDRs reflects accident status but also technical status of the vehicle (fuel consumption, airbag functionality), but they will also register and describe (directly or indirectly) driver's behaviour in a dynamic way (e.g., brake fluid pressure at the beginning and the end of braking, vehicle speed, including that during braking, engine speed, throttle percentage, using or not usingsafety belts) [24].

Based on agreements with mobile service providers, EDRs are linked to onboard communication systems, which transmit the relevant information to a remote location when the event occurs. A collision notification system (or in-vehicle emergency call system) can, therefore, be activated automatically or manually and provide data to emergency services. Initiatives have been launched in the US and in the EU to promote the implementation of such systems and to enforce standards across the different transport sectors and applications. Section 31406 of Senate Bill 1813 has stated EDR as mandatory and it must be installed in all cars in USA starting with 2015, and outlined civil penalties against violators.

European Commission determined a quantitative target: 50% reduction of the number of road fatalities by the 2020, starting from 2010 [21]. EC decided (Recommendation of 8 September 2011) to equip all cars with an on-board system and ensure the implementation, , of the mechanism serving the eCall reporting indicator by the mobile networks' operators in their networks by 31 December 2014 [2].

Minimum set of data means information that must be sent to the entry point for reports of accidents in accordance with EN 15 722 standard. Successful implementation in the entire EU of a harmonized interoperable eCall service requires automatic transfer of voice and audio connection and the minimum data set of the accident, generated by the on-board system, to appropriate public accidents reporting exchange.

Member States should commit its national authorities to notify the Commission, by the end of March 2012, about the measures that were applied in response to this recommendation. Furthermore, according to European Parliament resolution of 27 September 2011 on road safety in Europe for the years 2011 to 2020, there should be following legislative document developed [4] by 2013 - proposals for legislative changes, assuming that each new vehicle must be equipped with a system reminding about fastening seat belts in the front and rear seats, operating on the basis of acoustic and visual signal.

According to EC statement, EDRs would reduce probability of deaths as well as for serious and light injuries [3], as a result of a collision, by 20%. For the reasons mentioned, the EC recommends eCall pan-European system for the EU [20].

The Research and Technological Development Framework of the European Union launched a large number of EDR projects which have been finalized or are still carried out with a view to enhance the road safety. Another study based on available practical experiences of EDR concludes that a reduction in the number of accidents by 20% would generate a reduction of 26, 1% of lightly injured, of 36, 9% of seriously injured and of 50, 4% of killed road users [1].

There is evidence that drivers who know that their cars have black boxes drive more cautiously. Case studies from Europe and the US show that the number of crashes can be reduced by 20 to 30 percent [1]. Crash severity is also reduced. Berlin Police Department reports that deaths, as well as serious and light injuries are reduced by 20% in the crashes of vehicles with EDRs. As a result, all Berlin Police radio patrol cars use EDRs. A similar trial in Vienna led to the use of EDRs in all the city's police cars. In mid-nineties, a Europe-wide program studied the effect of different types of EDRs in fleets in Great Britain, the Netherlands and Belgium. The overall crash rate fell by 28% and costs by 40%.

Last year in November German parliament, the Bundestag backed a proposal to the European Commission to put black boxes, which gather information from vehicles involved in accidents, in all of the country's new cars from 2015 on [25]. German ministry for traffic, the Verkehrsministerium, lent its support to the proposal after details of the motion emerged this week. German council for road safety – which advises the ministry for traffic – is now in the process of setting up a taskforce to evaluate the black boxes.

Berlin police's experience by bears this out: since the city's police cars had the boxes installed in 2010, accidents involving the vehicles dropped by 35 percent.

FUNCTIONAL REQUIREMENTS OF THE RECORDER

The device will receive data from selected circuits of the vehicle via digital and analogue input ports. Sensors may be possible to be connected e.g. to the doors, lights, turn indicators or brakes.

EDR should be installed in protected place. This kind of place is vehicle cabin, but optimal place is the area under driver's seat (Figure 2).

Event Data Recorder should be in line with all specifications (environmental, physical and electromagnetic compatibility) determined by EU directives and standards defined by CEN, ISO and ETSI. The main terms are the following:

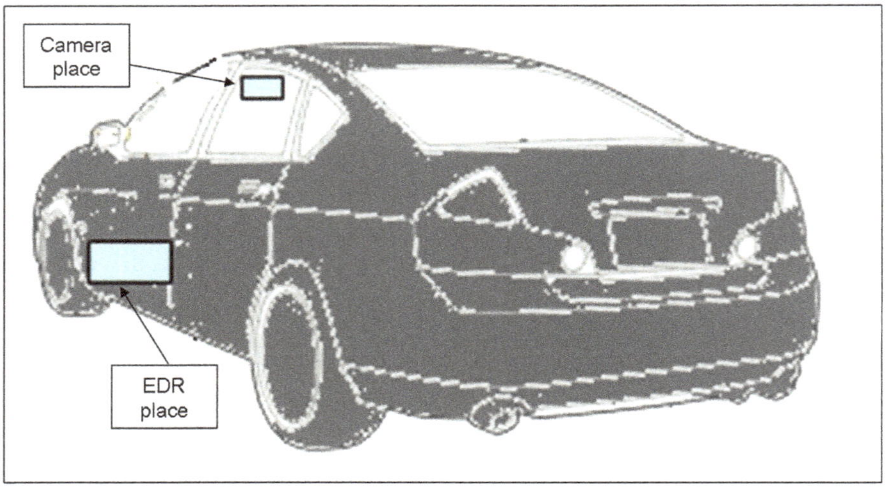

Fig. 2. Installation place of EDR and camera [Based on own study and work]

- Council Directive 73/23/EEC of 19 February 1973 on the harmonization of the laws of Member States relating to electrical equipment designed for use within certain voltage limits. (Official Journal L 077, 26/03/1973 P. 0029 – 0033). Electrical equipment means any equipment designed for use with a voltage rating of between 50 and 1 000 V for alternating current and between 75 and 1 500 V for direct current;
- Council Directive 89/336/EEC of 3 May 1989 on the approximation of the laws of the Member States relating to electromagnetic compatibility. (Official Journal L 139, 23/05/1989 P. 0019 – 0026);
- Directive 1999/5/EC of the European Parliament and of the Council of 9 March 1999 on radio equipment and telecommunications terminal equipment and the mutual recognition of their conformity. (OJ L 91, 7.4.1999, p. 10–28). Telecommunications terminal equipment means a product enabling communication or a relevant component thereof which is intended to be connected directly or indirectly by any means whatsoever to interfaces of public telecommunications networks (that is to say, telecommunications networks used wholly or partly for the provision of publicly available telecommunications services). Radio waves means electromagnetic waves of frequencies from 9 kHz to 3000 GHz, propagated in space without artificial guide;
- PN-ETS 300 135:1997/A1:1999. Radio Equipment and Systems. Angle-modulated Citizens' Band radio equipment. Technical characteristics and methods of measurement. This standard is based upon CEPT Recommendation T/R 20-02 (1972, with subsequent amendments): "Low-power radio transmitter-receivers intended to provide voice radio communications in the 27 MHz band (PR 27 Radio Equipment), and CEPT Recommendation T/R 20-07 (1982, with subsequent amendment): "Free circulation for use in different countries, of low-power mobile and portable transmitter-receivers in the 27 MHz band;
- PN-ETS 300 673:2005. Radio Equipment and Systems (RES). Electromagnetic Compatibility (EMC) standard for 4/6 GHz and 11/12/14 GHz Very Small Aperture Terminal (VSAT) equipment and 11/12/13/14 GHz Satellite News Gathering (SNG) Transportable Earth Station (TES) equipment;
- PN-ETSI EN 300 433-2 V1.1.1:2003. Electromagnetic compatibility and Radio spectrum Matters (ERM), Land Mobile Service, Double Side Band (DSB) and/or Single Side Band (SSB) amplitude modulated citizen's band radio equipment, Part 2: Harmonized EN covering essential requirements under article 3.2 of R&TTE Directive;
- PN-ETSI EN 301 489-1 V1.6.1:2006. Electromagnetic compatibility and Radio spectrum Matters (ERM); Electromagnetic Compatibility (EMC) standard for radio equipment and services; Part 1: Common technical requirements;
- PN-ETSI EN 301 489-13 V1.2.1:2003. Electromagnetic compatibility and Radio spectrum Matters (ERM); Electromagnetic Compatibility (EMC) standard for radio equipment and services; Part 13: Specific conditions for

Citizens' Band (CB) radio and ancillary equipment (speech and non-speech);
- PN-EN 60950-1:2007/A1:2011. Information technology equipment – Safety – Part 1: General requirements/ Application of a safety standard is intended to reduce the risk of injury or damage due to the following: electric shock; energy related hazards; fire; heat related hazards; mechanical hazards; radiation; chemical hazards;
- IEEE P1616. Motor Vehicle Event Data Recorders Institute of Electronic and Electrical Engineers (IEEE), 2010. This standard specifies minimal performance characteristics for onboard tamper- and crash-proof memory devices for all types and classes of highway and roadway vehicles. This international protocol will help manufacturers develop what is commonly called "black boxes" for autos, trucks, buses, ambulances, fire trucks and other vehicles. It includes a data dictionary of 86 data elements and covers device survivability [7, 11];
- J1698 – J1698-1. Vehicle Event Data Recorder Interface – Vehicular Output Data Definition Society of Automotive Engineers, SAE, 2003. The scope of the SAE J1698 standards development effort is to develop common data output formats and definitions for a variety of data elements that may be used for analyzing vehicle "events," most notably crashes. The standards are intended to govern data element definitions and data extraction methodology as applicable for light-duty (less than 8500 lbs GVW) original equipment applications. Further, the standard will specify common connectors and network communications protocols to facilitate the extraction of such data [7, 22].
- SAE J1939-71 2013. Vehicle Application Layer is the SAE J1939 reference document for the conventions and notations that specify parameter placement in PGN data fields, the conventions for ASCII parameters, and conventions for PGN transmission rates. This document previously contained the majority of the SAE J1939 data parameters and messages for information exchange between the ECU applications connected to the SAE J1939 communications network [15].

The electronic system will record and remember the course of acceleration during the collision and remember the result of changing the vehicle speed. In addition, it will remember certain information prior to the accident and immediately after the accident.

The device will interpret digital information transmitted on the CAN bus and the FMS-CAN to record the following parameters (range, depending on the type of vehicle and equipment, such as truck equipped with a tachograph): speed, engine rpm, the position of the brake pedal, clutch pedal position, accelerator pedal position, the state of the cruise control, fuel level, mileage, total fuel consumption, tachograph – operation mode, tachograph – speed, tachograph functioning, dealing with the event status, the mileage remaining until the next tests, engine hours, coolant temperature, the axis load.

CAN bus (Controller Area Network) refers to vehicle bus standard, developed by Robert Bosch, which had quickly gained acceptance in the automotive and aerospace industries. CAN is a serial bus protocol to connect individual systems and sensors as an alternative to conventional multi-wire looms. It allows automotive components to communicate on a single or dual-wire networked data bus up to 1Mbps. The protocol was Publisher in 1986 at the Society of Automotive Engineers (SAE). Bosch published the CAN 2.0 specification in 1991. CAN bus is one of five protocols used in the OBD-II (On-board diagnostics) standard. The OBD-II standard has been mandatory for all cars and light trucks produced in the United States since 1996, and the EOBD (European on Board Diagnostics, regulations are the European equivalent of OBD-II). EOBD is a system for warning the vehicle driver that there is a fault which may cause the emission levels to exceed those allowed by the European directive. EOBD standard is mandatory for all M1 vehicles and petrol vehicles produced in European Union since 2001 and all diesel vehicles produced since 2004.

Fleet management system (FMS) – the mission of six major European truck manufacturers was the creation of a standard to make the surveillance of a fleet over the internet possible. The main problem was the mixture of trucks from different manufacturers in a fleet. FMS can be seen as an interface between truck and internet data transmission. FMS uses a physical layer according to ISO 11898-2 (250 Kbit/s), an application layer according to SAE J1939/71 and a data link layer according to SAE J1939/21. The physical connectors are not yet standardized.

Additional sensors will be able to monitor: the time of release (reaction), a longitudinal, lateral acceleration, vehicle speed (the counter), the engine throttle (gas pedal), brake status (enabled or disabled), supply voltage, the position of the ignition, cushions signalling, the number of events, the time between events, horn, light switches, traffic lights, parking lights, turn indicators, the change in the car deflection (car rotational speed with respect to the vertical axis), the driver's seat belt status, events registration time. The actual sensors in car were presented in Figure 3.

TECHNICAL REQUIREMENTS OF THE RECORDER

One element of accident reconstruction the recreation of time-space relationships of the event participants. Motion reconstruction process is based on the analysis of records of the parameters characterizing the motion of the car body. The forward motion is recorded as standard in a form of linear acceleration components (components: longitudinal, lateral, and vertical).

The device is designed for installation in all types of vehicles (passenger cars, trucks, buses) to record the driving parameters such as speed, acceleration, braking, use of direction indicators, etc. Such information can be extremely helpful in identifying those responsible for road accidents and will allow reconstructing the accident. They are also to replace the witnesses who are not always reliable.

The recorder will have a small size, will be much smaller than the car radio, and made of durable materials, and the place specially protected in it should be the SD card casing, on which the data will be stored. The version 1 of device will be equipped with the following elements (Figure 4):

- Microprocessor module – the element controlling the operation of all other components.
- GPS – to receive the geographical coordinates.
- Module „black box" – memory chips, including SD card, capable of recording the driving and operating parameters.
- Digital inputs module – this module allows connection of digital signals.
- Analogue inputs Module – this module allows connection of analogue sensors.
- CAN inputs module – to connect CAN bus in the FMS standard.

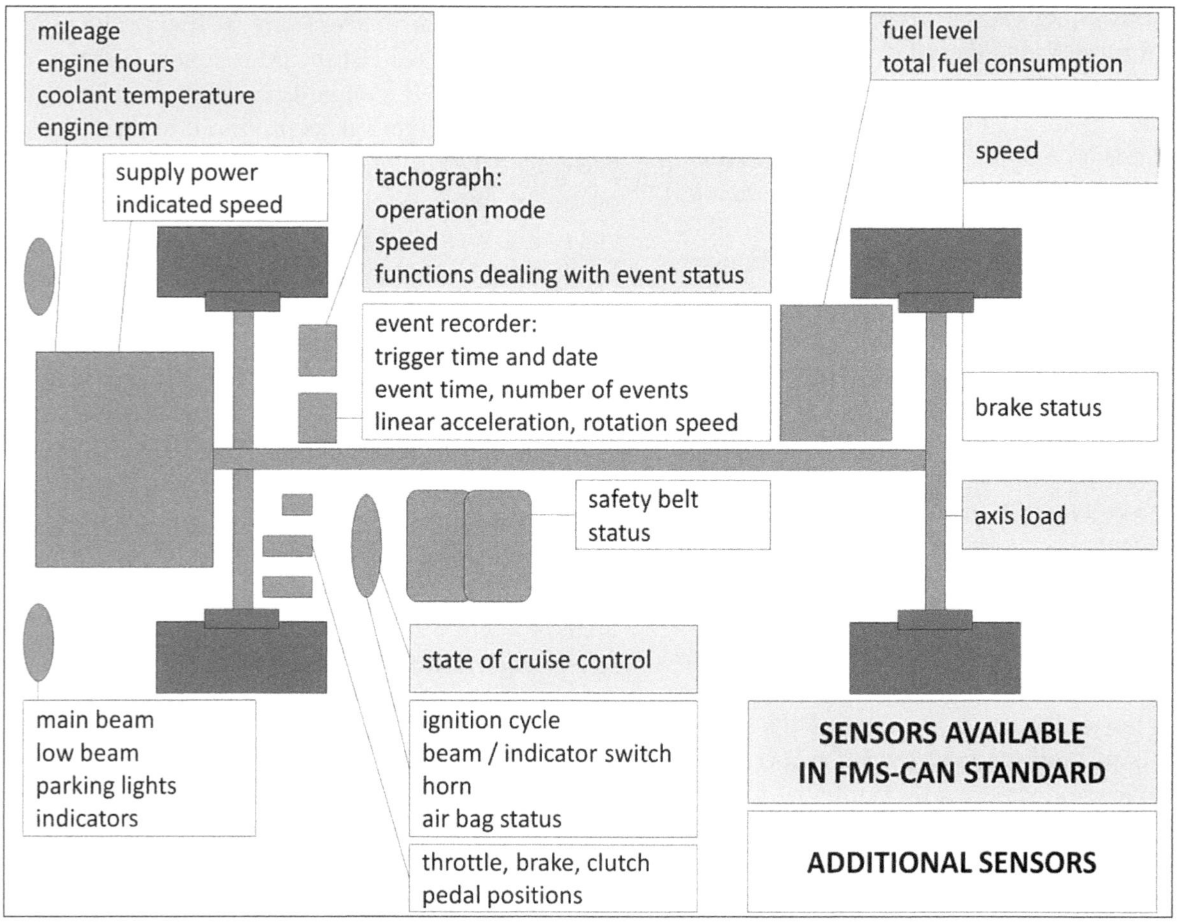

Fig. 3. Monitoring sensors in vehicle [8]

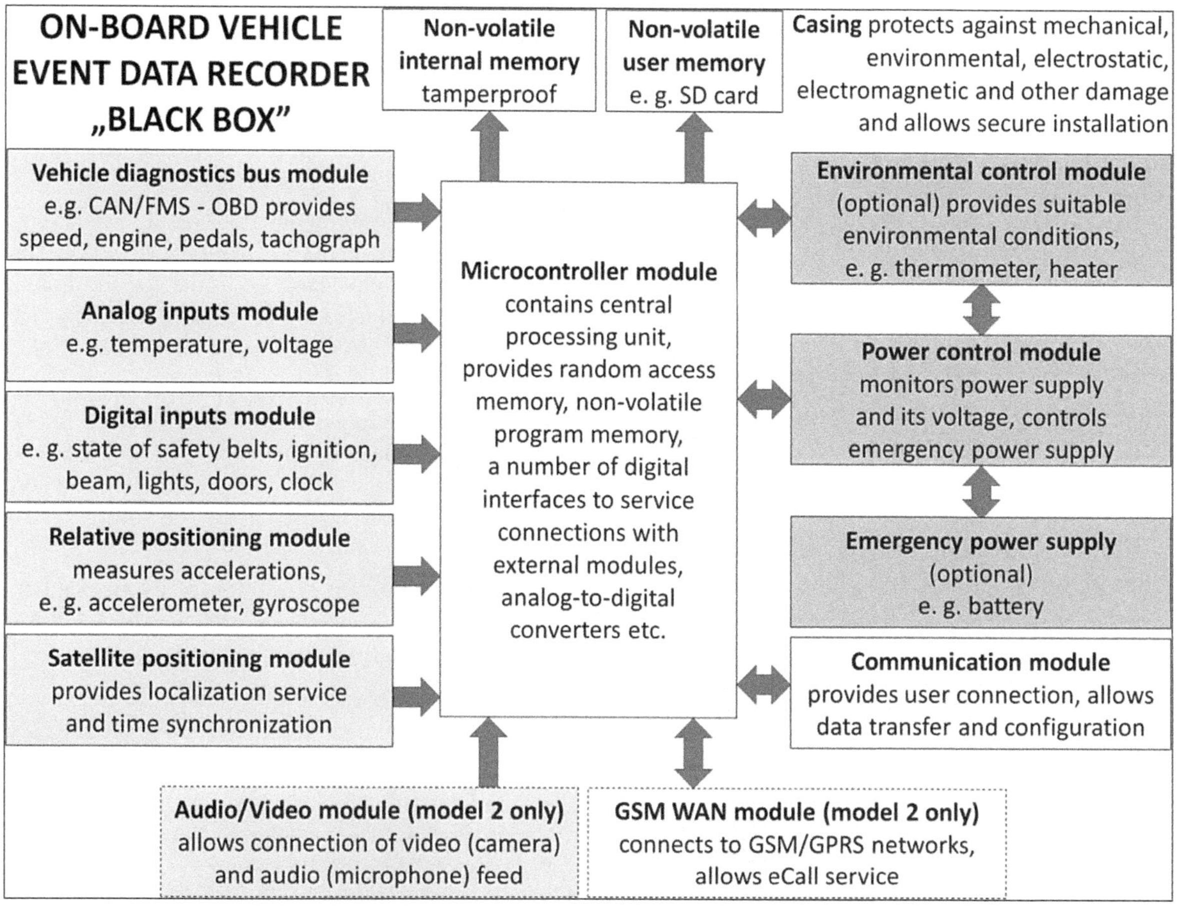

Fig. 4. On-board vehicle, event data recorder functional structure [G. Nowacki, A. Niedzicka]

- Battery backup module.
- Housing.

Devices will be connected to the CAN bus and to the selected electrical circuits of the car. The design may not allow for manipulation or external sensors, such as their exclusion. It will be deprived of the possibility of changing the stored data, connected directly to the accident.

The device must be designed so as to minimize the reconstruction error of the motion parameters, in particular trajectory of the movement. This error should be the smallest degree dependent on the vehicle load.

The unit will have an airtight case ensuring resistance to a short-term immersion in water and service liquids (fuel, oil, hydraulic fluid). It will withstand exposure to direct flame and high overload for a period of several milliseconds.

Equipment should be protected against shock and vibrations in accordance with the standard defined in the EN normative documents (environmental conditions, electromagnetic compatibility). All mobile devices should withstand the following exposure: a one-time shock and falls from a height of 1 m. The devices should be able to withstand vibration, both sinusoidal and random.

For safe and reliable operation of all electrical and electronic equipment in the car, it is necessary to ensure the electromagnetic compatibility of the recorder – the black box. The device should be compatible with all environmental specifications, physical and compatibilities defined in the CEN, ISO and ETSI standards. It should meet all requirements, relating to this group of products, of the EU Council Directives, European standards and national legal regulations.

CONCLUSIONS

The end result of the conception will be turning out two devices: 1 – the economic, universal simple event data recorder 'black box', 2 – the economic, universal event data recorder 'black box' for all types of vehicles, taking into account the eCall reporting.

Recorder – a car black box can be used to record data concerning the technical condition of the vehicle, the driving technique, and the driver's compliance with the traffic regulations and maintaining the road traffic safety in all motor vehicles.

European Commission is currently considering the implementation of legislation in this area, prescribing mandatory installation of black boxes in all vehicles. Psychological impact of the black box will revolutionize road safety. Drivers will be more cautious, knowing that their every manoeuvre may be recorded, so in the event of an accident they will not be able to make false statements.

The device, connected to vehicle monitoring sensors, will be installed behind the dashboard or under the driver's seat. Each sudden change of speed or opening of the airbag will activate it, so that also the collisions involving pedestrians are be recorded. In order not to violate privacy, car black boxes will store the data recorded for 30 seconds before the accident and 15 seconds after it. The machine will automatically alert the emergency road services about the accident. In Britain, black boxes are standard equipment in many privileged vehicles. When in 1999 the London police installed them in a 3.5 thousand of company cars, within 18 months, the costs of road accidents fell by 2 million pounds. The devices are also placed in some newer car models.

In the U.S., black boxes are quite commonly used, and right now they belong to a standard equipment of over two thirds of new cars. U.S. Senate approved the bill, under which from the 2015 on all new vehicles must be equipped with digital driving parameters recorders, known as black boxes. The failure install such equipment will result in punishment.

The studies conducted in the U.S. and the UK have shown that drivers who drive with black boxes, were 20% less likely to have participated in the fatal cases, the failure rate and repair bills for their cars fell by 25 percent. In Poland, the annual cost of road accidents alone are 5 billion, so if that gets reduced by about 20%, one will get the savings for the state – amounting to 1 billion annually.

Acknowledgement

The paper was prepared as a part of own research project 6207/CZT realized by Motor Transport Institute in the years 2012–2013 in the framework of statutory activity financed with the means of the Ministry of Science and Higher Education (Poland).

REFERENCES

1. Bach, H., The use of drive data recorders and accident data recorders and their socio-economic evaluation, Technische Universität Dresden, 2000. www.roadpeace.org
2. Commission Recommendation of 8 September 2011 on support for an EU-wide eCall service in electronic communication networks for the transmission of in-vehicle emergency calls based on 112 ('eCalls'). 2011/750/EU) L 303/46, Brussels, 2011.
3. Cost-benefit assessment and prioritization of vehicle safety technologies in particular 142 seq. (144) and p. 8). European Commission Directorate Energy and Transport. ww.ec.europa.eu/transport/roadsafety_library/publications/vehicle_safety_technologies_final_report.pdf.
4. European Parliament resolution of 27 September 2011 on European road safety 2011-2020. P7_TA-PROV(2011)0408. Brussels, 2011.
5. Guzek M., Car 'black boxes' in road accident reconstruction analysis. Transport No 82, Warsaw, 2012.
6. Kowalick T.M., Protects the security, integrity and authenticity of vehicle crash data. Georgetown University Law Center 600 New Jersey Ave NW, Washington, D.C. 2011.
7. Kowalick T.M., Motor vehicle EDR' global standardization and related issues. National Research Council / National Academies. Keck Center, 500 Fifth St., NW, Washington, D.C. 2010.
8. Merkisz J., Inteligentne rozwiązania systemów informacyjnych w pojazdach osobowych i ciężarowych (Intelligent solutions of information systems in HGV and cars). No 602 Transport and Logistics Problems, 12, 2010, 169-172.
9. Merkisz J., Tarkowski S., Selected aspects of using deck recorders in automotive vehicles. Reliability and Maintenance, 2, 2011.
10. Messerschmidt W., Smith B., Messerschmidt K., Minimizing the Risk of Losing Valuable Forensic Data When Downloading the Electronic Control Modules (ECMs) of Heavy Commercial Vehicles Technical White Paper. Messerschmidt Safety Consultants Birmingham, 2011.
11. Motor Vehicle Event Data Recorders. Institute of Electronic and Electrical Engineers (IEEE), IEEE P1616 Standards Committee, February, 2005. http://www.fmcsa.dot.gov/facts-research/research-technology/report/vehicle-data-recorders-dec05/chap1.htm
12. NHTSA 49 CFR Part 563 Issued 2006, amended 2008. Effective Sept. 2012 New NHTSA EDR Part 563 rule for Sept 2012 standardizes EDR minimum data set, insures public access to data, and that data survives in severe crashes. http://www.nhtsa.gov/EDR.

13. Plíhal J., Pípa M., Black box - project V&V. National Highway Traffic Safety Administration. MD ČR, Praha 2010.

14. Report on accidents, Care & national data. European Commission, Brussels 2012.

15. SAE J1939-71 Standard. Revised: 2013-09-03. Vehicle Application Layer, http://standards.sae.org/j1939/71_201309.

16. SAMOVAR - Safety Assessment Monitoring On-Vehicle with Automatic Recording Project. 1995, https://gforge.inria.fr/projects/samovar.

17. Schmidt-Cotta, R. R., Vehicle Event Recording based on Intelligent Crash Assessment. Final Report, VERONICA II. 06.10.2009.

18. Spek, A., Bot, H., Accuracy of freeze frame data and EDR data in full scale crash tests. International Conference, European Associiation for Accidnt Research and Analisis, 2012, http://www.evuonline.org/en/component/content/article/1066.

19. Symon, E., Prophylactic and Analyze Group of BRD. KGP, Warsaw 2012.

20. The EU eCall: Time for Deployment, Brussels, 21.8.2009, COM (2009) 434 final; http://ec.europa.eu/information_society/activities/esafety/ecall/index_en.htm.

21. Towards to the European Road Safety Area: Policy orientations on road safety 2011-2020. EC 2010.

22. Vehicle Event Data Interface – Vehicular Output Data Definition Society of Automotive Engineers (SAE). J1698 Standards Committee – J1698-1 Standard, SAE, December, 2003.

23. Westbrook, R., Evaluation of electronic data recorders for incident investigation, driver performance, and vehicle maintenance project Bd549-50. Final report. University of South Florida, September, 2009.

24. Working Paper. Event Data Recorders (EDR) on Vehicles. Privacy and data protection issues for governments and manufacturers. EDR: Definitions and facts. 49th Meeting, Montreal (Canada) 4-5 April 2011.

25. ZDNet article. Germany's cars to get teched-up as parliament proposes black boxes for vehicles by Jung, J. 2012. http://www.zdnet.com/germanys-cars-to-get-teched-up-as-parliament-proposes-black-boxes-for-vehicles-7000003488.

6

INTEGRATION OF SHIP HULL ASSEMBLY SEQUENCE PLANNING, SCHEDULING AND BUDGETING

Remigiusz Romuald Iwańkowicz[1]

[1] Faculty of Maritime Technology and Transport, West Pomeranian University of Technology, Al. Piastów 41, 71-065 Szczecin, Poland, e-mail: iwankow@zut.edu.pl

Abstract

The specificity of the yard work requires the particularly careful treatment of the issues of scheduling and budgeting in the production planning processes. The article presents the method of analysis of the assembly sequence taking into account the duration of individual activities and the demand for resources. A method of the critical path and resource budgeting were used. Modelling of the assembly was performed using the acyclic graphs. It has been shown that the assembly sequences can have very different feasible budget regions. The proposed model is applied to the assembly processes of large-scale welded structures, including the hulls of ships. The presented computational examples have a simulation character. They show the usefulness of the model and the possibility to use it in a variety of analyses.

Keywords: assembly sequence planning, schedulling, shipbuilding, budgeting, resource consumption, ship hull.

INTRODUCTION

Assembly is one of the most common operations in the production processes. It involves combining components in a separable and inseparable way. As a result, a complex product called the assembly is created. The specificity of the ship's structure is connected with a large number of elements [10]. The production of the ship's hull consists of a combination with welding methods in an inseparable way [13, 14]. The hull, as the final assembly, constitutes a unified whole. It is a distinctive feature differentiating the hull assembly from the mechanical assembly. In case of the shipbuilding process, the component may be the so-called subassembly, i.e. the structure assembled in advance (Figure 1).

The components can be assembled in a different order called a sequence. Each sequence of the assembly leads to the same result – the final structure. However, the sequences differ in terms of value of many criteria, for example: the difficulty of implementation, energy costs, safety of employees, the required qualifications of welders, the total execution time, etc. The selection of the best sequence is therefore a multi-objective optimisation problem. The whole class of this type of problems and methods of their solving is called the assembly sequence planning (ASP).

The ASP methods are developed mainly in the area of mechanical assembly, however, in recent years also the assembly of hulls becomes the subject of theoretical studies.

The article proposed the views on ASP considering the resource budgeting. The proposed approach clearly shows that every sequence of assembly is accompanied by the space of allowed schedules and budgets. Integration of planning the assembly sequence with the scheduling allows to take phenomenon into consideration as early as at the stage of designing the structure.

DESCRIPTION OF THE PROBLEM

Scheduling problems of the assembly processes of the seagoing ship hulls are included in the models of assembly sequence planning (ASP) in a

Fig. 1. Subassemblies of ship hull

simplified manner. The dominant approach here is the minimisation of the total time of the process.

The construction of the hull is a complex and lengthy process. This process is characterised by a variable intensity of work. There are three main stages: machining, prefabrication and final assembly of the hull. The prefabrication stage is characterised by the highest density. As a result, a typical graph of the labour budget of the hull's construction has the shape of a stretched letter S [12]. Prefabrication and assembly of the hull mostly include the welding works [1]. In practice, not only the work as the resource is considered, that is a subject to budgeting. One should also take the resources of steel materials, welding materials, production area, and many other factors into consideration. Most often the planning of the new structure assembly takes place when the yard has already a fixed timetable of other works. This means that the demand for the resources is already fixed in a certain time perspective. The new assembly should be planned in order to take the existing timetable of works and the maximal intensity of the resource consumption into account, which the yard is able to secure.

REVIEW OF THE LITERATURE

For the construction of the hull so far the issue of ASP considering the resource budget has not been developed. Wei [15] considered the problem of outfitting the hull in terms of the distribution of workload over time. However, the issue of outfitting is more similar to the classical mechanical assembly and differs from the assembly of the welded structures.

Zhong., et al. [16] states that the methods of planning the hull assembly require a different approach. Models existing in this area are based on the analysis of construction in terms of connections and components occurring in it [5]. The considered criteria are mainly the similarity to the previously implemented solutions [2, 3, 4, 6, 8, 11, 16]. Authors very often propose the use of case based reasoning methods.

The general project budgeting problem on the basis of the fixed sequence was considered by Lin [7]. He proposed the method of generating the so-called feasible budget region (FBR) based on the critical path analysis.

In the assembly planning the acyclic graphs are used. Arcs of the graph are the organised neighbourhood relations between activities. For the fixed number of connections many assembly sequences can be generated. Robinson [9] provided the dependence of the number of directed acyclic graphs on the number of nodes. Here we can observe the effect of explosion of the number of solutions.

FEASIBLE BUDGETS REGION OF FIXED ASSEMBLY SEQUENCE

There are n connections considered, which should be performed during the assembly process. Every connection is attributed with a separate action. For every i-th activity there is a determined time of performance t_i and the amount of resources r_i used in the time unit. For simplicity, it is assumed that both of these values are expressed in the whole numbers. The overall resource consumption during the assembly is:

$$ORC(\mathbf{t},\mathbf{r}) = \sum_{i=1}^{n} t_i \cdot r_i \qquad (1)$$

The sequence of assembly is presented as the acyclic directed graph G, whose nodes are the steps of the performed calls. In the further part of determination, the sequence and graph of the assembly are treated as clear.

It is important that the overall resource consumption (1) is constant and independent of the choice of assembly sequence. The collection of acyclic graphs with the fixed number of nodes n is determined with $\Gamma(n)$.

For the fixed sequence $G \in \Gamma(n)$ and the fixed time vector $\mathbf{t}$, based on the analysis of the critical path, for each activity there can be determined the terms of:

- the earliest beginning $eb_i = eb_i(G, \mathbf{t})$,
- the earliest ending $ef_i = ef_i(G, \mathbf{t})$,
- the latest beginning $lb_i = lb_i(G, \mathbf{t})$,
- the latest ending $lf_i = lf_i(G, \mathbf{t})$.

Assuming that the assembly is started in the term 0, the ending term of the process is as follows:

$$t_f(G,\mathbf{t}) = \max_{i=1,\dots,n} \{ lf_i(G,\mathbf{t}) \} \qquad (2)$$

If at least one actions has a non-zero time supply ($lb_i \neq eb_i$), then in the time period $[0, t_f]$ different schedules can be realised.

The set of allowable assembly schedules for the sequence G and times $\mathbf{t}$ is defined as follows:

$$S(G,\mathbf{t}) = \{ \mathbf{s} = (s_1,\dots,s_n) \mid \forall i = 1,\dots,n : eb_i(G,\mathbf{t}) \le s_i \le lb_i(G,\mathbf{t}) \} \qquad (3)$$

where the schedule $\mathbf{s}$ is the vector of terms of beginning the activity. It can be observed that the number of all allowable schedules is:

$$|S(G,\mathbf{t})| = \prod_{i=1}^{n} \left(lb_i(G,\mathbf{t}) - eb_i(G,\mathbf{t}) + 1 \right) \qquad (4)$$

For each fixed schedule $\mathbf{s}$ there is determined the use of resources in subsequent time units 1, 2, ..., t_f:

$$rc_j(G,\mathbf{t},\mathbf{s},\mathbf{r}) = \sum_{i \in N(G,\mathbf{t},\mathbf{s},j)} r_i, \quad j = 1,\dots,t_f(G,\mathbf{t}); \quad \mathbf{s} \in S \qquad (5)$$

where $N(G, \mathbf{t}, \mathbf{s}, j)$ is the set of indices of activities realised in the j-th time unit of the schedule $\mathbf{s}$.

The budget of the assembly is defined as the function of the schedule and time:

$$B(G,\mathbf{t},\mathbf{s},\tau) = \sum_{j=1}^{\tau} rc_j(G,\mathbf{t},\mathbf{s},\mathbf{r}), \quad \tau = 1,\dots,t_f(G,\mathbf{t}) \qquad (6)$$

At the beginning of the process, the use of the resource is zero. Therefore, it is assumed:

$$\forall \mathbf{s}: \; B(G,\mathbf{t},\mathbf{s},0) = 0 \qquad (7)$$

The budget determines how the use of the determined amount is used in time $ORC(\mathbf{t}, \mathbf{r})$ (see equation (1)). It can be seen that for the final term: $B(G, \mathbf{t}, \mathbf{s}, t_f) = ORC(\mathbf{t}, \mathbf{r})$ for any $\mathbf{s}$. The earliest budget of the sequence is the one in which all connections are used the earliest it is possible ($\mathbf{s} = (eb_1, \dots, eb_n)$). Similarly, the latest budget of the sequence is the one in which $\mathbf{s} = (lb_1, \dots, lb_n)$. Lin [7] calls the area included between the graphs of these budgets the feasible budgets region (FBR).

The way of determining FBR of the assembly sequence is presented in example 5 of the connections and the sequences presented in Figure 2.

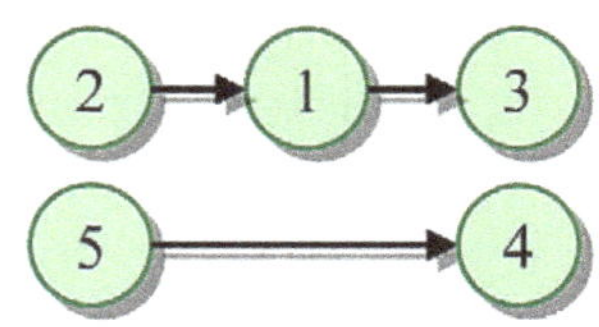

Fig. 2. An exemplary assembly sequence for 5 connections

Let the execution times and the required resources be as follows: $\mathbf{t} = (7, 5, 9, 1, 12)$, $\mathbf{r} = (10, 6, 9, 8, 7)$. The overall consumption of resources during the assembly is: $ORC(\mathbf{t}, \mathbf{r}) = 7 \cdot 10 + 5 \cdot 6 + 9 \cdot 9 + 1 \cdot 8 + 12 \cdot 7 = 273$.

Based on the determined earliest and latest terms it can be observed that the beginning of the performance of the connection no. 4 can take place in 9 different terms: 12, 13, ..., 20. Similarly, connection 5 can be started in 9 terms: 0, 1, ..., 8. Therefore, there is considered $9 \cdot 9 = 81$ variants of schedules. Each variant has a different resource budget. Figure 3 presents 2 budgets creating FBR and 3 intermediate budgets.

It can be observed that even such a simplified assembly sequence (only 5 connections) allows to generate a diverse set of schedules and budgets.

Attention should be paid to the fact that for the determined amount of connections there can be

Table 1. Results of the analysis of the critical path for the exemplary sequence

Connection /activity i	Earliest terms		Latest terms	
	beginning eb_i	ending ef_i	beginning lb_i	ending lf_i
1	5	12	5	12
2	0	5	0	5
3	12	21	12	**21**
4	12	13	20	**21**
5	0	12	8	20

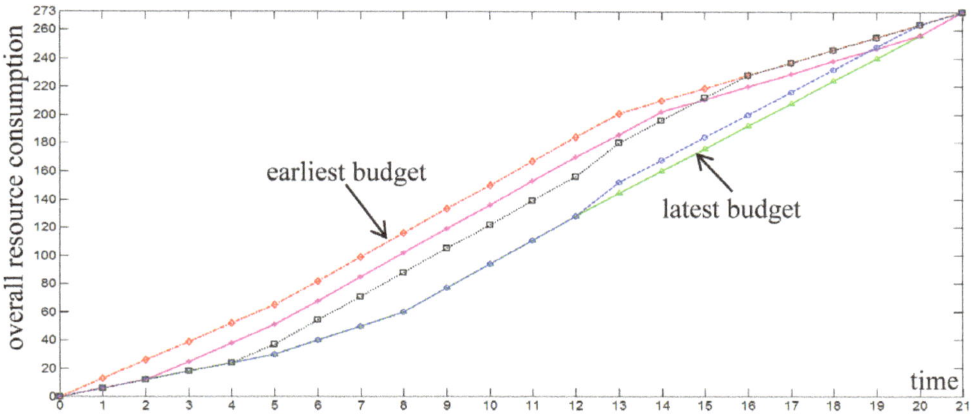

Fig. 3. FBR and the selected allowable budgets for the considered assembly sequence

generated many assembly sequences, which differ in the FBR shape. For the determined durations and times of using the resources all sequences have the same determined value *ORC*(**t**, **r**). This means that the height of the FBR is the same. However, these regions may differ in the dimension along the time axis, what results directly from differences in the duration of the sequence variants. This will be illustrated by the example of three assembly sequences for 5 connections (Figure 4).

It is assumed that the time vector **t** and the resource vector **r** are the same as in the previous example. FBR of sequence 1 was shown in Figure 3. For the sequence 2 and 3 the terms determined with the critical path method are presented in Tables 2 and 3. It is visible that three basic variants of the assembly sequence differ in the final terms. These are, respectively: 21, 26 and 28. FBR for the considered three assembly sequences are presented in Figure 5.

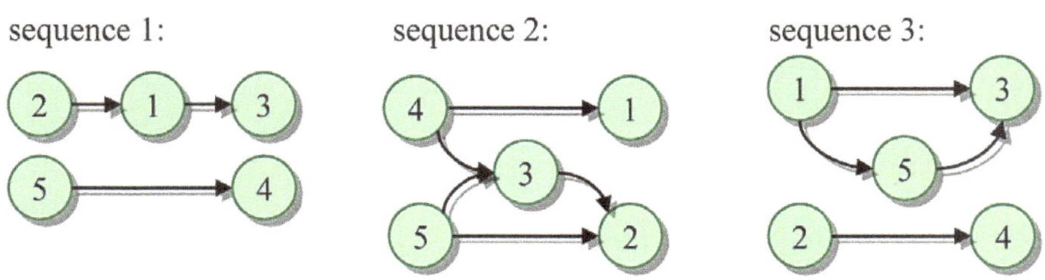

Fig. 4. Exemplary assembly sequences for 5 connections

Table 2. Results of the critical path analysis for sequence 2

Connection /activity	Earliest terms		Latest terms	
	beginning eb	ending ef	beginning lb	ending lf
1	1	8	19	**26**
2	21	26	21	**26**
3	12	21	12	21
4	0	1	11	12
5	0	12	0	12

Table 3. Results of the critical path analysis for sequence 3

Connection /activity	Earliest terms		Latest terms	
	beginning *eb*	ending *ef*	beginning *lb*	ending *lf*
1	0	7	0	7
2	0	5	22	27
3	19	28	19	**28**
4	5	6	27	**28**
5	7	19	7	19

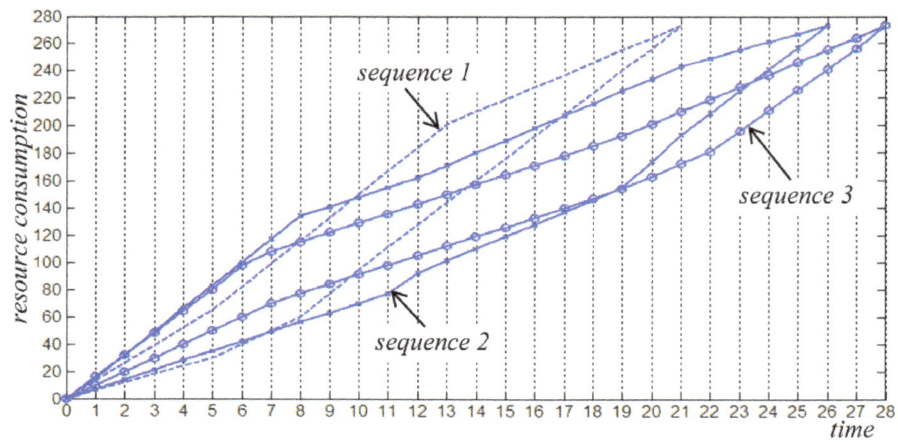

Fig. 5. Feasible budget regions for three exemplary assembly sequences

Based on the analysis of feasible budgets regions it can be observed: differences in the duration of the process and in the available scope of control of the speed of the resource consumption. Sequence 2 in every time unit has the widest scope and as a result the greatest FBR surface.

CRITERIA FOR THE SELECTION AND LIMITATION OF THE ASSEMBLY SEQUENCE

The network model and the FBR of the every considered assembly sequence variant allows to implement limitations and criteria to the ASP process. From the point of view of the work organisation of the production system it is important to consider the level of available production capacities of the system. They are inherently limited. The level of the free production capacity can be decreased by simultaneous implementation of other processes. One should also take into consideration the fact that the excess of the contracted completion term of the construction to the client or the co-operating party may result in the financial penalties or disorganisation of the production system work.

The peak resource consumption and the duration of the process are the basic measures characterising the assembly sequence. The time of the process $t_f(G, \mathbf{t})$ (see equation) is the number clearly defined for a given sequence G. The peak resource consumption $rc_{peak}(G, \mathbf{t}, \mathbf{r})$ requires the analysis of all schedules $\mathbf{s} \in S(G, \mathbf{t})$ and search for the variant with the lowest peak (see equation (5)):

$$rc_{peak}\left(G,\mathbf{t},\mathbf{r}\right)=\min_{\mathbf{s}\in S(G,\mathbf{t})}\left\{\max_{j=1,\dots,t_f(G,\mathbf{t})}\left\{rc_j\left(G,\mathbf{t},\mathbf{s},\mathbf{r}\right)\right\}\right\} \tag{8}$$

The sequence analysis for 5 connections was conducted for the data similar to the ones in chapter 4: $\mathbf{t} = (7, 5, 9, 1, 12)$, $\mathbf{r} = (10, 6, 9, 8, 7)$. There was generated the whole set $\Gamma(n)$, and then for every sequence there was determined:

- time of realisation of the assembly process t_f,
- all possible schedules and their corresponding peak resource consumption rc_{peak},
- schedule with the lowest peak consumption.

Consequently, a 29 281-element set $\Gamma(n)$ was obtained. Figure 6 presents the durations of all sequences.

It is visible that for the adopted data the assembly can last from 12 to 34 time units. There are no sequences lasting for 23, 30, 31 and 32 time units. Sequences with the longest time are serial. Their time is the sum of the times of all activities. Sequences with the shortest time have

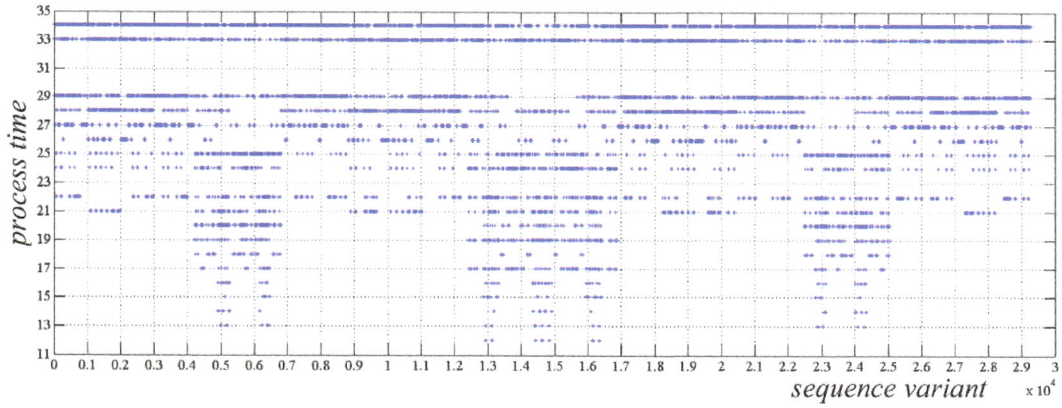

Fig. 6. Durations of the generated assembly sequences

the parallel character. All others are realised during the longest activity.

Figure 7 presents the peak levels of the resource consumption achievable for every sequence. The peak resource consumption fits into the range [10, 32]. The highest peak values (32) have 4 sequences shown in Figure 8. The lowest level of the peak consumption is equal to the demand of the most resource-absorbent activity.

Figure 9 presents the dependency between the studied variables. It can be observed that the total number of different variants equals 55. It is a very small value compared to the number of the analysed sequences. As expected, it can be observed that all serial sequences, with the longest duration, have the same peak resource consumption level. Sequences with the highest peak consumption, shown in Figure 8, have the same duration. It is the lowest and equals 12. Also other sequences have the same duration, which peak resource consumption is lower.

In the figure the points creating the pareto-optimal set have been marked due to the criteria

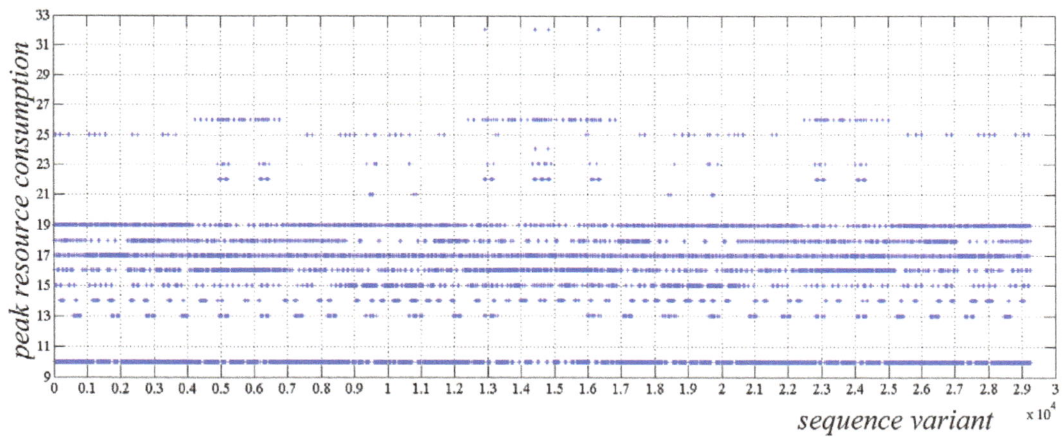

Fig. 7. Peak resources consumption for the generated assembly sequences

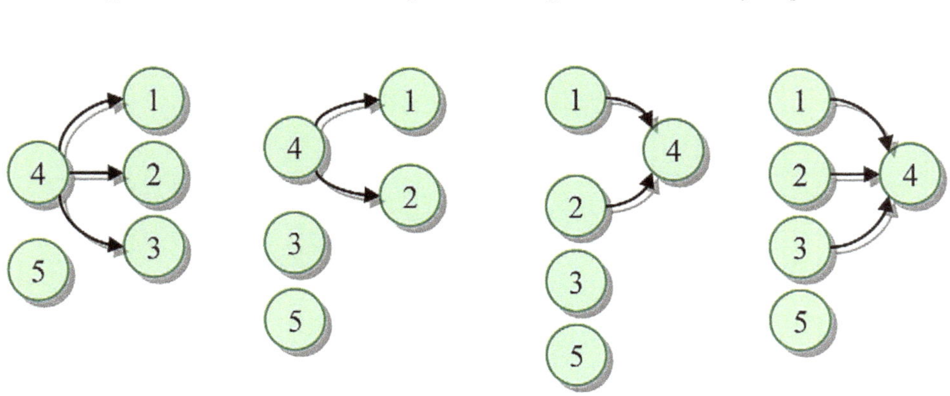

Fig. 8. Sequences with the highest peak resource consumption

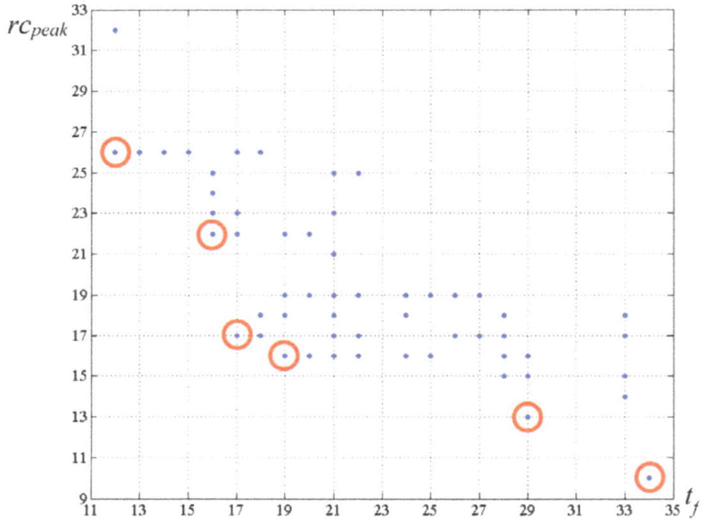

Fig. 9. Dependence of the peak resource consumption level on the process time

of time minimisation and the peak resource consumption. In many cases this set is searched in order to find a rational solution.

Time boundary values of the assembly completion t_F and the peak resource consumption rc_{Peak} can be adopted as the parameters dividing the set $\Gamma(n)$ into subsets of the acceptable and unacceptable sequences. As a result two conditions are obtained, which must be met by the acceptable sequence G with the fixed vectors $\mathbf{t}$ and $\mathbf{r}$:

$$t_f(G,\mathbf{t}) \le t_F \qquad (9)$$

$$rc_{peak}(G,\mathbf{t},\mathbf{r}) \le rc_{Peak} \qquad (10)$$

For the studied case 5 connections determined the number of the acceptable sequences depending on the values of the limiting parameters t_F and rc_{Peak} (Figure 10).

The analysis of the number of the set of the acceptable sequences may be significant, when during the planning of the assembly sequences other criteria and limitations not connected with the schedule are also considered. Insufficient sequences acceptable at the stage of the schedule analysis may lead to an empty set of solutions after considering other limitations. On the other hand, a very large set of solutions will require the use of the randomised methods of search and the long computation time.

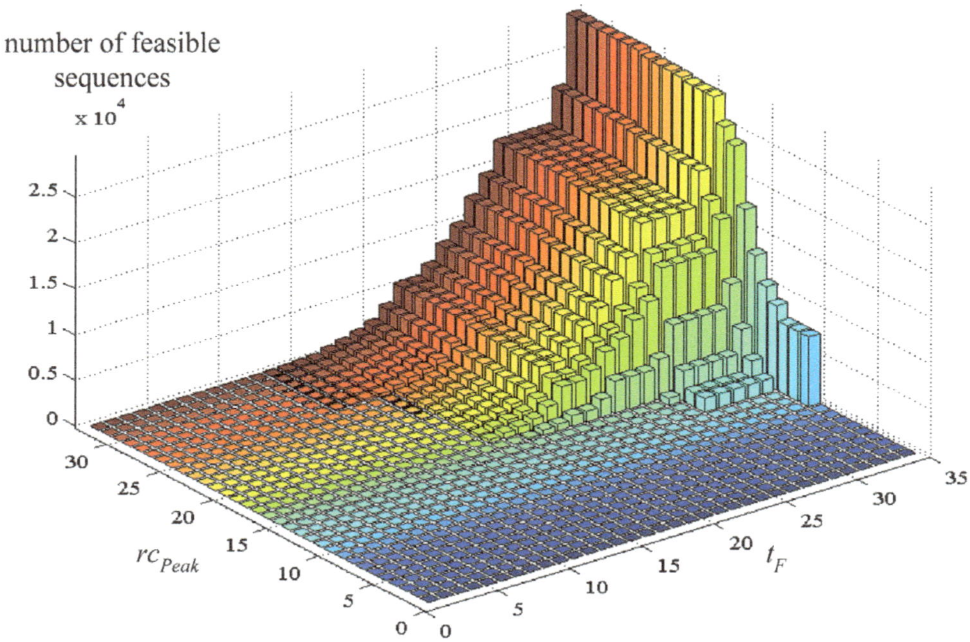

Fig. 10. The number of acceptable sequences depending on the limitation values

CONCLUSIONS

The presented theoretical model allows to account for scheduling issues and budgeting during planning of the assembly sequences. Such an approach is necessary in case of the assembly of large-scale structures. This type of processes are long and consume significant amounts of resources. Yards planning the production must synchronise many assembly processes implemented at the same time. Resource budget management in time is therefore crucial for them.

The proposed method to determine the peak resource consumption for the determined assembly sequence of course requires the use of a computer. The analysis should be performed on a huge amount of variants, which is strongly increasing with the increase of the number of connections. In practice, the shipbuilding number of the analysed connections is reduced by the use of the subassembly.

Two approaches to the assembly scheduling analysis were shown. In the first one, the process time and the peak resource consumption are the basis for the formulation of optimisation criteria. The way of determining the pareto-optimal solutions was shown. In the second approach, conditions limiting the set of acceptable solutions are formulated due to the time and consumption of the resource.

The examples of the calculations were conducted for 5 connections and randomly selected vectors of data. Vast possibilities of the assembly sequence analysis in terms of the schedule and resources budget were presented.

REFERENCES

1. Batista H.S., Graczyk T., An analysis of the welding-induced deformation of ship panels using a statistical tool. Polish Maritime Research, 20 (2), 2013, 74–81.
2. Chi Z., Jun S., Intelligentized work-preparation for ship hull construction with Optimized Assembly Planning System. Proceedings of the IEEE International Conference on Systems, Man and Cybernetics, Istanbul, Turkey, 10-13 October 2010, 2740–2744.
3. Cho K.-K., Lee S.-H., Chung D.-S., An automatic process-planning system for block assembly in shipbuilding. Annals of the CIRP, 45 (1), 1996, 41–44.
4. Cho K.-K., Sun J.-G., Oh J.-S., An automated welding operation planning system for block assembly in shipbuilding. International Journal of Production Economics, 60-61, 1999, 203–209.
5. Iwańkowicz R., Optimization of assembly plan for large offshore structures. Advances in Science and Technology Research Journal, 6(16), 2012, 31–36.
6. Iwańkowicz R., Application of multi-dimensional grouping to building steel stiffened shell structures. Polish Maritime Research Vol. 20, 1(77), 2013, 48–58.
7. Lin E.Y.H., The role of the activity adjacency matrix in the critical path method. APL Quote Quad, 21 (3), 1991, pp. 1419.
8. Qu S., Jiang Z., Tao N., An integrated method for block assembly sequence planning in shipbuilding, The International Journal of Advanced Manufacturing Technology, 69 (5-8), 2013, 1123–1135.
9. Robinson R.W., Counting labeled acyclic digraphs. [In:] New Directions in Graph Theory (Ed. F. Harary). Academic Press, New York 1973.
10. Sekulski Z, Ship hull structural multiobjective optimization by evolutionary algorithm. Journal of Ship Research, 58 (2), 2014), 45–69.
11. Seo Y., Sheen D., Kim T. , Block assembly planning in shipbuilding using case-based reasoning. Expert Systems with Applications, 32, 2007, 245–253.
12. Storch R.L., Hammon C.P., Bunch H.M., Moore R.C., Ship production, 2nd ed. Cornell Maritime Press, Centreville, Maryland 1995.
13. Urbański T., The influence of welding deformation types of hybrid node on assembly suitability. Advances in Science and Technology Research Journal, 6 (16), 2012, 42–47.
14. Urbański T., Taczała M., Experimental investigation of deformations of hybrid structural joint. [In:] 4th International Conference on Maritime Structures (MARSTRUCT 2013), Espoo, Finland, 25-27 March 2013, Taylor&Francis Group, London 2013, 441–446.
15. Wei Y., Automatic generation of assembly sequence for the planning of outfitting processes in shipbuilding [Ph.D. thesis]. VSSD, Delft, the Netherlands 2012.
16. Zhong Y., Xue K., Shi D., Assembly unit partitioning for hull structure in shipbuilding. Computer-Aided Design, 45 (12), 2013, 1630–1638.

7

EFFECT OF THE NARROWING ANGLE IN BEVELIOD GEAR ON THE TEMPERATURE PROFILE ON THE ACTIVE FLANK OF TOOTH

Mariusz Sobolak[1]**, Piotr Strojny**[1]

[1] Department of Mechanical Engineering, Rzeszow University of Technology, al. Powstańców Warszawy 12, 35-959 Rzeszów, Poland, e-mail: msobolak@prz.edu.pl; pstrojny@prz.edu.pl

ABSTRACT

This paper contains a description of the test bench to research the temperature distribution in gears made of a plastic polymer. The paper presents the research results of the change and the temperature distribution on the side surface of the Beveloid tooth gear. The research were performed for a series of Beveloid gears with different values of narrowing angle.

Keywords: Beveloid gear, temperature profile, narrowing angle.

INTRODUCTION

The gears are very often used in the transmission of torque. They are used in almost all industries. One of the areas in which they play a huge role is the electrical engineering industry and especially in production of household appliances, where most gears of nearly all types and varieties of wheel teeth are applied.

Nonetheless, gears in household appliances are mostly made of plastics i.e. materials which are far more susceptible to heat. The problem of overheating the gears of plastics is one of the biggest problems household appliances designers meet upon construction.

There have been numerous efforts made to reduce the excessive heating up of the gears for example, by introducing additional components to trigger a forced air flow inside the device or achieved through an appropriate gears geometry (less susceptible to heat during work).

RESEARCH

The preliminary research shows that one of the gears varieties that are less susceptible to heating are the Beveloid gears. Beveloid gears are shaped by a gear tool (with constant width of a tooth face) on the rotation surface of the truncated roller. These gears are in the mid-section of the toothed-wheel rim and have a geometry identical to the spur gear (Figure 1).

Variable thickness of the teeth geometry is a result of using a tool with a constant tooth space width (Figure 2).

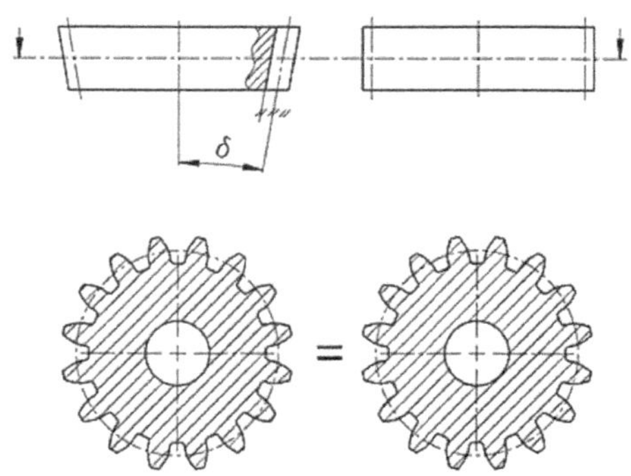

Fig. 1. Comparison of Beveloid gear geometry with a spur gear

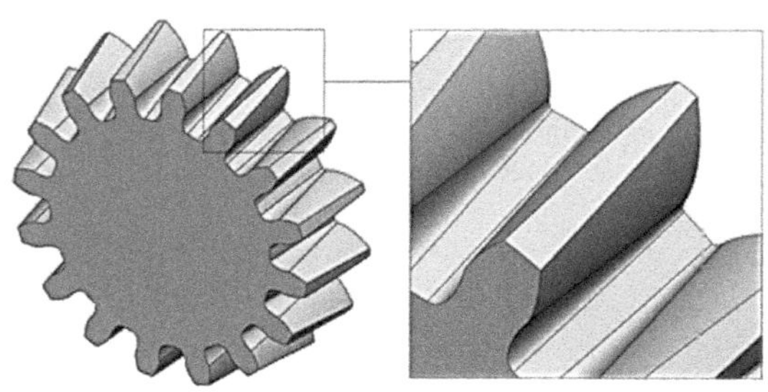

Fig. 2. The shape of the tooth space in Beveloid gear

With this solution 3 variants of the gear mate and the possibility to a certain extent regulation of the angle between the gearing is obtained:

- Beveloid gears mate with axes parallel (Figure 3),
- Beveloid gear mate with spur gear – we can obtain an angle between the gears axes (Figure 4),
- Beveloid gears mate with angled axes (Figure 5).

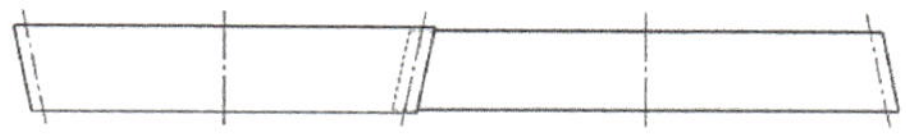

Fig. 3. Beveloid gears with parallel axes

$\Sigma = \delta$

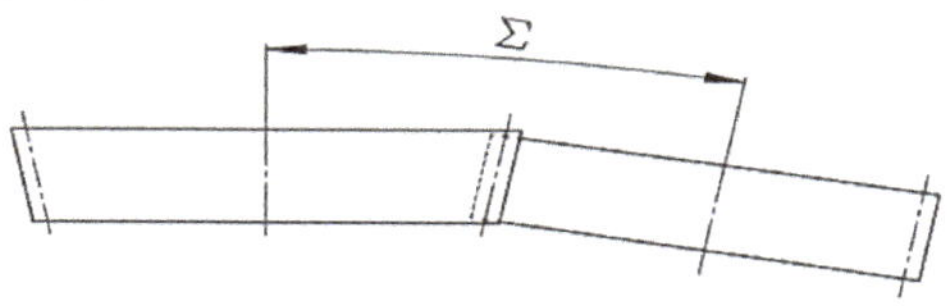

Fig. 4. Beveloid gear with spur gear

$\Sigma = 2\delta$

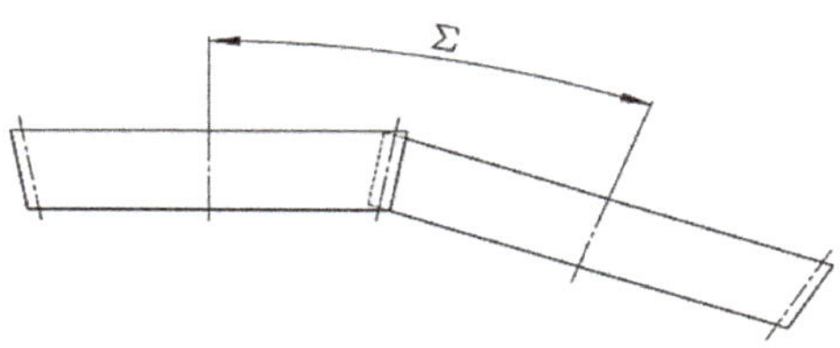

Fig. 5. Beveloid gears with angeled axes

On the basis of the conducted analyses, the following relation has been developed, which allows to determine the maximum allowable narrowing angle of Beveloid gear (equ. 1).

$$\delta_{max} = arctan\frac{2,5 + z(\cos(\alpha) - 1)}{\psi z} \quad (1)$$

where: z – number of teeth,
α – pressure angle,
ψ – factor face width of gear.

It can be observed that the smaller number of teeth and the lower width of the toothed-wheel rim, the greater the narrowing angle value is (Figure 6). This information will help in the future to produce gears for different variants of arrangements of gearing, such as atypical planetary gear.

One of the criteria of quality of gears, is the quantities of heat emission during gearing. In order to conduct such an analysis, a deducted test stand has been designed and manufactured (Figure 7).

The test stand consists of a speed controller (1) to control the electric motor (2) slide guides (3) for fine adjustment gear mate (4), the disc brake (5) to cause the counter-torque, high-precision thermal imaging camera FLIR SC5000 type (6) (Figure 8), shield with aluminum insert (7) to isolate the influence of temperature of the engine.

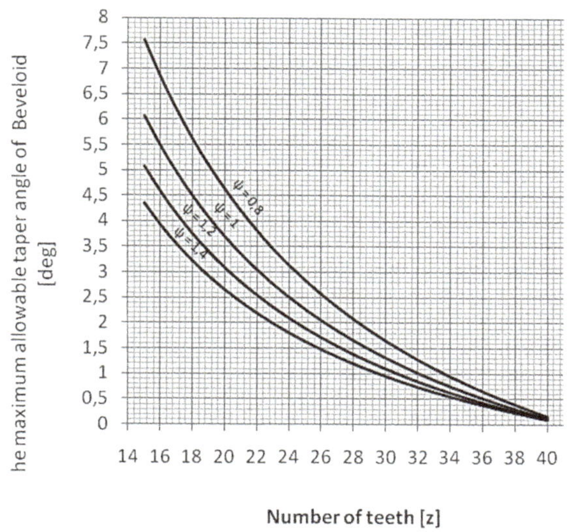

Fig. 6. Maximum allowable narrowing angle in Beveloid gears [deg]

Fig. 7. Description of the test stand

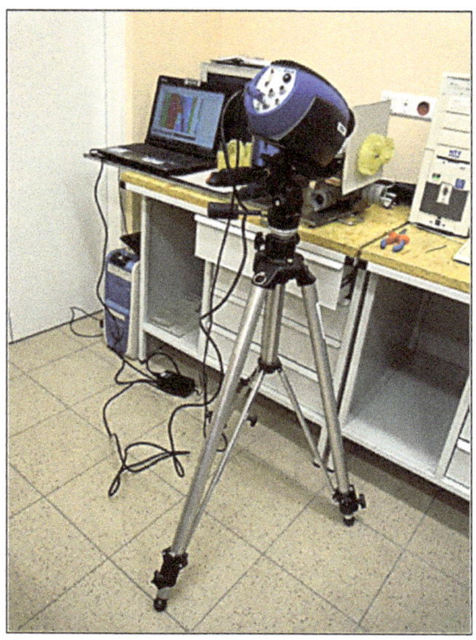

Fig. 8. Thermal imaging camera type FLIR SC5000

Additionally, the device is equipped with a positioner (8) (Figure 9) which provides the temperature measurements that are always at the same gear mate position. There have been five pairs of Beveloid gears tested, each of different narrowing angle (0 °, 1.25 °, 2.5 °, 3.75 °, 5 °) (Figure 10).

The results of measurements have been developed using the "Altair" software version 5.90.001 (Figure 11).

CONCLUSIONS

Studies suggest that Beveloid gears possess decreased tendency to get overheat during mate than the spur gears (Figure 12). With the narrowing angle increase, the temperature measured on the surface of the gear mate was decreasing.

Fig. 9. The positioner of fine adjustment gears

Fig. 10. Beveloid gears used in the experiment

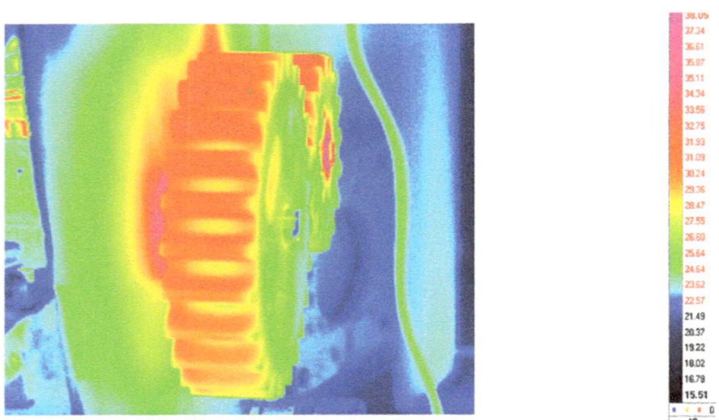

Fig. 11. The results of heat measurement

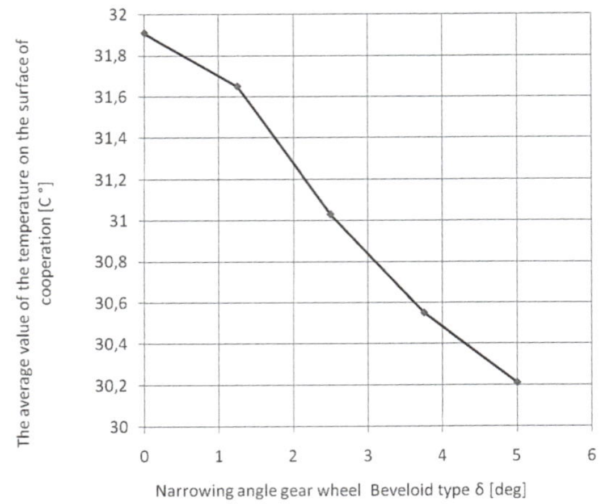

Fig. 12. The results of measurements, for the various gears, measured as the average surface temperature of gear mate

It may be conditioned by the geometry of Beveloid gear by which they are characterized with higher number of tooth contact ratio and smoother entrance into the engagement.

Nevertheless, the test stand itself needs to be modified, as it has been observed that the engine warm-up brings an unfavourable impact on the temperature measurement on the surface of the teeth, what consequently hinders the temperature reading on the pinion.

REFERENCES

1. Brecher C., Röthlingshöfer T., Gorgels C. Manufacturing simulation of beveloid gears for the use in a general tooth contact analysis software. MACHINE TOOL, German Academic Society for Production Engineering (WGP) 2008.
2. Chia-Chang L., Chung-Biau T. Contact characteristics of beveloid gears. Pergamon, Department of Mechanical Engineering, National Chiao Tung University, Hsinchu, 30010, Taiwan, November 2001.
3. Sobolak M. Analiza i synteza współpracy powierzchni kół zębatych metodami dyskretnymi. Oficyna Wydawnicza Politechniki Rzeszowskiej, Rzeszów 2006.
4. Sobolak M. Bezpośrednia numeryczna symulacja kształtowania uzębienia kół przekładni stożkowych typu Gleason o kołowo-łukowej linii zęba. Archives of Mechanical Technology and Automatization, Poznań 2001.
5. Sobolak M. Metody numeryczne (dyskretna i bezpośrednia) symulacji kształtowania uzębienia kół przekładni stożkowych typu Gleason o kołowo-łukowej linii zęba. Gears G2001, Zeszyty naukowe Politechniki Rzeszowskiej, Mechanika z. 57, Rzeszów 2001.

8

DETECTION OF CHANGES OF THE SYSTEM TECHNICAL STATE USING STOCHASTIC SUBSPACE OBSERVATION METHOD

Andrzej Puchalski[1]

[1] University of Technology and Humanities in Radom, Malczewskiego 29, 26-600 Radom, Poland, e-mail: andrzej.puchalski@pr.radom.pl

ABSTRACT

System diagnostics based on vibroacoustics signals, carried out by means of stochastic subspace methods was undertaken in the hereby paper. Subspace methods are the ones based on numerical linear algebra tools. The considered solutions belong to diagnostic methods according to data, leading to the generation of residuals allowing failure recognition of elements and assemblies in machines and devices. The algorithm of diagnostics according to the subspace observation method was applied – in the paper – for the estimation of the valve system of the spark ignition engine.

Keyworda: system diagnostic, subspace method, vibroacoustic.

INTRODUCTION

The methods of identification of dynamic models by means of subspace, known from the control theory [1–4]: N4SID (Numerical Algorithms for Subspace State Space System IDentification), MOESP (Multivariable Output-Error State Space), CVA (Canonical Variate Analysis) or the method of principal components analysis (PCA) are also used for diagnostics of technical objects.

The idea of identification by subspace methods is applied in structural health monitoring (SHM) [5]. The SHM range, aimed at improving a broadly considered safety and at decreasing maintenance costs, encompasses low-frequency methods originated from vibroacoustic diagnostics of machines, as well as high-frequency methods (the so-called non-destructive inspections).

The condition monitoring is used in the automotive and aircraft industry, power industry (mainly rotating machines) and in civil engineering [6–10]. The result of the vibroacoustic analysis is often a modal model in a form of a natural frequencies set, damping ratios and mode shapes used in various detection procedures of structural defects based on these parameters.

Algorithms of vibroacoustic diagnostics of mechanical objects by subspace methods lead to the determination of the signature (pattern) of the system in a serviceable state [11–14]. This approach requires neither knowledge of modal parameters nor finite elements method (FEM) in relation to the identification of the system state.

Both direct measuring data and covariances of these data or impulse characteristics of investigated systems can be used in constructing subspace matrices. As the result of the decomposition of the subspace observation matrix the vectors of residuals – assessed as 'null' in a nominal (serviceable) state – are obtained.

BASIC PROBLEMS OF STOCHASTIC FILTRATION OF EMPIRICAL DATA

The assessment method of vectors of residuals, by means of the theory of local approach to the statistic identification of changes in dynamic systems, is used in the hereby paper [15–17].

The vector of random variables with the observation space Y= $\{Y_1,\ldots,Y_N\}$, $Y_k \in R^m$, encompassing sampling periods k = 1,…,N in a form of measurement data collection (time window)

and parameters space Θ generating residual $R_N(\theta \in \Theta)$ – is considered.

The subset of the parameters space in a nominal state is marked as Θ_0, while the subset of the parameters space in a damage state as Θ_1. In addition: $\Theta_0 \cap \Theta_1 = 0$ and $\Theta_0 \cup \Theta_1 = \Theta$.

It is assumed that the distance between the investigated models or experiments, represented by the error vector:

$$\delta\theta = \frac{const}{\sqrt{N}} \quad (1)$$

is sufficiently small when the observation data record N is large. The local approach to the identification of the nominal model means that the deviation of the error vector – determined by equation (1) – is of the order of $\frac{1}{\sqrt{N}}$.

The statistics of diagnostics parameters constructed on the basis of observation of the system in a serviceable state constitutes, in turn, grounds for formulating the zero hypothesis, which is tested against the alternative hypothesis, corresponding to small deviations of parameter values.

These assumptions allow to transfer the problem of detection of changes in the system into the plane of investigating the expected value of the residual in a form of random vector of the normal Gaussian distribution in a serviceable state as well as during further monitoring.

Applying generalized likelihood ratio (GLR) the rejection criterion of the verified hypothesis, in a form of the chi-square test with the assumed critical level for the given confidence interval, is obtained.

CONSTRUCTING THE SIGNATURE OF THE VIBRATING STRUCTURES

The model of the linear mechanical system – given by equations [18] – is considered.

$$M\ddot{q}(t) + D\dot{q}(t) + Kq(t) = f(t) \quad (2)$$

$$y(t) = Lq(t) \quad (3)$$

where: *M, D, K, L* are matrices of mass, damping, stiffness and monitored states filter – respectively, *q(t)* is the displacement vector, *f(t)* is the external forces vector (modelled in a general case as the non-stationary Gaussian white noise) and *y(t)* is the vector of the monitored states.

Modes of the system μ (representing natural frequencies and damping ratios corresponding to them) are the solution of the equation:

$$\det(M\mu^2 + D\mu + K) = 0 \quad (4)$$

while mode shapes:

$$\psi_\mu = L\varphi_\mu \quad (5)$$

satisfy the formula:

$$(M\mu^2 + D\mu + K)\varphi_\mu = 0 \quad (6)$$

Assuming forcing in a white noise form and discretising the above model with a sampling time T, the system description is obtained by means of differential equations of state and observations:

$$x_k = Ax_{k-1} + w_{k-1} \quad (7a)$$

$$y_k = Cx_k + v_k \quad (7b)$$

where:

$$x_k = \begin{bmatrix} q(kT) \\ \dot{q}(kT) \end{bmatrix}, \ A = e^{\mathcal{A}T}, \quad (8)$$

$$\mathcal{A} = \begin{pmatrix} 0 & I \\ -M^{-1}K & -M^{-1}D \end{pmatrix}, \ C = [L \quad 0],$$

$x_k \in R^n$ – state vector, $y_k \in R^m$ – response (observation) vector in the k-ts sampling moment, $A \in R^{nxn}$, $C \in R^{mxn}$, w_k and v_k represent white Gaussian noises of mean values being zero and covariance matrices Q_k and R_k – respectively.

Discrete eigenvalues and eigenvectors $(\lambda, \varphi_\lambda)$ of matrix *A* complying with the equations:

$$\det(\lambda I - A) = 0, (\lambda I - A)\varphi_\lambda = 0 \quad (9)$$

represent the modal description for a continuous time, as follows:

$$e^{T\mu=}\lambda, \psi_\mu = C\varphi_\lambda \quad (10)$$

Equations (9) and (10) indicate that the matrix pair (*A, C*), and thus also the observability matrix of the system described by equations (7), allows to calculate modal parameters of the object.

The problem of detection of defects in a mechanical system on the bases of vibration monitoring is thus equivalent to the problem of analysing changes in the structure of its linear model of state variables with an unstable process noise.

Defining various ways of signature determination and generation the investigated system residual will be the main subject of further considerations.

When the modal model is available, the signature of the nominal system $S^T(\theta_0)$ corresponding to the vector of parameters $\theta_0 \stackrel{\text{def}}{=} \begin{pmatrix} \Lambda \\ vec\,\Phi \end{pmatrix} \in R^{n(m+1)}$ ($\Lambda \in R^n$ is the vector of eigenvalues λ of matrix *A*, $\Phi \in R^{mxn}$ is the matrix, whose columns are vectors $C\varphi_\lambda$ and φ_λ is the matrix eigenvector corresponding to eigenvalue λ) is constructed as the left null space of the observability matrix $\mathcal{O}_{p+1}(\theta)$ [19]:

$$\mathcal{O}_{p+1}(\theta) = \begin{pmatrix} \Phi \\ \Phi\Delta \\ \vdots \\ \Phi\Delta^p \end{pmatrix} \in R^{m(p+1)xn}, \quad (11)$$

where $\Delta = diag(\Lambda)$.

In case of diagnostics of systems for which the modal model is not available, the observability matrix can be determined by means of singular value decomposition (SVD) of the Hankel matrix $\mathcal{H}_{p+1,q}$ of covariances of experimental data obtained for the serviceable system.

In order to do this, the series of covariances of the output signals vector Y_k: $R_i \stackrel{\text{def}}{=} \boldsymbol{E}(Y_k Y_{k-i}^T)$ will be considered and the Hankel matrix of data covariances, for the selected constants p and q, will be defined as follows:

$$\mathcal{H}_{p+1,q} = Hank(R_i) \stackrel{\text{def}}{=} \begin{pmatrix} R_1 & R_2 & \cdots & R_q \\ R_2 & R_3 & \cdots & R_{q+1} \\ \vdots & \vdots & & \vdots \\ R_{p+1} & R_{p+2} & \cdots & R_{p+q} \end{pmatrix} \quad (12)$$

The empirical Hankel matrix $\hat{R}_i$ is determined on the basis of the covariance estimator in a form:

$$\hat{R}_i \stackrel{\text{def}}{=} \frac{1}{N-i} \sum_{k=i+1}^{N} y_k y_{k-i}^T \quad (13)$$

Then the SVD of the Hankel matrix of data covariance is considered:

$$\mathcal{H}_{p+1,q} = U\Sigma V^T \approx (U_1 \quad U_2)\begin{pmatrix} \Sigma_1 & 0 \\ 0 & 0 \end{pmatrix}\begin{pmatrix} V_1^T \\ V_2^T \end{pmatrix} = U_1\Sigma_1 V_1^T \quad (14)$$

where $U \in R^{(p+1)x(p+1)}$ and $V \in R^{qxq}$ are orthogonal matrices, and $\Sigma_1 = diag[\sigma_i]$ is squared diagonal matrix of n – singular values σ_i of the Hankel matrix $\mathcal{H}_{p+1,q}$.

It is known, from the theory of realisation by subspace methods, that the observability matrix and the Hankel matrix are linked by the equation:

$$\mathcal{H}_{p+1,q} = \mathcal{O}_{p+1}\mathcal{Z}, \quad (15)$$

where $\min(pm, qm) \geq n$ and $\mathcal{Z}$ is the matrix of a form resulting from the assumed identification algorithm. In addition the observability matrix of the system is determined by:

$$\mathcal{O}_{p+1} = U_1 \Sigma_1^{\frac{1}{2}} \quad (16)$$

Matrices $\mathcal{H}_{p+1,q}$ and $\mathcal{O}_{p+1}$ have the same left null space. The left null space S^T determined for the observability matrix $\mathcal{O}_{p+1}$ satisfies asymptotically – under serviceable conditions – the equation:

$$S^T(\theta_0)\mathcal{H}_{p+1,q} = 0 \quad (17)$$

for the empirical Hankel matrix $\mathcal{H}_{p+1,q}$.

DIAGNOSTICS PROCEDURE

The described procedure can be presented in a form of the algorithm of the dynamic system diagnostics, based on the observed vibrations output signals:

- Step 1. To calculate of the averaged Hankel matrix of data covariance, on the basis of J data sets recorded under various working and environmental conditions of the serviceable system:

$$\hat{\mathcal{H}}_{p+1,q}^{0} \stackrel{\text{def}}{=} \frac{1}{J}\sum_{j=1}^{J} \hat{\mathcal{H}}_{p+1,q}^{0,j} \quad (18)$$

- Step 2. To perform SVD of the matrix $\hat{\mathcal{H}}_{p+1,q}^{0}$ according to equation (14).
- Step 3. To determine the observability matrix according to equation (16).
- Step 4. To determine the left null space S_0^T of the observability matrix $\mathcal{O}_{p+1}$ of the data obtained in the serviceable state:

$$S_0 = null(\mathcal{O}_{p+1}^T) \quad (19)$$

- Step 5. To determine the residual vector in a form:

$$R_N \stackrel{\text{def}}{=} \sqrt{N} vec[S^T(\theta_0)\hat{\mathcal{H}}_{p+1,q}] \quad (20)$$

for the empirical matrix $\hat{\mathcal{H}}_{p+1,q}$ obtained for the current experimental data

- Step 6. To determine the estimator $\hat{\Sigma}$ of the covariance matrix of the residual vector R_N, after dividing data into K segments of the same dimension l acc. to:

$$\hat{\Sigma} = \frac{1}{Kl}\sum_{k=1}^{K} R_N R_N^T \quad (21)$$

- Step 7. To calculate the index of defects detection χ^2 in a form of the chi-square test:

$$\chi^2 = R_N^T \hat{\Sigma}^{-1} R_N \quad (22)$$

The algorithm is illustrated in Figure 1 and verified on the example of data sets obtained in measurements of the acceleration of the IC engine head vibrations, at cylinder 1 in a horizontal direction.

The analysis was performed on the basis of the data originated from road tests of 4-cylinder IC engine of a cubic capacity 1.4 l and a mileage of 400 000 km described in details in [20, 21].

Data collections formed from the recorded and properly transformed vibration responses

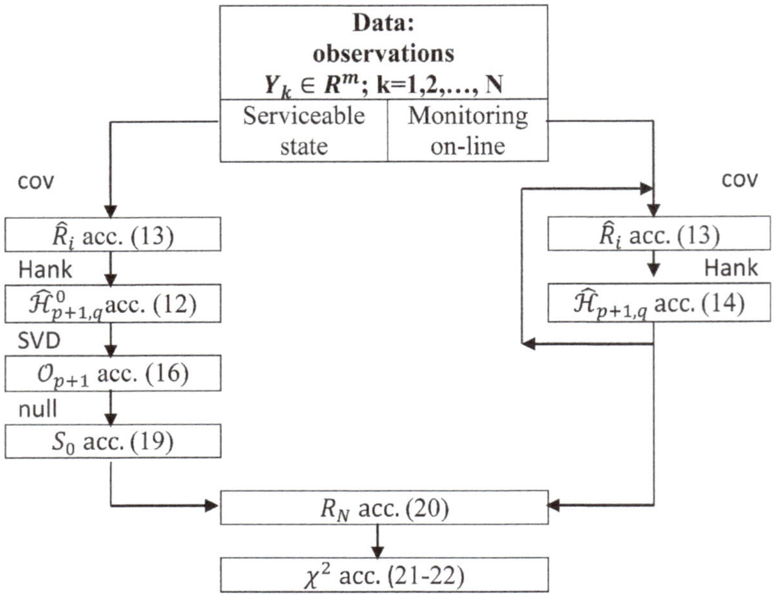

Fig. 1. Algorithm of the index χ^2 determination for the monitored object

for closing the exhaust valve were utilised. Each observation contained 26 signal samples, recorded at a constant rotational speed of 3000 rpm, at states:

- Non-defective valve, a proper valve clearance (0.25 mm),
- Defective valve type I (valve head cut 3 mm), a proper valve clearance.

Values of detection indices in a form of chi-square test of the discussed defects for the successive observation collections of non-defective and defective systems are presented in Figure 2.

CONCLUSIONS

The presented example of a non-parametric diagnostics method of the IC engine valve system, on the basis of data subspace, belongs to algorithms of filtration realised according to statistic properties. Original results of the acceleration vibrations signal, obtained in several road tests for various maintenance states of the valve system, proved the algorithm efficiency.

The analysis of results indicates the possibility of applying this method for fault detection and isolation. However, the method does not allow for defect identification, i.e. for the determination of its size and discriminating the reason of such a state.

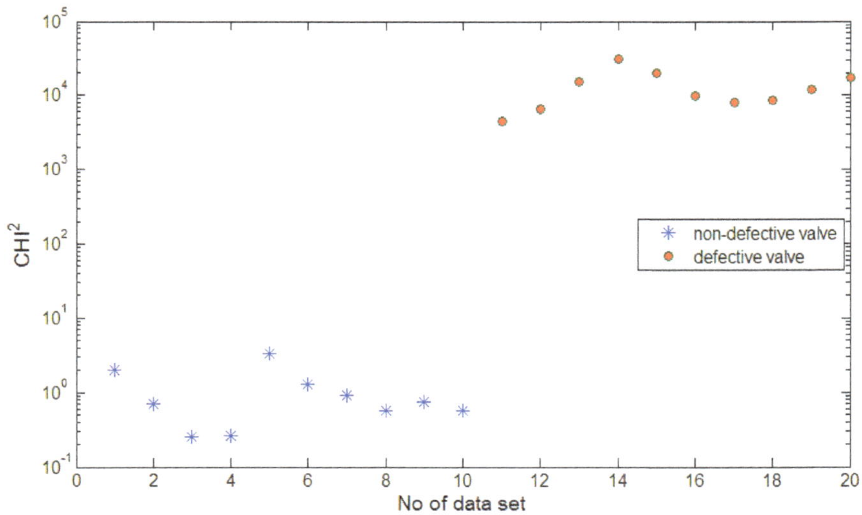

Fig. 2. Chi-square statistic values of the diagnosed valve system as a function of successive observation collections

REFERENCES

1. Van Overshee P., De Moor B., Subspace Identification for Linear Systems. Theory – Implementation – Applications. Kluwer Academic Publishers, Boston - London - Dordrecht 1996.
2. Katayama T., Subspace methods for system identification. Springer-Verlag, London 2005.
3. Huang B., Kadali R., Dynamic modeling, predictive control and performance monitoring. A data drive subspace approach. Springer-Verlag, London 2008.
4. Isermann R., Münchhof M., Identification of Dynamic Systems. Springer-Verlag, Berlin 2011.
5. Farrar Ch.R., Worden K., An introduction to structural health monitoring. Phil. Trans. R. Soc. 365, 2006, 303-315.
6. Radkowski S., Wibroakustyczna diagnostyka uszkodzeń niskoenergetycznych. Wyd. ITE, Radom 2002.
7. Uhl T., The use and challenge of modal analysis in diagnostics. Diagnostyka, 30, 2004, 151-161,
8. Mendrok K., Uhl T., Maj W., Paćko P., SHM system based on modal filtration. Key Engineering Materials, 518, 2012, 289-297.
9. Dybała J., Vibrodiagnostics of gearboxes using NBV-based classifiers: A pattern recognition approach. Mechanical Systems and Signal Processing, 38, 2013, 5-22.
10. Zimroz R., Bartkowiak A., Two simple multivariate procedures for monitoring planetary gearboxes in non-stationary operating conditions. Mechanical Systems and Signal Processing, 38, 2013, 237-247.
11. Basseville M., Mevel M., Goursat M., Statistical model-based damage detection and localization: subspace-based residuals and damage-to-noise sensitivity ratios. Journal of Sound and Vibration, 275, 2004, 769-794.
12. Mevel L. at al., Input/output versus output-only processing for structural identification-application to in-flight data analysis. Journal of Sound and Vibration, 295, 2006, 531-552.
13. Ding S.H. at al., Subspace method aided data-driven design of fault detection and isolation systems. Journal of Process Control, 19, 2009, 1496-1510.
14. Mevel L., Hermans L., Van Der Auweraer H., Application of a subspace-based fault detection method to industrial structures. Mechanical Systems and Signal Processing, No.mssp.1999.1247, available online at http://www.idealibrary.com
15. Pollard D., Some thoughts on LeCam's statistical decision theory. Cornell University Library, 19 Jul 2011, arxiv.org/abs/1107.3811
16. Benveniste A., Basseville M., Moustakides G.V., The Asymptotic Local Approach to Change Detection and Model Validation, IEEE Trans. on Automatic Control, 32(7), 1987, 583-592.
17. Benveniste A., Basseville M., Goursat M., Mevel L., The local approach to change detection, diagnosis, and model validation: application to vibration mechanics, 2006, www.irisa.fr/distribcom/benveniste/pub/Ljung2006_birthday.html
18. Awrejcewicz J., Drgania deterministyczne układów dyskretnych. WNT, Warszawa 1996.
19. Basseville M., Abdelghani M., Benveniste A., Subspace-based fault detection algorithms for vibration monitoring. Automatica, 36, 2000, 101-109.
20. Komorska I. Vibroacoustic diagnostic model of the vehicle driving system. Wydawnictwo ITE-PIB, Radom 2011.
21. Puchalski A., Komorska I.: Application of vibration signal Kalman filtering to fault diagnostics of engine exhaust valve. Journal of Vibroegineering, 15(1), 2013, 194-200.

9

MEASUREMENT UNCERTAINTY ANALYSIS OF DIFFERENT CNC MACHINE TOOLS MEASUREMENT SYSTEMS

Leszek Semotiuk[1], Jerzy Józwik[1], Kuric Ivan[2]

[1] Department of Production Engineering, Mechanical Engineering Faculty, Lublin University of Technology, 36 Nadbystrzycka Str., 20-816 Lublin, Poland, e-mail: l.semotiuk@pollub.pl; j.jozwik@pollub.pl

[2] Department of Automation and Production Systems, Faculty of Mechanical Engineering, University of Žilina, 1 Univerzitná Str., 010-26 Žilina, Slovakia, e-mail: ivan.kuric@fstroj.utc.sk

ABSTRACT

In this paper the results of measurement uncertainty tests conducted with a Heidenhain TS 649 probe on CNC machine tools are presented. In addition, identification and analysis of random and systematic errors of measurement were presented. Analyses were performed on the basis of measurements taken on two different CNC machine tools with Heidenhain control system. The evaluated errors were discussed and compensation procedures were proposed. The obtained results were described in tables and figures.

Keywords: measurement uncertainty, inspection probe, control system, CNC machine tool.

INTRODUCTION

Application of inspection probes in computer numerical control machine tools is becoming increasingly popular. The spectrum of their uses includes zero point measurement at workpiece surface and, to an increasing extent, inter-operational dimension control. In order to use it as a measuring device, the knowledge on its metrological characteristics should be required. However, the metrological characteristics of this meter circuit, which a computer numerical control machine tool with a mounted probe, are not known. Admittedly, manufacturers of the inspection probes provide unidirectional repeatability (2s) as the parameter characterising measurement uncertainty, however, it is only one of many components of uncertainty budget of the whole measuring system. Other components that characterise inspection probe include: geometric and kinematic accuracy of the machine tool, probe calibration accuracy as well as kinematics of the measurement process. Multiple research centres, both domestic and foreign, use probes in their research. Research works aim, *inter alia*, at integrating control and technology design processes in order to increase the efficiency of production processes (CLM-closed-loop machining) [1, 10-13, 18], the possibility and ways of utilizing inspection probes in control processes [17], comparing accuracy of measurement conducted on CMM and CNC machines [9] or assessing positioning accuracy of 5-axix CNC machine tools using measurements taken with inspection probes and based on a theoretical mathematical model [10, 14, 15].

At the Department of Production Engineering of the Mechanical Engineering Faculty at Lublin University of Technology a series of tests aiming at determining the manner of testing metrological characteristics of the measuring system CNC machine – inspection probe were conducted. The tests were focused on determining one-dimensional and two-dimensional measurement uncertainty [2, 3], repeatability of probe fix (a component of the measurement uncertainty budget) [5], the correlation between CNC machine tool spindle bearing failure and measurement uncertainty [6] as well as the influence of thermal deformations resulting from processing on inaccuracy of inspection probe measurement [4]. Moreover, potential application of inspection probes for automatic control and correction of dimensional inaccuracies of the workpiece [7] as well as uncertainty of laser probes measurement were investigated [8, 16].

The development of interfaces and software, such as PC-DMIS NC GAGE and STEP-NC proves that measurement control systems used on CNC machine tools are in constant progress. Not only do they aid taking measurement but also allow preparing reports. The user is equipped with modern software for planning control procedures and machining allowances based on three-dimensional models.

The aim of this paper was to determine one-dimensional measurement uncertainty of machining centre DMC 635V with a Heidenhain TNC 620 control system equipped with a DMG TS649 probe as well as to compare test results with the measurement system: machining centre DMC 635V with a FANUC 0iMC control system – OMP60 touch trigger probe.

EXPERIMENTAL SETUP

The research consisted of determining one-dimensional measurement accuracy (repeatability and accuracy) for each axis of the machine tool using end gauges. This required measurement-taking of gauge blocks of length equal to 50 mm, 75 mm and 100 mm along each machine tool axis (X, Y, Z).

The tests were conducted in The State School of Higher Education in Chełm. The measurement system consisted of a DMC 635V machining centre with the Heidenhain TNC 620 control, equipped with DMG TS 649 touch probe. The maximum travel of the table and machine tool spindle are as follows: for the X axis: 0 – 635 mm, for the Y axis – 510 mm, and for the Z axis – 460 mm. The probe technical specifications are presented in Table 1.

In order to program the measuring motions, measurement cycles implemented in the control system of the machining tool were used. The length of the gauge block along X and Y axes was measured using measuring cycle 426. Gauge block measurement parallel to the Z axis was indirect. The surface of the gauge block clamped to the machine table, in order to eliminate the influence of the table's geometric features, was adopted as the measuring base. The gauge block was connected with an identical gauge block already fixed on the table to form a stack. The 427 measuring cycle was used for measurement. An example of the 426 measuring cycle (50 mm long gauge block) is presented in Figure 1. and the process of measurement is presented in Figure 2.

Prior to the measurement, gauge blocks were positioned at the machine table and left for 24 hours with the aim of temperature equalization. In accordance with the Central Office's of Measures directions, measurements of each gauge block were repeated thirty times [19].

Table 1. Technical specifications of DMG TS 649 probe

Sense directions	±X, ±Y, ±Z
Signal transmission	infrared radiation $0^0 - +30^0$
Probe accuracy with a standard gauge plunger	≤ ± 5 μm
Unidirectional repeatability 2σ with sampling speed equal to 1 m/min	≤ 1 μm
Stylus trigger force: XY +Z	 1 N 8 N
Maximum measuring speed	5 m/min

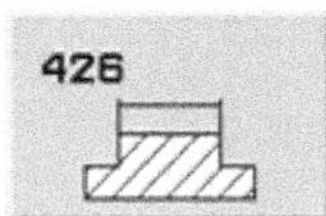


```
TCH PROBE 426 POMIAR MOSTKA ZEWN.
 Q263=+50    ;1.PKT POMIAROW 1.OSI
 Q264=+4.5   ;1.PKT 2.OSI
 Q265=+0     ;2-GI PUNKT W 1. OSI
 Q266=+4.5   ;2-GI PUNKT W 2. OSI
 Q272=+1     ;OS POMIAROWA
 Q261=+5     ;WYSOKOSC POMIARU
 Q320=+5     ;BEZPIECZNA WYSOKOSC
 Q260=+30    ;BEZPIECZNA WYSOKOSC
 Q311=+50    ;ZADANA DLUGOSC
 Q288=+52    ;MAKSYMALNY WYMIAR
 Q289=+48    ;MINIMALNY WYMIAR
 Q281=+1     ;PROTOKOL POMIARU
 Q309=+0     ;PGM-STOP JESLI BLAD
 Q330=+0     ;NARZEDZIE
```

Fig. 1. Structure of 426 measuring cycle of 50 mm gauge block measurement

RESULT ANALYSIS

The conducted tests allowed calculating: arithmetic mean, standard uncertainty u_a, combined standard uncertainty u, expanded uncertainty U as well as correction P_{Ex}. Measurement results for individual machine tool axes are presented in Table 2.

The calculations were conducted by the following formulas in accordance with EA-4/02 [19]:

- arithmetic mean $\overline{L}$ of a series of measurements

$$\overline{L} = \frac{1}{n}\sum_{i=1}^{n} L_i \tag{1}$$

where: L_i – measured length of gauge block.

a)  b) 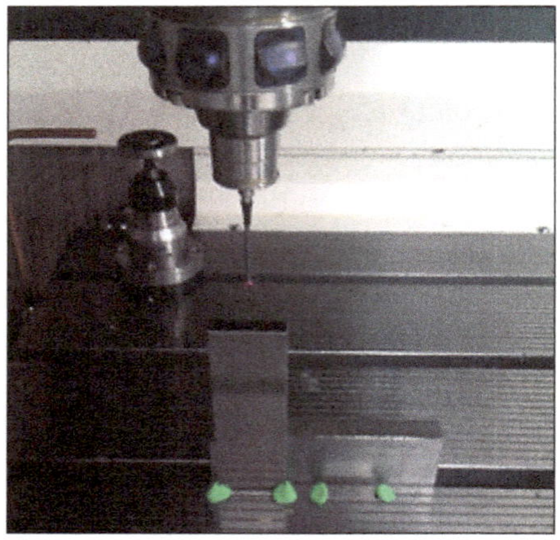

Fig. 2. Measurement of gauge block length: a) X axis, b) Z axis

- systematic indication error E_x:

$$E_x = \overline{L} - L_N \quad (3)$$

where: L_N – nominal length of gauge block.

- correction P_{Ex}:

$$P_{Ex} = -E_x \quad (4)$$

- standard uncertainty u_a:

$$u_a = \sqrt{\frac{\sum_{i=1}^{n}\left(L_i - \overline{L}\right)^2}{n(n-1)}} \quad (5)$$

where: n – number of measurements,
L_i – length of gauge block in i – this measurement.

- combined standard uncertainty u:

$$u = \sqrt{u_a^2 + u_L^2 + u_R^2} = \sqrt{u_a^2 + \frac{te^2}{6} + \frac{r^2}{3}} \quad (6)$$

where: u_a – standard uncertainty,
u_L – is the uncertainty of the gauge length,
u_R – is the resolution uncertainty,
t_e – limit deviation of the gauge block length,
r – variation of indications.

The triangular probability distribution was assumed, with the centre in the nominal dimension of the gauge block. The measurement system resolution equalled 0.001 mm. The variability of indications was estimated at r = ±0.0005 mm, assuming the rectangular probability distribution. The values of t_e for the gauge blocks used in the tests (for L_N nominal values) had the following values: t_e = ±0,4 mm (for L_N = 50 mm), t_e = ±0,5 mm (for L_N = 75 mm), t_e = ±0,6 mm (for L_N = 100 mm).

- expanded uncertainty U:

$$U = k \cdot u \quad (7)$$

where: k – coverage factor,
u – combined standard uncertainty. The coverage factor is commonly set to 2. This corresponds to expanded certainty at the trust level of 95%.

The analysis of results presented in Table 2 indicates that uncertainty of measurement using inspection probe DMG TS 649 is much lower than the systematic indication error determined for each analysed controlled axis. It is furthermore noteworthy that the standard uncertainty is lesser than the one-dimensional repeatability 2σ as declared by the manufacturer.

The combined standard uncertainty and extended uncertainty analysis proves that regardless of the length of the gauge block measured or the type of controlled axis, their variation range does not exceed 0.0002 mm, which makes it an exceptionally good result. At the same time, those values are considerably lower than measurement accuracy declared by the probe manufacturer.

When analysing systematic indication error E_x what should be highlighted is that for X axis its value equalled ±2 mm, for Y axis 2.5 mm, and for Z axis it produced the highest values, approximately 11 mm. High values of systematic indication error for Z axis (when compared to values of this error for X and Y axes) may indicate poor calibration of probe length, which furthermore seems to be corroborated by the fact that P_{Ex} correction value was practically identical in the case of Z axis, regardless of the length of the analysed gauge block. Test results are presented in the form of graphs (Figures 3–5).

The main goal of this article was to compare measurement uncertainty of two different measuring systems. One of them is DMC 635V machine tool with DMG TS 649 touch probe, and the

Table 2. Measurement results analysis of gauge block lengths on numerically controlled vertical milling machines

Measuring system	DMC 635V - Heidenhain TNC620. DMG TS 649			FV 580A - Fanuc 0iMC OMP 60
	X Axis			
Gauge block dimension. mm	**50**	**75**	**100**	**100**
Mean. mm	49.9984	74.9981	100.0014	100.011
Standard uncertainty u_A. mm	0.0001	0.0001	0.0001	0.0004
Combined standard uncertainty u. mm	0.0003	0.0004	0.0004	0.0006
Extended uncertainty U. mm	0.0007	0.0007	0.0008	0.0112
Systematic indication error E_x mm	-0.0016	-0.0019	0.0014	0.011
Correction P_{Ex}. mm	0.0016	0.0019	-0.0014	-0.011
	Y Axis			
Gauge block dimension. mm	**50**	**75**	**100**	**100**
Mean. mm	49.9979	74.9956	99.9948	100.009
Standard uncertainty u_A. mm	0.0001	0.0001	0.0001	0.0004
Combined standard uncertainty u. mm	0.0003	0.0004	0.0004	0.0005
Extended uncertainty U. mm	0.0007	0.0007	0.0008	0.0010
Systematic indication error E_x mm	-0.0021	-0.0040	-0.0050	0.009
Correction P_{Ex}. mm	0.0021	0.0040	0.0050	-0.009
	Z Axis			
Gauge block dimension. mm	**50**	**75**	**100**	**100**
Mean. mm	50.0110	75.0111	100.0114	100.041
Standard uncertainty u_A. mm	0.0001	0.0001	0.0001	0.0001
Combined standard uncertainty u. mm	0.0003	0.0004	0.0004	0.0004
Extended uncertainty U. mm	0.0007	0.0007	0.0008	0.0008
Systematic indication error E_x. mm	0.0110	0.0111	0.0114	0.041
Correction P_{Ex}. mm	-0.0110	-0.0111	-0.0114	-0.041

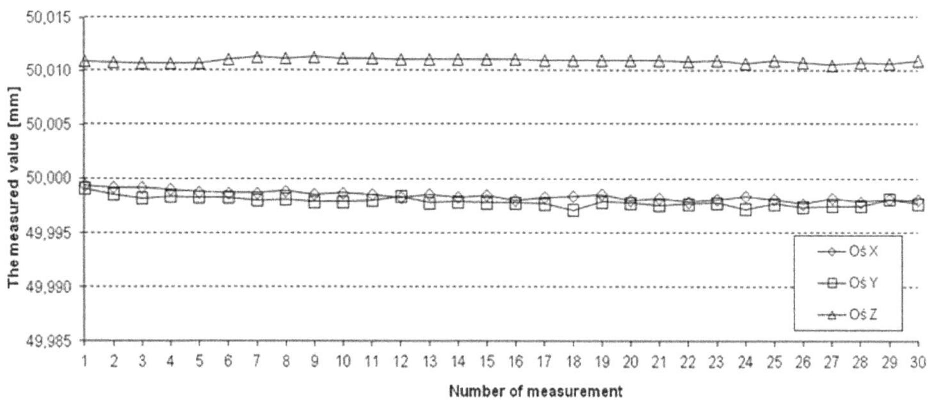

Fig. 3. Measurement results of 50 mm gauge block on DMC 635V machine tool with a DMG TS 649 touch probe

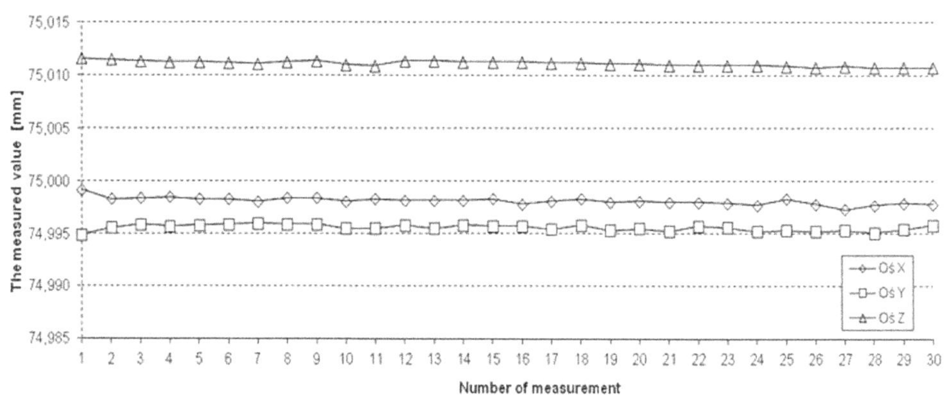

Fig. 4. Measurement results of 75 mm gouge block on DMC 635V machine tool with DMG TS 649 touch probe

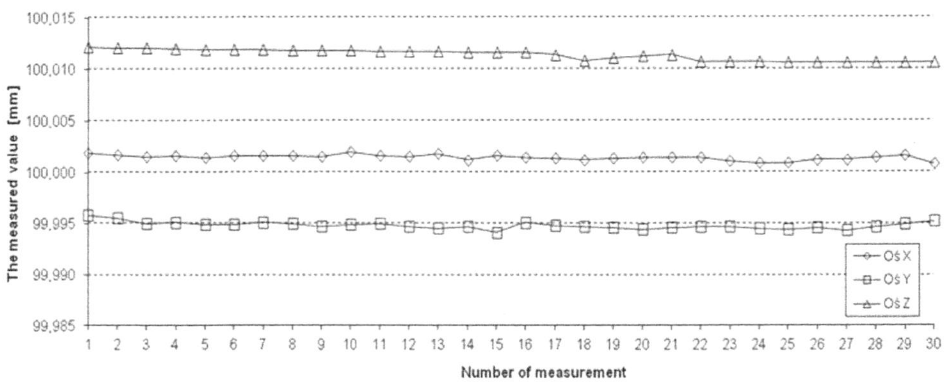

Fig. 5. Measurement results of 100 mm gouge probe on DMC 635V machine tool with DMG TS 649 touch probe

other one is FV 580A machine tool with OMP60 touch probe. Measurement uncertainty analysis was conducted on a 100 mm gauge probe. Graphic representation of the measurement results is shown in Figures 6–8.

What should be unambiguously stated when analysing Figure 6. and 7. is that the scatter of results is greater for FV 580A machine tool. This may indicate inaccurate machine tool spindle positioning. This situation may be a consequence of worse geometric and kinematic accuracy resulting from machine tool wear (over 700 hours of operation). DMC 635V is a new machine tool. It may be, therefore, assumed that measurement accuracy depends largely on the operation life of the machine tool.

Figure 8. shows that scatter of results along Z axis is similar for both machine tools. However, a considerable difference can be noticed in systematic indication errors. As mentioned previously, this can result from poorly calibrated probe length in both of the analysed instances.

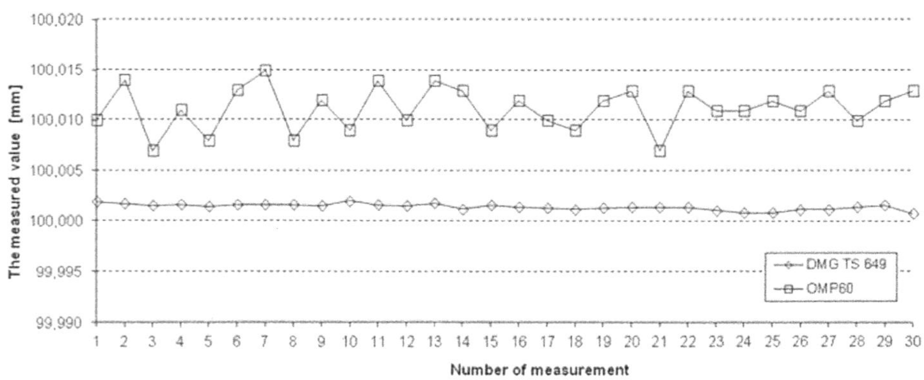

Fig. 6. Comparison of measurement results of 100 mm gauge block along X axis on DMC 635V machine tool with DMG TS 649 touch probe and FV 580A machine tool with OMP60 touch trigger probe

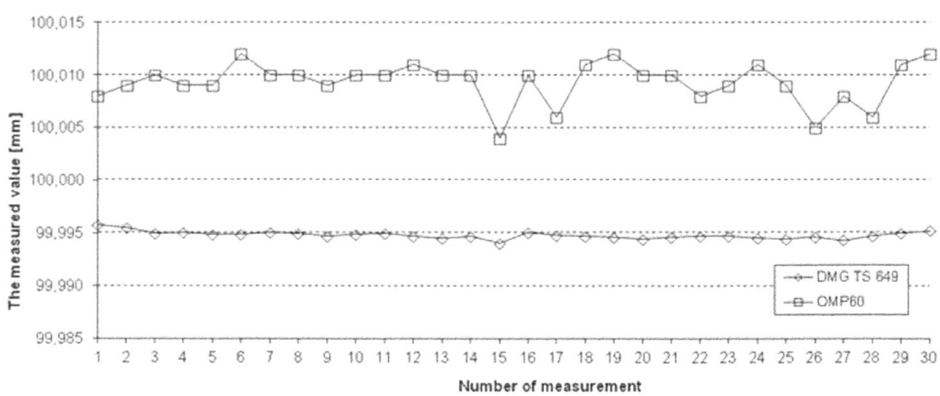

Fig. 7. Comparison of measurement results of 100 mm gauge block along Y axis on DMC 635V machine tool with DMG TS 649 touch probe and FV 580A machine tool with OMP60 touch trigger probe

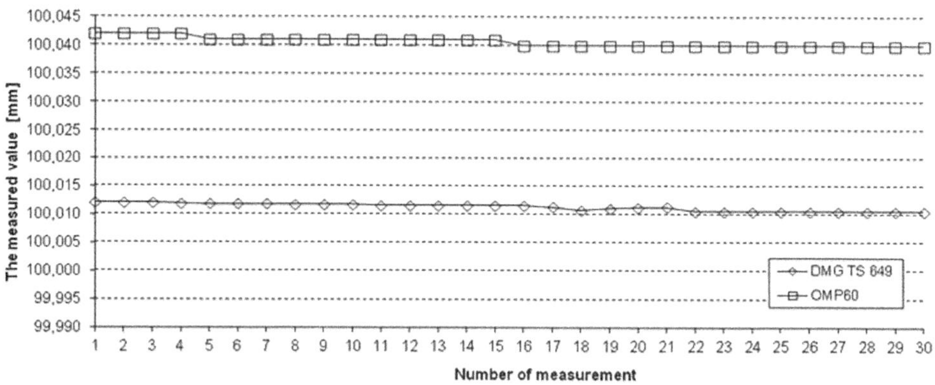

Fig. 8. Comparison of measurement results of 100 mm gauge block along Z axis on DMC 635V machine tool with DMG TS 649 touch probe and FV 580A machine tool with OMP60 touch trigger probe

CONCLUSIONS

Based on the measurement results for two different machine tools with different exploitation time, the following observations concerning measurement uncertainty of both systems can be made:

1. Exploitation time is a key factor influencing geometric and kinematic accuracy of the machine tool, which influences uncertainty of measuring systems measurements with the application of inspection probes. CNC machine tools require constant control of the machine condition, especially with the occurrence of collision.
2. Regular scatter of results analysis with end gauges can prove beneficial to diagnosing technical condition of machine tools, in particular the accuracy of spindle positioning. This method does not seem to demand extensive amount of labour and its cost is considerably low. An increase in scatter of results may be a signal for detailed diagnosis with the application of specialist diagnostic equipment (e.g. Ballbar QC20).
3. Probe calibration cannot be disregarded though. Measurement results show that careless calibration may cause systematic errors which lower the 'quality' of measurements. Z axis provides an ample evidence here. For this reason, calibration accuracy tests of the probe should be conducted. This would allow eliminating factors causing incorrect correction data entering into the machine tool control system.
4. The application of inspection probes on CNC machine tools should induce constant measurement accuracy diagnostics. This type of error can influence workpiece geometric features forming, especially the ones that are produced in a number of fixings. Determining zero point of the workpiece in each fixing accumulates errors arising from subsequent measurements. Consequently, the accuracy of geometric features location of the workpiece is affected.

REFERENCES

1. Duan X., Wang X., Liu W., Liu H, Huo Y.: Development of on-line inspection system for machine center based on CAD. International Conference on Automation and Logistics. Qingdao, China 2008.
2. Jacniacka E., Semotiuk L., Babkiewicz M.: Wyznaczenie dwuosiowej niepewności pomiaru wewnątrzobrabiarkowego systemu pomiarowego z zastosowaniem sondy OMP60. Pomiary Automatyka Robotyka, 12, 2012.
3. Jacniacka E., Semotiuk L., Pieśko P.: Niepewność pomiaru wewnątrzobrabiarkowego systemu pomiarowego z zastosowaniem sondy OMP60. Przegląd Mechaniczny, 6, 2010.
4. Jacniacka E., Semotiuk L.: Odkształcenia cieplne a niedokładność pomiaru sondą przedmiotową. Pomiary Automatyka Kontrola, 9, 2011, 985-988.
5. Jacniacka E., Semotiuk L.: Powtarzalność mocowania jako składnik budżetu niepewności pomiaru sondą przedmiotową na obrabiarkach CNC. Mechanik, 5-6, 2012, 456-459.
6. Jacniacka E., Semotiuk L.: Wpływ eksploatacyjnego zużycia łożysk wrzeciona frezarki sterowanej numerycznie na niepewność pomiaru sondą inspekcyjną, Przegląd Mechaniczny, 3, 2012, 22-27.
7. Józwik J., Gęca T.: Wykorzystanie tensometrycznej sondy pomiarowej OMP400 do automatycznej kontroli międzyoperacyjnej i korekcji błędów wymiaru na obrabiarkach CNC. Mechanik, 1, 2012.
8. Józwik J.: Niepewność wyniku pomiaru średnicy frezów laserową sondą narzędziową NC4. Pomiary Automatyka Robotyka, 1, 2009.

9. Kamieńska-Krzowska B., Semotiuk L., Czerw M.: Analiza możliwości zastosowania sondy przedmiotowej do kontroli czynnej na pionowym centrum obróbkowym FV580A. Acta Mechanica at Automatica, 1(2) 2007.

10. Kumara S., Nassehia A., Newmana S.T., Allenb R.D., Tiwaric M.K.: Process control in CNC manufacturing for discrete components: A STEP-NC compliant framework. Robotics and Computer-Integrated Manufacturing, 23, 2007, 667–676.

11. Kwon Y., Tseng T.L., Ertekin Y.: Characterization of closed-loop measurement accuracy in precision CNC milling. Robot Computer-Integrated Manufacturing, 22(4), 2006, 288–296.

12. Kwona Y., Jeongb M.K., Omitaomu O.A.: Adaptive support vector regression analysis of closed-loop inspection accuracy. International Journal of Machine Tools & Manufacture, 46, 2006, 603–610.

13. Lei W.T., Hsu Y.Y.: Accuracy test of five-axis CNC machine tool with 3D probe-ball. Part I: Design and modeling. International Journal of Machine Tools & Manufacture, 42, 2002, 1153–1162.

14. Lei W.T., Hsu Y.Y.: Accuracy test of five-axis CNC machine tool with 3D probe-ball. Part II: Errors estimation. International Journal of Machine Tools & Manufacture, 42, 2002, 1163–1170.

15. Lei W.T., Hsu Y.Y.: Error measurement of five-axis CNC machines with 3D probe–ball. Journal of Materials Processing Technology, 139, 2003, 127–133.

16. Sydor P., Józwik J: Ocena niepewności standardowej i rozszerzonej pomiaru laserową sondą narzędziową obrabiarki CNC. Международная научно-техническая конференция – Прогрессивные направления развития машино-приборостроительных отраслей и транспорта, Издательство Сев НТУ, Ukraina, Sewastopol 2008.

17. Wang W., Lu K., Chen Z.: Study on edge detection based on sampling in CNC non-contact measuring system. International Technology and Innovation. Conference, Hangzhou, China 2006.

18. Zhao F., Xu X., Xie S.: STEP-NC enabled on-line inspection in support of closed-loop machining. Robotics and Computer-Integrated Manufacturing, 24, 2008, 200–216.

19. Dokument EA-4/02. Wyrażanie niepewności przy wzorcowaniu. GUM, Warszawa 1999.

10

THE TECHNIQUE OF FLOW CYTOMETRY IN DIAGNOSTIC RESEARCH

Arkadiusz Macheta[1], Katarzyna Radko[1], Agnieszka Szymczyk[2], Monika Podhorecka[1]

[1] Department of Hematooncology and Bone Marrow Transplantation, Medical University of Lublin, Al. Racławickie 1, 20-059 Lublin, Poland, e-mail: arek.macheta@gmail.com; katarzyna.radko@gmail.com; monika.podhorecka@onet.pl

[2] Students Scientific Association at Department of Hematooncology and Bone Marrow Transplantation Medical University of Lublin, Al. Racławickie 1, 20-059 Lublin, Poland, e-mail: agusia_szymczyk@wp.pl

ABSTRACT

Flow cytometry is a technology that simultaneously counts and then examines multiple physical and/or chemical characteristics of single particles, usually cells, as they flow in a fluid stream through a beam of light. The properties measured include a particle's relative size, relative granularity or internal complexity, and relative fluorescence intensity. These characteristics are determined using an optical-to-electronic coupling system that records how the cell scatters incident laser light and emits fluorescence. One of the most significant applications is immunophenotyping of cells - the most important tool in diagnosis and monitoring haematological disorders, such as acute and chronic leukemia, lymphoma, monoclonal gammopathy, myelodisplastic and myeloproliferative diseases.

Keywords: flow cytometry, immunophenotype, hematooncology.

INTRODUCTION

Flow cytometry technique dates back to the 1930s, when *Science* magazine published an article on the possibility of using a photoelectric sensor in counting the cells flowing through a capillary tube. In fact, though, ten or more years had passed from when the idea was conceived to when the design works began [9, 16]. Originally, cytometric tests were focussed on detecting traces of biological warfare toxins, which explains the commitment of the American Army to develop this technique during the Second World War. The prototype of a modern-day cytometer was composed of a basic dark-field flow chamber, with the brightest lamp available at that time – the Ford headlight – providing the source of visible light [15]. Further design and methodological research focussed on numerous areas of perfection and the use of flow cytometry. The most interesting is the fact that following the recent terrorist attacks history is making a full circle – now the American Army is again interested in the usefulness of the technology discussed for the detection of microorganisms in the air [2].

THE FLOW-CYTOMETRY TECHNIQUE

Flow cytometry is an advanced technology involving the quantitative and qualitative multiparameter assessment of individual cells [3]. The analysis consists of the identification of surface or intracellular antigens by means of specific monoclonal antibodies conjugated to fluorescent compounds, the so called-fluorochromes. Cells marked with these compounds are irradiated with a laser beam while they flow through the measurement chamber, as a result of which they disperse the light and emit it through activated fluorochromes. By means of a system of photomultipliers, light impulses are converted into electrical impulses, which in turn are subjected to analysis using a computer system compatible with the cytometer [3, 7, 17]. Coordination with the main construction systems of the flow cytometer, that is with the hydraulic, optical and electronic systems, ensure the great measurement potential of this complex measuring apparatus. This technique is a rapid and sensitive method for the diagnosis and monitoring of the clinical course of diseases of the haematopoietic system [3].

The Hydraulic System

The role of the fluid-transfer system is to transport the suspension of the examined cells to the measurement chamber. When the suspension flows through the chamber it is surrounded by an isotonic sheath fluid and prevents the cells from adhesion. Measurements are conducted during the transfer of the compounds suspended in a stream of fluid through a tapering microcanal. In line with the principle of hydrodynamic focussing, the construction of the cytometer enforces the laminar flow of the analysed cells and their concentration, which is low enough to allow them to flow one by one into the area irradiated by a focussed light beam, exactly in the middle of the hydraulic canal (Fig. 1) [3, 11].

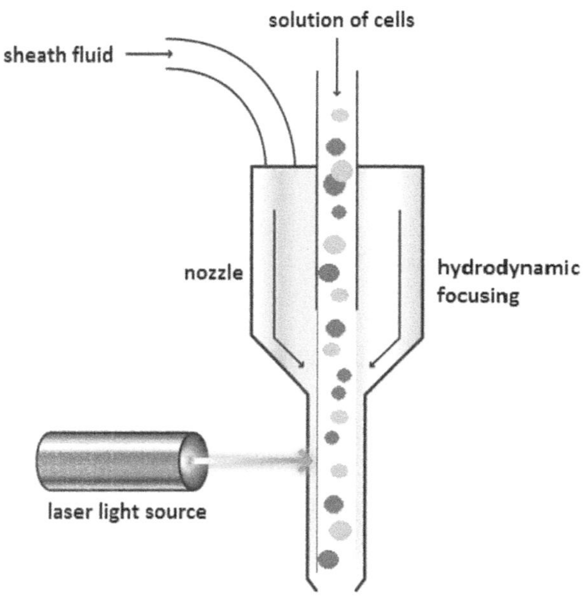

Fig. 1. The hydraulic module in flow cytometry

The Optical System

The role of the optical system is to generate and collect light signals – i.e. dispersion and fluorescence. Initially, the beam of light is pointed at the cells flowing through the chamber, perpendicularly to their flow direction. The type of the source of light used depends on the manufacturer - these are in the majority fixed-wavelength lasers, e.g. argon ion (488 nm), krypton, or helium-neon lasers (Fig. 2.) [7, 11, 13]. The mercury-vapour lamps that were commonly used in older types of cytometers (ICP, Phywe) for the production of light beams are nowadays supported by laser technology. Increasingly more devices have more than one source of light [8, 13].

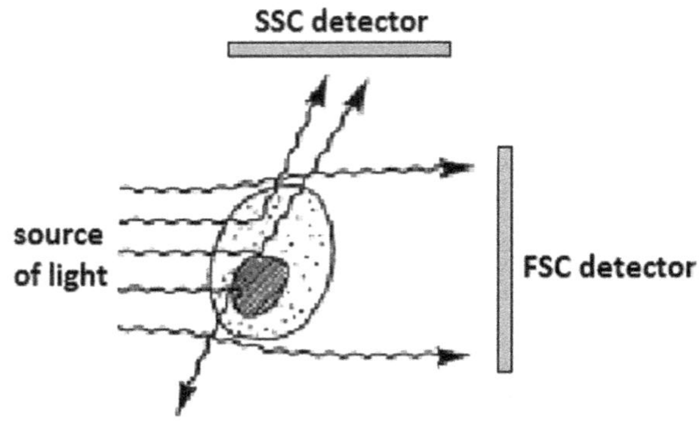

Fig. 2. The optical module in flow cytometry

The Electronic System

The electronic system in flow cytometry converts light signals into electronic impulses, which in turn will be converted into graphic signs (PMT). It comprises devices for converting signals into values, i.e. into digital images, and a computer system for storing and processing the resultant data. It contains a set of photomultipliers, signal amplifiers and an analogue-digital converter. The reception of signals is by means of forward scatter (FSC) detectors, side scatter (SSC) detectors and fluorescence detectors, because the cells diffract and disperse the light when they cross the laser light beam. The intensity of light dispersed at the edges of the analysed cell in the same direction as the laser beam enables the FSC detector to divide the cells according to their size. Based on the diffraction of light at an angle of 90 degrees towards the laser beam on the cell membrane and in the intracellular structures, the SSC detector divides the cells according to their shape and granularity [10]. After marking the cells with fluorescent dyes the intensity of the fluorescence can also be measured. The laser light beam forces a brief transfer of electrons to an orbit of the atom with lower energy, i.e. with a wavelength greater than the length of the light wave. The phenomenon of fluorescence is the result of electrons transferring back to their original orbit. It is important to employ fluorochromes with absorption maxima within the range of the wavelength of the laser in our cytometer in the tests [10, 13]. Subsequently, light intensity is measured using appropriate detectors and the resultant signals are amplified by the photomultipliers, formed and send to the computer.

The Computer Presentation of the Results

The data analysis module allows comprehensive data processing, with several types of charts available for the visualisation of results:

- one-dimensional charts (histogram)
- two-dimensional charts (dot plot (Fig. 3), density plot, contour plot)
- three-dimensional charts (perspective charts)

During the analysis, it is possible to gate a group of cells, to subject the group of cells to quantitative assessment in every area and the calculation of various statistical values, which means that they are repeatable and accurate, collected in a short period of time and based on the analysis of several thousand cells [6].

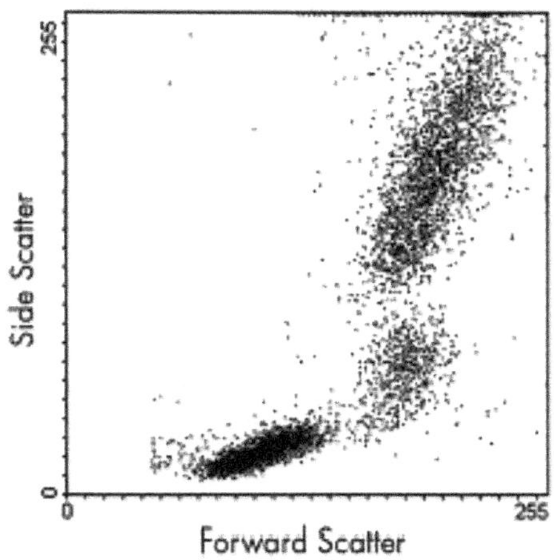

Fig. 3. Two-dimensional dot plot for peripheral blood

THE APPLICATION OF FLOW CYTOMETRY IN HAEMATOLOGY

Flow cytometry enables the assessment of the morphological features of cells with the determination of their maturity level and diversity. In clinical diagnostics, this method is mainly applied in the test concerning diseases of the haemopoietic system – thanks to developed panels of antibodies, flow cytometry helps not only to assess the immunophenotype of proliferating cells but also to establish their origin. Medicine employs a panel of monoclonal antibodies enabling the assessment of the affiliation of cells to a line and stage of haematopoietic development. The classification of affiliation to the line and the recognition of the differentiation stage are based on establishing the co-expression of some antigens and the phenomenon of the disappearing group of antigens with others emerging in the subsequent stages of normal cell maturation. The earlier stage of differentiation and maturity is disrupted, the more difficult it becomes to diagnose the cell origin is only on the basis of morphology [4, 5, 6, 19] (Table 1).

Table 1. Basic immunophenotype markers of peripheral blood cells [12]

CD45	All leucocytes
CD19, CD20	B lymphocytes
CD3, CD7	T lymphocytes
CD4	Helper T lymphocytes
CD8	Suppressor T lymphocytes
CD56	NK cells
CD25, CD71, CD69	Lymphocyte activation
CD13, CD33	Myeloid differentiation
CD15	Granulocytes
CD14	Monocytes
CD34	Stem cells (progenitor cells)

The technique of flow cytometry has many advantages, including the possibility to perform examination using a small amount of biological material, even without their initial treatment, and the concurrent identification of surface and intracellular markers [3, 14]. It enables testing in the *in vitro* setting, whereas it is possible to run analysis only of the cells in the suspension. Therefore, the most convenient materials are blood or body-fluid cells. However, the analysis of solid-tissue cells is also viable after precise fragmentation and obtaining single and undamaged cells. It is also crucial to prepare a sample properly, especially in the case of assessing cellular functions. The most reliable results of analyses are obtained after testing freshly-taken material and, if it is necessary to store samples, cells should be fixed and stored in conditions preventing them from degradation; otherwise non-specific bindings of damaged dead cells with monoclonal antibodies may occur and, consequently, false results may be generated. Thanks to its broad applicability, flow cytometry is considered to be the basis for diagnostics and disease monitoring. Moreover, it is useful in assessing phenotypes in physiology, as it facilitates quantitative and qualitative assessment of antigen expression in various cell populations, even in very heterogeneous ones. The basis for routine cytometric diagnostics includes the identification of the presence of particular subpopulations and their physical features, assessing the absolute amount of population cells tested and the percentage of particular populations. There are conditions where assessing immunophenotype is helpful and sometimes even essential to recognise or monitor a disease, as exemplified by:

- proliferative disorders (especially lymphomas and leukemias),
- immunodeficiency syndromes (congenital and acquired),
- organ transplantations,
- autoimmune diseases,
- diseases treated with immunosuppression,
- cancers [2, 19].

The employment of this technique is not only limited to determining immunophenotype, as exemplified by its application in medicine involving the testing of platelets of the whole blood, which helps observe accurately the status of thrombocytes in haemorrhagic diathesis, or hypercoagulability or assessing the effectiveness of antiaggregation treatment, or assessing the functions of phagocytes through the assessment of phagocytosis, chemotaxis, or producing reactive oxygen species, or identifying chromosomal mutations along with the *in situ* hybridsation technique (FISH) [1, 2].

This technique is unquestionably of a great importance in the diagnosis of lypmhoma and leukemia. Cell subpopulations are characterised in detail and classified based on cell differentiation, surface and cytoplamsatic antigens. This brings unique possibilities for describing cancer and its biology on the basis of molecule expression in the process of neoplasia and provides opportunities to identify the stage of the development of the cell on which the cancer is growing, and the degree of its maturity, and helps determine the functional features, such as cytotoxicity or the production of immunoglobulins. As regards of lymphomas and leukemias, flow cytometry is for:

- providing detailed diagnosis,
- confirming/excluding residual disease,
- as the method of choice for monitoring anticancer therapies,
- controlling cell transplants,
- assessing the regeneration of the bone-marrow microenvironment [18].

CONCLUSIONS

The flow cytometry technique developing is the result of cooperation between many research teams, especially biologists and engineers. The first devices enabled cellular analysis with regard to size and granulation. At present, the modern cytometer is even equipped with several dozen detectors for collecting data, and several lasers with various light lengths and complex optoelectronic systems. Contemporary equipment provides possibilities for several simultaneous, multicoloured testing of up to 10 fluorescent markers and two parameters of dispersion in a stream of cells transported at the speed of up to 70000/sec. So, we are involved in significant analytical options burdened with the risk of error and difficulties in comparing the obtained data. Particularly important are the parameters of the flow cytometer itself. Its model and type of software used should be included in the methodology. The efficiency of laser lines and filters of emissions used with the appropriate fluorochromes should be also employed. The use of many dyes increases the risk of disruptions resulting from overlapping spectra. This can be avoided thanks to the so-called compensation, i.e. narrowing the spectrum of detection to a very narrow wave range. This function is set into a cytometer during the collection of data and it should be included in the description of an experiment, e.g. the size of samples, statistical data and the minimum detection threshold. The lack of possibility to assess the topographical origin of cells is an impediment to the diagnostics and is a form of disadvantage for flow cytometry. Sometimes the necessity to deliver live and undamaged cells is also problematic. The technique of flow cytometry is becoming more and more useful in diagnostic research in haematology and other fields of medicine, although it is also commonly applied in many various areas of biological research.

REFERENCESS

1. Baj Z. Wykorzystanie cytometrii przepływowej w badanich płytek krwi. Central European J. of Immunology, 21, 1996, 119-127.
2. Baran J. Nowa epoka cytometrii przepływowej – przewodnik po współczesnych cytometrach i ich zastosowanie. Post. Biol. Kom., 24, 2008, 3-15.
3. Brown M, Wittwer C. Flow cytometry: principles and clinical applications in hematology. Clin. Chem., 46, 2000, 1221-1229.

4. Elghetany M.T., Patel J, Martinez J, Schwab H. CD87 as a Marker for Terminal Granulocytic Maturation: Assessment of its Expression During Granulopoiesis. Cytometry B, 51B, 2003, 9-13.
5. Fujimoto H, Sakata T, Hamaguchi Y, Shiga S, Tohyama K, Ichiyama S, Wang F, Houwen B. Flow Cytometric Method for Enumeration and Classification of Reactive Immature Granulocyte Populations. Cytometry, 42, 2000, 371-378.
6. Kaczmarek A, Osawa T, Leporowska E, Mackiewicz A. The role of flow cytometry in clinical diagnosis. Współcz. Onkol., 6, 2002, 366-373.
7. Kawiak J., Skierski J. S.: Cytometria przepływowa w badaniach komórek i niektóre jej zastosowania lekarskie. Post. Biol. Kom., 19, 1992, 2339-2354.
8. Mandy F.F., Bergron M, Minkus T. Principles of flow cytometry. Transfus. Sci. 16, 1995, 303-314.
9. Moldavan A. Photo-electric techniques for counting microscopical cells. Science, 88, 1934, 188-189.
10. Nunez R. Flow cytometry: principles and instrumentation. Curr. Issues Mol. Biol., 3, 2001, 39-45.
11. Owens M.A., Loken M.R. Flow cytometry principles for clinical laboratory practice. Quality assurance for quantitative immunophenotyping. Wiley-Liss, Inc. 1995.
12. Pituch-Noworolska A. Zastosowanie cytometrii przepływowej do diagnostyki białaczek i chłoniaków nieziarniczych. Central European J. of Immunology, 21, 1996, 138-146.
13. Rytwiński K, Wnuk A. Cytometria przepływowa, zasady działania, zastosowanie – ze szczególnym uwzględnieniem pediatrii. Medycyna Wieku Rozwojowego, 2, 1999, 281-290.
14. Sędek Ł, Mazur B. The application of monoclonal and policlonal antibodies in flow cytometry. Post. Biol. Kom., 24, 2008, 17-34.
15. Stopa P.J. The flow cytometry of Bacillus anthracis spores revisited. Cytometry, 41, 2000, 237-244.
16. Tucker F.T, O'Konsky C.T., Pickard H.B., Pitts J.N. A photoelectric counter for colloidal particles. J. Am. Chem. Soc., 69, 1947, 2422-2431.
17. Weiss D.J. The application of flow cytometric techniques to veterinary clinical hematology. Vet. Clin. Pathol., 31, 2002, 72-82.
18. Żeromski J, Sikora J, Dworacki G. Utility of flow cytometry in diagnostics and monitoring of lymphomas. Post. Biol. Kom., 24, 2008, 113-140.
19. Żydowicz G, Mazur B. Cells immunophenotype in normal hematopoiesis. Post. Biol. Kom., 24, 2008, 35-44.

11

DAMAGES TO INJECTORS IN DIESEL ENGINES

Piotr Ignaciuk[1], Leszek Gil[2]

[1] Institute for Transport, Internal Combustion Engines and Ecology, Lublin University of Technology, Nadbystrzycka 36, 20-618 Lublin, Poland, e-mail; p.ignaciuk@pollub.pl

[2] Department of Transport and Computer Science, University of Economics and Innovation in Lublin, Projektowa 4, 20-209 Lublin, Poland, e-mail: leszek.gil@wsei.lublin.pl

ABSTRACT

The article describes damages to high pressure injectors used in common rail injection systems. The conducted analysis of their causes includes the diagnosis of injectors on a test bench and the results of microscopic research of damaged components. The tribological damages of high pressure injectors are local and cavitations pitting. The place of cavitations pitting are mainly check valves, where the reduction in the quantity of injected fuel is forming.

Key words: tribological damages, fuel injectors, common rail, compression ignition engine.

INTRODUCTION

The development of the modern diesel engines is now determined by the limits of toxic exhaust emissions. One of the limited toxic components is the particulate matter (PM). Reduced emission of particulates is promoted by very good fuel atomization and high temperature of combustion process. Obtaining the proper atomization of the fuel is only possible using small gaps (holes) in the ends of the injectors in conjunction with high-pressure injection. Currently used common rail injection systems are characterized by high injection pressures of up to 2200 bar. The amount of fuel per one cycle is usually divided into 3 to 5 or even 7 parts. Fragmentation of injected fuel dose allows for controlling the combustion process. Therefore, the components of such injectors perform more moves (opening – closing) than in classic injection systems with no fragmentation of injected fuel. This type of work is conducive to reduce wear and durability of the injectors. Currently durability of injectors is estimated at more than 60 million engine cycles [4, 5].

The injector is therefore a critical part of modern injection system, determining the correctness of its work. There are two types of injectors: solenoid injectors (Figure 1), which allow to divide the injected dose of fuel into 5 parts and piezoelectric injectors (Figure 2), in which the dose can be divided into up to 7 parts. Injectors controlled with solenoid valves have check valves and spring arrangement, which is the cause of some inertia in their work. Piezoelectric injectors can be made in versions without any springs and check valves. Reduction in the number and weight of moving parts (up to 75%) reduces the inertia of the injectors and allows for the increase in the accuracy and speed of fuel delivery [5].

RESEARCH OBJECT

For the research, electromagnetic and piezoelectric Bosch injectors were used. The injectors were installed in Mercedes-Benz Sprinter LCV. The construction of the surveyed injectors is shown in Figures 1 and 2.

The injectors have been removed from damaged engines, and then diagnosed. Table 1 summarizes the observed signs of damage and vehicle mileages for the given type of type injectors.

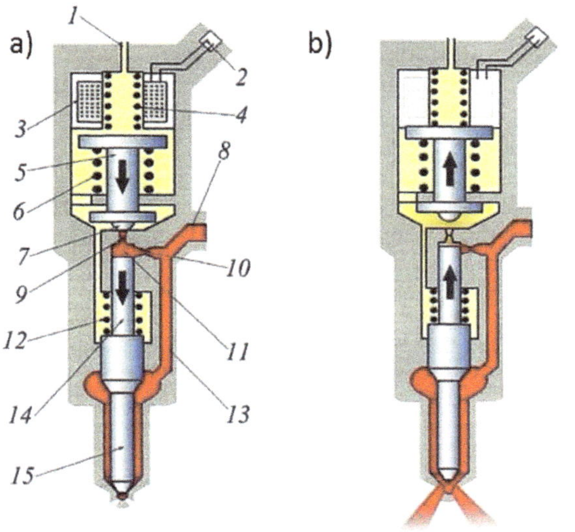

Fig. 1. Cross section of an electromagnetic Bosch injector used in common rail systems of I and II generation; a) injector closed, b) fuel injection into combustion chamber: 1 – drain channel of excess fuel, 2 – electrical connector, 3 – coil, 4 – upper plunger spring, 5 – piston, 6 – lower plunger spring, 7 – ball valve, 8 – high pressure fuel supply channel, 9 – drain gland, 10 – inlet gland, 11 – valve control chamber, 12 – spray nozzle spring, 13 – nozzle fuel supply channel, 14 – valve control plunger, 15 – nozzle needle [3]

Table 1. Investigated injectors and their failure symptoms

No.	Injector type	Mileage till damage	Failure symptom
1.	Electromagnetic	423 000 km	Inability to start the engine
2.	Piezoelectric	168 000 km	Engine hard to start
3.	Piezoelectric	224 000 km	Uneven engine work at idle

DIAGNOSTIC MEASUREMENTS

Diagnostic measurements of electromagnetic injectors was done on Bosch EP 200 tester. Diagnostic measurements included measurement of fuel doses. The obtained values should fall within the proper range for a given type of injector. In addition, the injectors were checked for leaks. The entire diagnostic process takes place automatically. Piezoelectric injectors were tested using Bosch EPS 945 tester. The diagnostic unit checks injector tightness, and then collects and uploads the dosing characteristics into the memory of injector (this is a new IMA code). It should be emphasized that piezo injectors are generally unrecoverable (there are no possibilities to exchange components). Only their regeneration is possible consisting of the above-mentioned recording of new dosage characteristics in the memory of the injector. Any leaks eliminate the possibility of further operation of the injector [1, 2, 5]. In cases where the injector diagnostic test indicated that no further use is possible, the injector was disassembled, and the components of precision pairs were subjected to microscopic observation.

MICROSCOPIC OBSERVATION

Microscopic examination was carried out with optical microscope Zeiss Neophot 2. The microscopic observations were done on the external surfaces of the pistons, needles, tips and check valves of the damaged injectors. Figure 3 shows a typical electromagnetic injector pistons damage.

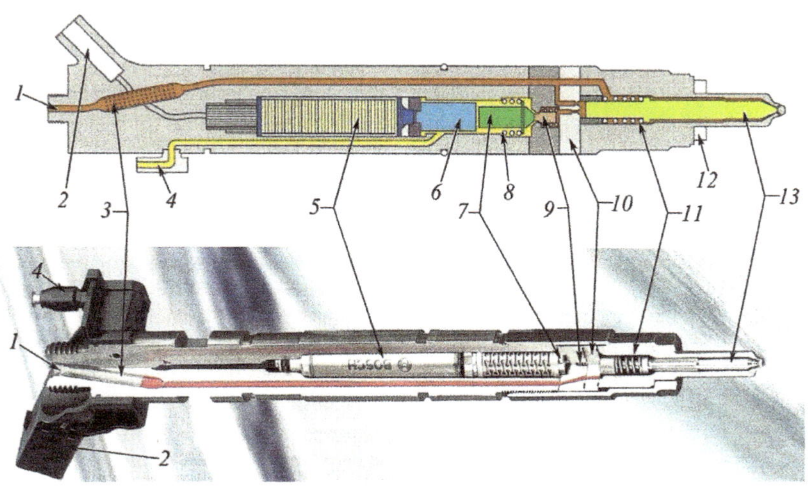

Fig. 2. Cross section of a piezoelectric Bosch injector used in common rail systems: 1 – fuel supply port, 2 – electrical connector, 3 – fuel filter, 4 – fuel return port, 5 – piezoelectric stack, 6 – intermediate piston, 7 – valve plunger, 8 – valve plunger spring, 9 – switching valve, 10 – throttling plate, 11 – nozzle needle, 12 – copper gasket in the housing of the injector head, 13 – needle [3]

That are usually small grooves and indents on the surface caused by the presence of mechanical impurities. As can be seen, that are local damages not covering the entire surface of the element. Damage to the piezoelectric injector needles also have a similar local character (see Figure 4). In the presented case these are minor scratches and pitting cavitation.

Another type of damage is characteristic for check and switching valves. In Figure 5 one can see distinctly the process of cavitation which started on the edge of valve opening in a piezoelectric injector. The consequence of this phenomenon is deterioration in the injector leak tightness, what with fuel at high pressure causes excessive flow through the overflow port of the

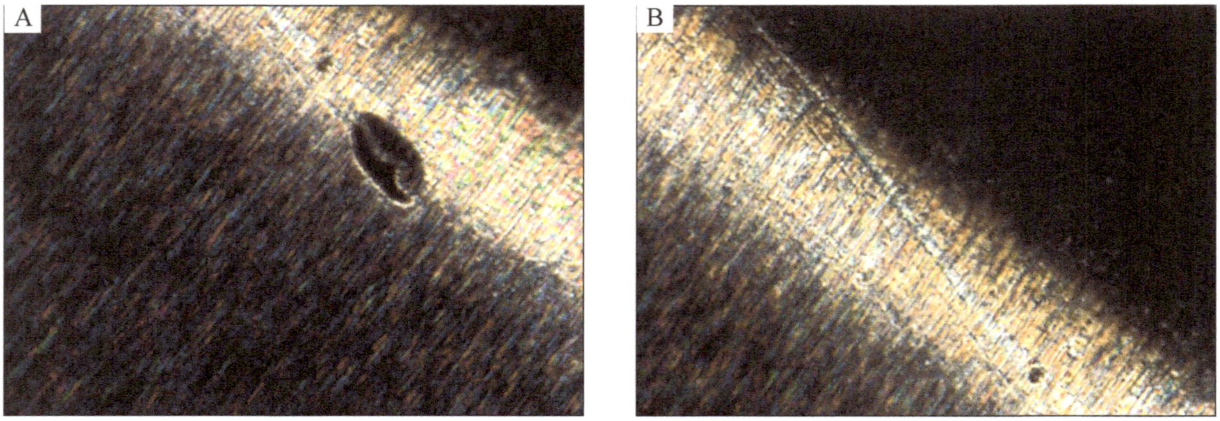

Fig. 3. Damages of the piezoelectric injector plunger: a) indentation, b) groove with a visible imprint of the contamination

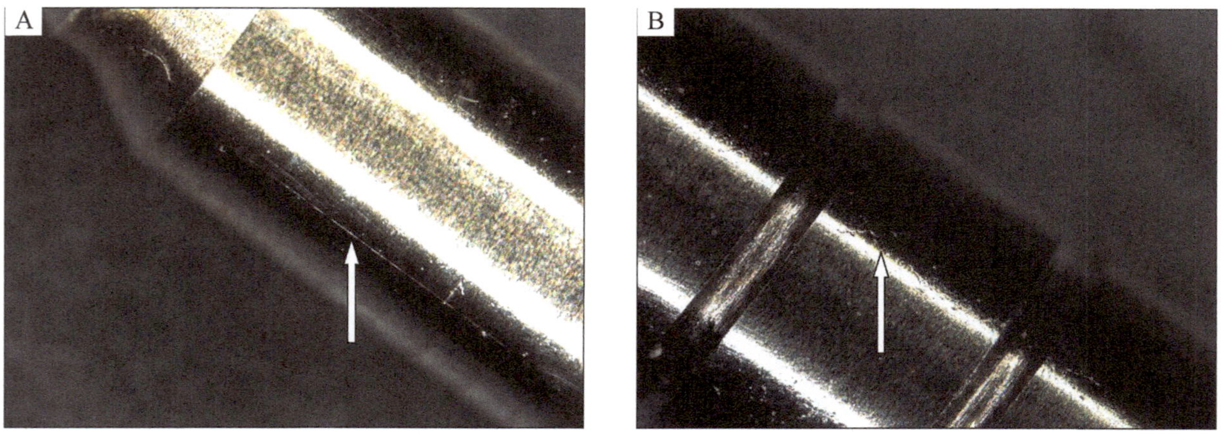

Fig. 4. Damages of the piezoelectric injector needle: a) longitudinal scratch, b) cavitation pitting

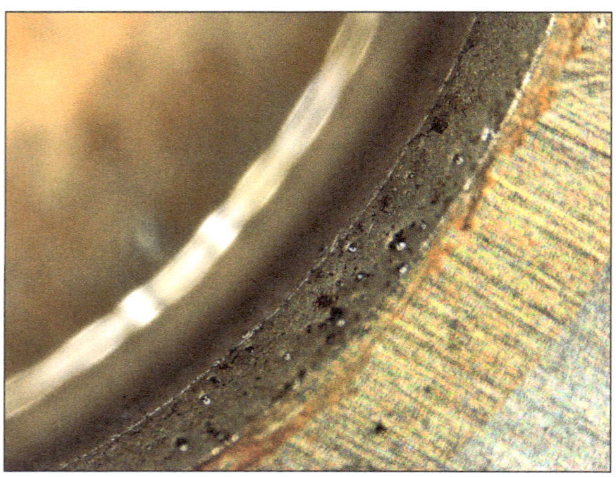

Fig. 5. Cavitation erosion on the edge of the pinhole overflow valve in a piezoelectric injector

Fig. 6. Developed cavitation on the edges of the pinhole overflow valve in a solenoid injector

injector. Such types of damage result in difficult starting of the engine and uneven idling. Figure 6 shows cavitation damage of the pinhole overflow valve in a solenoid injector. Also in the case of this type of damage the injector is not able to give a sufficiently large amount of fuel, resulting in difficult start-up and a non-uniform work of the engine. Reduction of the dose of injected fuel results in lean fuel-air mixture. Increasing the amount of oxygen relative to the amount of fuel is accompanied by the increase in the speed and temperature of combustion. This is confirmed by the observations – some of the damaged injectors had clear signs of overheating. Figure 7 shows overheated tip of a piezoelectric injector. Figure 8 shows clearly visible thermal discoloration on the needle of the above injector.

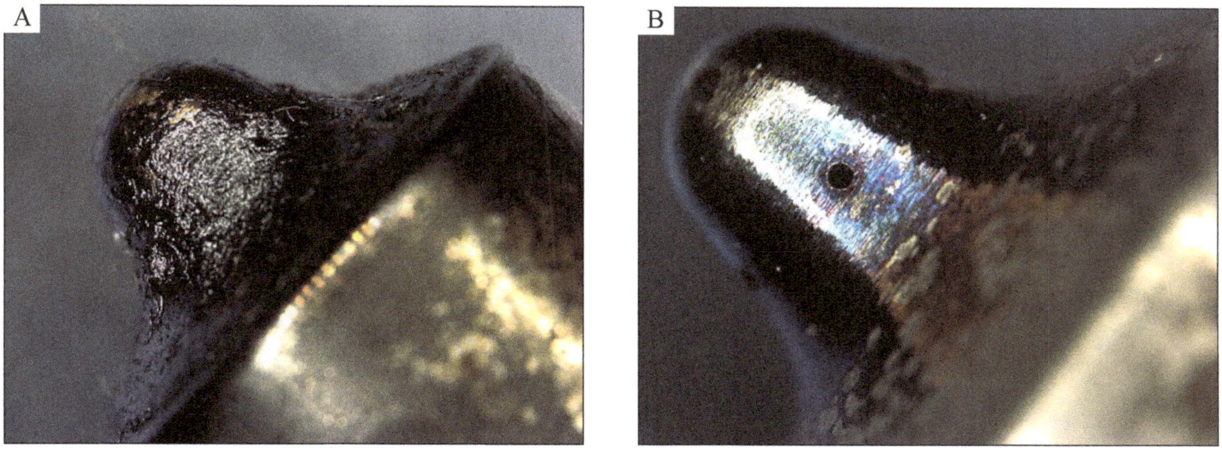

Fig. 7. Tip of the damaged piezoelectric injector: a) appearance after removing b) appearance after clearing – visible discoloration caused by the impact of high temperatures

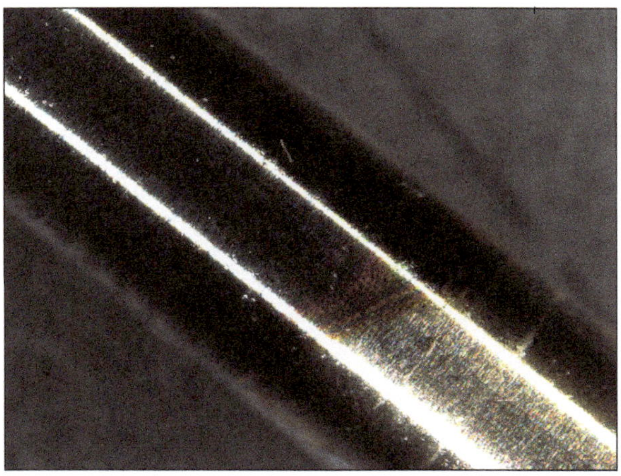

Rys. 8. Thermal discoloration of the needle of the piezoelectric injector

CONCLUSIONS

- Typical damage symptoms of the injectors in common-rail injection systems are uneven idling and the difficulty in starting the engine.
- Damage to the injectors may be either: tribological wear, pitting, cavitation and thermal damage.
- The tribological damages to the common-rail injectors are local and can be caused even by a small amount of impurities.
- The place of cavitation pitting in the injectors are mainly check valves.
- Damage of the return valve causes a reduction in the quantity of fuel actually injected, resulting in a leaner air-fuel mixture. The combustion of lean mixtures causes overheating of injector tips.

REFERENCES

1. Günter H.: Układy wtryskowe Common Rail w praktyce warsztatowej. WKŁ, Warszawa 2010.
2. Günter H.: Diagnozowanie silników wysokoprężnych. WKiŁ, Warszawa 2002.
3. Poradnik serwisowy. Układy common rail w silnikach z zapłonem samoczynnym samochodów osobowych, nr 6/2011.
4. Rokosch U.: Układy oczyszczania spalin i pokładowe systemy diagnostyczne samochodów OBD. WKiŁ, Warszawa 2007.
5. Układ wtryskowy Common Rail. Informator techniczny Bosch. Warszawa. WKŁ. 2000.

12

TECHNICAL AND ORGANISATIONAL ASPECTS OF PRODUCTION PREPARATION ON THE EXAMPLE OF DOOR FACTORY

Anna Rudawska[1], Agnieszka Seń[2], Leszek Semotiuk[3]

[1] Department of Production Engineering, Faculty of Mechanical Engineering, Technical University of Lublin, 36 Nadbystrzycka Str., 20-618 Lublin, Poland, e-mail: a.rudawska@pollub.pl

[2] Graduate student of Faculty of Mechanical Engineering,, Technical University of Lublin, Poland

[3] Department of Production Engineering, Faculty of Mechanical Engineering, Technical University of Lublin, 36 Nadbystrzycka Str., 20-618 Lublin, Poland, e-mail: l.semotiuk@pollub.pl

ABSTRACT

The aim of the work is the analysis of the organization of production, and door-producing undertaking to propose technical solutions that could improve production efficiency and reduce costs. On the basis of an analysis of the existing organizational and technical production plant a proposal of changes was presented. These changes depend on replacing a group of machines with one CNC EGURKO automatic line.

Keywords: technical and organizational production preparation, factory.

INTRODUCTION

The production process is a structured set of operations, whose aim is to perform the specified products. The production process can be of reference specified product or production cells, in which these operations are slotted. Production processes are specific for each type of product. The main part of the production process is a technological process during which the input materials change their shape, dimension, appearance, physical-chemical properties or parts change the relative position (during assembly) [1, 5]. It is often necessary to customize the manufacturing process to changing environmental conditions. Frequently, production process must be adjusted to changing external conditions, including e.g. supply and demand shift, since the production process is geared towards realisation of a given aim or group of aims, which may, however, change [2, 6].

The aim of the work is the analysis of organization of production, and door-producing undertaking to propose technical solutions that could improve production efficiency and reduce costs.

The purpose of technical preparation of production is to prepare all the activities associated with the production and, in particular, to produce a new product or service. The scope of these activities includes:

- modernization of the existing production,
- continuous exploration and continuous substitution of materials,
- the introduction of new technologies.

Technical production preparation comprises the following stages [2, 7, 9]:

- preliminary study and research,
- design preparation,
- technological preparation,
- organisational preparation,
- production initiation.

Organisational production conducts to determination of the time and space of the qualitative and quantitative compounds occurring between man work. Specific features affecting the organizational production are:

- different sequence and duration of individual processing processes,
- frequent changes of work pieces,
- a relatively high number of possible technological solutions,
- frequent changes.

Organisational production preparation is strictly connected with the preceding stage as the accepted technology, de facto, imposes certain production organisation solutions. Therefore, organisational production preparation works should be in parallel with technological and design preparation works. Additionally, what should be considered an essential aspect of organisational production preparation is a prior training for the staff, in order to improve their qualifications.

Organisational production preparation involves developing the following [1, 7]:

- preparatory works procedures required to start new production,
- production stations layout procedures, localisation of machining cells and production hierarchy,
- management structures,
- pre-operative planning procedures, including model schedules and production documentation,
- production start-up and development procedures.

Work stations and machining cells layout is to certain extent dependent on a selected machining technology. Nevertheless, proper planning of production stations layout may significantly shorten transport time or facilitate the transport between subsequent stations. Each such plan should take into consideration: particular stations size, required wall and transport route clearances, necessary installations accessibility, work ergonomics as well as transport between stations.

Production start-up involves test run, design and technological production preparation quality control as well as production stations and product quality control. If necessary, technical organisational documentation is revised. The purpose of the production start-up phase is to ensure full production capacity of production line in the shortest possible time [1, 3].

One of the key tasks of technical production preparation is to sustain technical development of both the product and its manufacturing, another is to ensure the producibility of the design. It is achieved by minimising the number of non-standard parts, selection of material and its shape. The importance of technical production preparation is growing with the technological and organisational progress of a company. It, moreover, enables introduction of new products and modernisation of already existing ones in terms of the technology of production, which is, in turn, reflected in higher product market value.

Among the main indices of technical production preparation there are: productivity improvement and technological-economic indices of a company [2].

THE PRODUCTION CHARACTERISTICS

The production characteristic of the selected product is carried out on the basis of the enterprise, which since 1992 has been producing all types of doors. The company offers are: the doors and door sash, doors, exterior doors, entrance doors, edging with natural veneers. The company has a hall divided into four smaller rooms, one of which meets the storage function of semi-finished products, and the other three are individual workstations equipped with specialized machines and automatic lines for the processing of wood and laminates and other equipment, as well as the position of the manual.

The company has the following work stations:

- position the glazing,
- for foil,
- installation of handles, hinges, locks,
- dryer,
- hand station:
  - – preparation of muntin (design frame),
  - – preparation of fill (particle or honeycombed plate – fill in the frame).

An actual layout of work stations in a production line of the door is presented in Figure 1.

The existing organizational arrangements in the workplace are, inter alia:

1) serial production type,
2) repeatable (rhythmical), production.

Among the benefits of the serial production organization to be replaced:

- increase specialisation of positions and workers' skills,
- increase the possibility of mechanization and automation of production processes,
- the ability to invest in production and technology, and also in production lines,
- shorten the production cycle and reduce stocks of work in progress, as well as increasing the rotation of current assets,
- reducing the time of idle workers' and machines' rest.

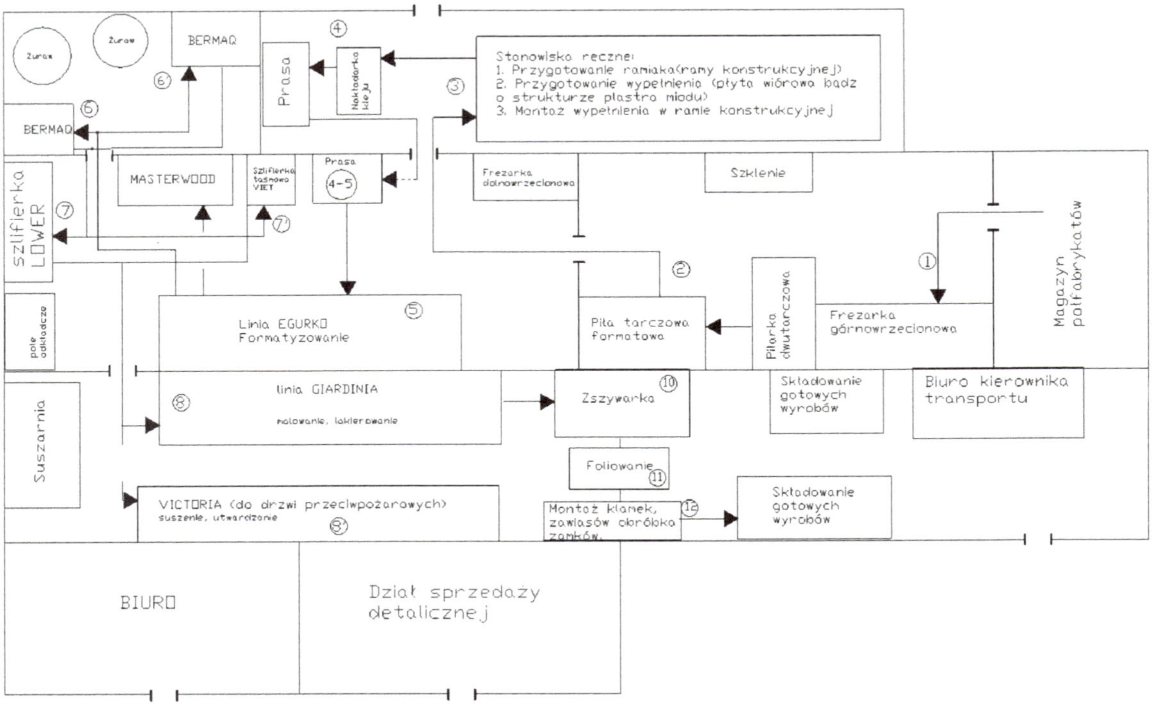

Fig. 1. Production stations layout of door (the original figure from [10])

While the repeatable production is characterized by:

- regular repetition of the same phenomena in equal intervals
- invariable adopted the technology, the production volume and the range of manufactured products,
- long-term break tasks into smaller units, carried out in shorter intervals,
- allocation of productive tasks to individual workstations to detail operation,
- fixed schedule during the production process.

The characteristics of the possibility of rational planning of production of certain products is presented, in order to plan the supply of materials, reduce storage costs and improve the use of posts.

THE TECHNICAL AND ORGANIZATIONAL MODIFICATION

The basis of an analysis of the organizational and technical solutions which existed in factory a proposal for change was presented, involving replacing a group of machines with one automatic line CNC EGURKO. It is proposed that machines can be replaced with one automatic line CNC EGURKO, the following:

- Jarome veneering machine,
- format the line EGURKO,

Fig. 2. The veneering machine in combination with dimension saw machine view [10]

- CNC milling machine EGURKO Bermaq,
- CNC milling machine MASTERWOOD,

In the prepared solution, it is proposed that the line will include the following stations: EGURKO CNC:
- glue spreader and veneering machine,
- press with a dryer,
- veneering machine in combination with dimension saw machine (Figure 2).

Between the lines the segments are mounted on a special automatic roller conveyor (Figure 3), thanks to which the material is moved from one location to another without human involvement (which saves 20% time and set the material transport industrial movement technology machine tools). The proposed scheme of CNC EGURKO automatic line is presented in Figure 4.

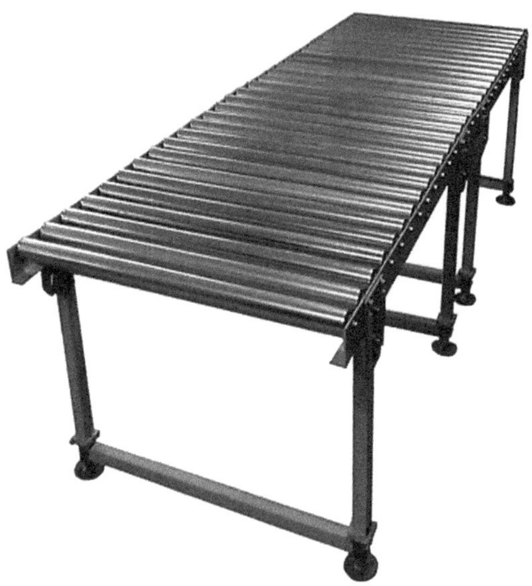

Fig. 3. The roller conveyor view [10]

ADVANTAGES AND LIMITATIONS OF THE PROPOSED SOLUTION

The advantages of the proposed changes include:
- time saving – 20% (no need for manual transport of materials between work stations),
- no manual setting and removing the material from the machine table- in the proposed dissolution of the machining time, one lot (10 pieces) is much shorter. On regular machines the time amounts to approximately 230 minutes, and in the automatic line it is estimated at about 130 minutes, and so it would be less by about 57%;
- the possibility reducing the number of qualified staff;
- 3 operators for 1 service line (while in the existing solution, the machines are operated by 6 people), in view of the fact that in the existing three shift system, employment reduction will reach 12 full time equivalents.

In addition, you can replace some of the production line, as the space-saving takes less space than 3 milling machines, veneering machine and dimension saw machine along with the spaces on the material before and after treatment. Another important advantage in view of the current volume production might include the improvement of ergonomics. However, the presented solution also has some disadvantages, among which one should mention that the line is dedicated to only one type of product (door), which affects some limitation in the implementation of the proposed solutions to the enterprise. In addition, an account should be taken of the need to reorganize the pro-

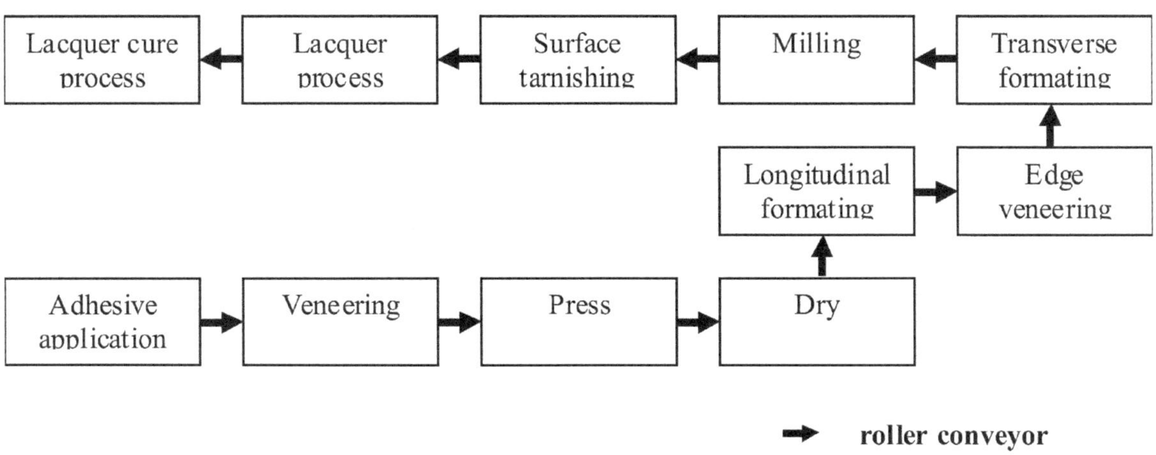

Fig. 4. The scheme of CNC EGURKO automatic line

duction establishment when changes are made in door manufacturing. This reorganization results in difficulties during the installation of automatic line production, especially in cases where the department was unable to make the breaks in the door production. One solution is to move the production of the door to another production hall or to limit the number of manufactured products at the time of the assembly line. Another limitation of the proposed solution is also a significant cost of purchase and installation of such a line, which should be considered in the financial analysis of the company.

SUMMARY

Superior purpose of the company is its survival and development, which is conditioned by gaining appropriate profit in longer periods of time. Proper preparation and organization of the production and the reduction of production cost and time are very important for the business. Despite the disadvantages of the proposed solution, the purchase of the automatic line will bring more benefits than losses. The cost of this line will return after 3–4 months of its work, and at the end of this time, the company will begin to make money on the investment.

REFERENCES

1. Pasternak K. Zarys zarządzania produkcją. PWE, Warszawa 2005.
2. Karpiński T. Inżynieria produkcji. WNT, Warszawa 2004.
3. Świć A. Elastyczne systemy produkcyjne. Wydawnictwo Politechniki Lubelskiej, Lublin 1998.
4. Feld M. Podstawy projektowania procesów technologicznych typowych części maszyn. WNT, Warszawa 2003.
5. Pająk E. Zarządzanie produkcją. PWN, Warszawa 2006.
6. Grabownik C., Skołuj B., Kalinowski K., Krenczyk D. Integracja technicznego i organizacyjnego przygotowania produkcji z zastosowaniem technik obiektowych. Przegląd Mechaniczny, 2009, 1: 36-41.
7. Dębski D. Ekonomika i organizacja przedsiębiorstw. Cz. I. WSiP, Warszawa 2006.
8. Szatkowski K. Przygotowanie produkcji. PWN, Warszawa 2008.
9. Gollinder-Tarajko M., Zaręba K. Projektowanie rozwoju produktu z wykorzystaniem technik komputerowego wspomagania. http://www.ptzp.org.pl/files/konferencje/kzz/artyk_pdf_2010 (maj, 2013)
10. Seń A. Techniczne aspekty przygotowania i organizacji produkcji na przykładzie wybranego przedsiębiorstwa produkcyjnego. Praca dyplomowa, Politechnika Lubelska, Lublin 2009.

13

STRENGTH OF ABS PARTS PRODUCED BY FUSED DEPOSITION MODELLING TECHNOLOGY – A CRITICAL ORIENTATION PROBLEM

Filip Górski[1], Radosław Wichniarek[1], Wiesław Kuczko[1], Przemysław Zawadzki[1], Paweł Buń[1]

[1] Chair of Management and Production Engineering, Poznań University of Technology, Piotrowo 3, 60-965 Poznań, Poland, e-mail: filip.gorski@put.poznan.pl; radoslaw.wichniarek@put.poznan.pl; wieslaw.kuczko@put.poznan.pl; przemyslaw.zawadzki@put.poznan.pl; pawel.k.bun@doctorate.put.poznan.pl

ABSTRACT

Numerous experimental studies, including experiments performed by the authors, have shown that the part orientation during layer deposition has a very strong influence on final strength of a product manufactured by additive Fused Deposition Technology. The paper presents the results of tensile, bending and impact strength tests performed on samples of various orientations, made out of ABS material using FDM technology. The results of these tests allowed discovering a unique phenomenon – with the changing orientation, not only the values of strength indexes change, but macroscopic material behavior under load as well. The transition between a "yield point" and "brittle" material usually happens in a certain range of orientation values, named a critical orientation by the authors. The paper indicates supposed ranges of critical orientation for various types of loads.

Keywords: ABS material, FDM technology, strength.

INTRODUCTION

Rapid development of the Additive Manufacturing Technologies (AMTs), also known as layered manufacturing technologies (or, in recent years, as 3D printing) allowed significant decrease of time needed for implementation of a new product. Additive manufacturing processes allow obtaining physical, three-dimensional shapes of nearly any complexity, directly from the digital representation of a product (usually a CAD model). There is no need of using any specialized tooling, besides the equipment of the manufacturing machine. Additive manufacturing technologies can be used for Rapid Prototyping, Rapid Manufacturing or Rapid Tooling. They are invaluable when there is a need of quick manufacturing of a physical prototype of a designed part [1, 2].

One of the most frequently used AMTs for the industrial purposes is the Fused Deposition Modelling technology, which can be used to obtain parts out of thermoplastic materials. The most widespread build material is acrylonitrile butadiene styrene (ABS), which ensures relatively good strength and acceptable thermal shrinkage. It also allows further processing of the obtained elements (by means of machining, coating or gluing). Machines for the Fused Deposition Modelling, in comparison with other additive manufacturing technologies, have small dimensions and are easy to maintain. They are also quiet and clean, which makes them available for use directly in design studios [1, 3]. FDM models can be recycled and there is not much waste, which makes the whole process ecologically friendly.

A final product manufactured using the Fused Deposition Modelling technology can be characterized by some coefficients, which are influenced by many factors [4]. Unlike in most manufacturing technologies, the values of parameters of the additive manufacturing process can be more significant than the properties of the part material – two different sets of process parameters applied to the same geometry can result in obtaining two products of entirely different properties, e.g. strength [4, 5] or accuracy [3]. Each set of the process parameters: orientation of the product in the working chamber, layer thickness and method of filling of the layer contour, will make the part structure look different, which will result in different values of coefficients such as strength, accuracy or surface quality.

So far, many scientists have investigated the influence of the manufacturing process parameters on mechanical properties of products made using FDM technology [4, 5]. Some researchers focused on optimization of a selected parameter in relation to a specific evaluation criterion, for example time of the process [6], accuracy of shape representation [7], surface quality [8] and mechanical properties [9]. The process parameter which influences values of the product properties in the most significant way is spatial orientation of the product in the working chamber during the manufacturing process [4, 5, 10, 11].

Mutual relations between the FDM process parameters and properties of the obtained products are not fully discovered yet. Attempts at their experimental determination have been made [4, 5, 12], but obtaining full characteristics of these relations is still an open research problem.

The aim of the present work was to summarize a set of experimental studies performed to obtain knowledge about an influence of the key parameter of the additive manufacturing with FDM technology – orientation of the product in the working chamber during layer deposition. During numerous studies, the authors have discovered that the FDM parts behave in two very distinct ways under load – the part can be either "brittle" or with "yield point" – it fails either via the thread fracture or the layer disjoint [13]. The transition between the two behaviors happens in a certain range of orientations [14]. The preliminary research, described in the paper, was aimed at finding out a general range of transition between material behaviors – this range was named a critical orientation. The critical orientation problem was defined on the basis of the tensile, bending and impact strength tests – their results are presented in this paper.

THEORETICAL INFORMATION

Fused Deposition Modelling (Fig. 1) is a process consisting in layered, linear deposition of a heated thermoplastic material supplied in a form of a wire by an extrusion head with nozzle of a certain external dimension (in the machine used for the described research, nozzle diameter is 0.4 mm). The head can move in two axes (X and Y). It deposes material on the model basis in a form of a contour and filling. After manufacturing one, complete layer, the model table moves in vertical (Z) direction to the distance equal to the selected layer thickness. Deposed material goes back to a solid state a moment after leaving the nozzle, bonding with the previously created geometry. For the more complex shapes, support structures are needed to prevent deformation by gravity. The extrusion head is usually equipped with two nozzles, one for the build material, the other one for the support material (usually similar types of thermoplastics are used, in some machines the soluble support material, such as PVA, is used for easier removal). A produced part is ready for use immediately after the support material is removed [1, 2].

The macrostructure of parts made by FDM technology consists of material threads deposed in alternate directions, creating layers bound together only adhesively. The manufactured elements behave in a specific way under any given load. Even a simple load (single force or torque) applied to non-complex geometries like a rectangular beam will result in a complex stress state inside the element. As such, it is very hard to predict whether a certain part will break under the applied load or not.

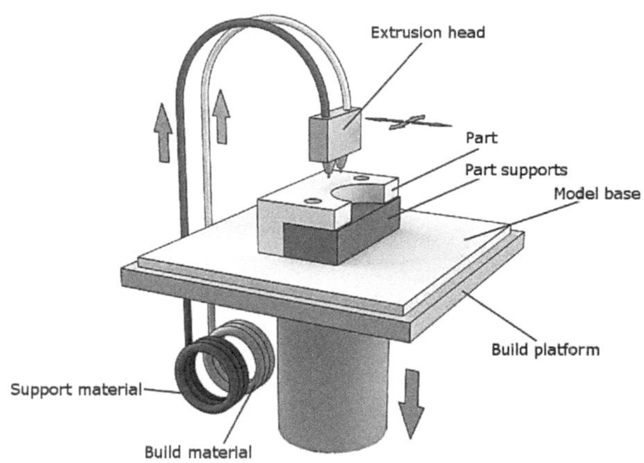

Fig. 1. FDM process scheme [2]

As the structure is largely dependent on the part orientation, prediction of the orientation influence on the properties of the finished product is of fundamental significance for evaluation of the possibilities of FDM technology application in small batch or piece production. Orientation in the working chamber can be intuitively defined as an angular difference between plane slicing the object into layers and a selected, base plane of the object (Fig. 2). Orientation may be therefore defined by the three rotation angles (rotation between the object and the machine coordinate system), where only two angles are relevant – the third angle (around the vertical axis, perpendicular to the layer slicing plane) has no influence on the manner of slicing the object into layers, so it does not affect product properties significantly.

STUDY DESCRIPTION

For the studies, samples made using FDM technology on the Dimension BST 1200 machine were used. The samples were manufactured out of ABS material supplied by the Dimension company in a form of a wire wound on a spool, sealed hermetically in the cartridge, what ensures proper levels of humidity required for the process.

Three types of samples were manufactured during the course of the study (Fig. 3). Shape of the samples was compatible with guidelines of specific standards, namely:

- PN-EN ISO 527 – procedures and parameters of tensile tests of polymer materials,
- PN-EN ISO 178 – procedures and parameters of bending tests of polymer materials,
- PN-EN 10045-1 – procedures and parameters of Charpy impact tests of polymer materials.

The samples were manufactured in different orientations. Two orientations in the X axis were considered – 0° and 90°, named Flat and Side, respectively. The Y axis orientations were the main variable in the presented studies, differing in 15° for the tensile tests, 30° for the bending tests (with additional 45° orientation samples) and 45° for the impact strength tests. An ID of a given sample is composed of the name of the X orientation and the value of the Y orientation (e.g. Flat-0 means that orientation of the is 0° in both axes), except for samples manufactured in the vertical direction, that is 90° in Y axis – for these samples, the X orientation value does not affect the process of slicing into layers, so this sample was simply given the "Vert" ID. A summary of the samples produced for tests is presented in Table 1.

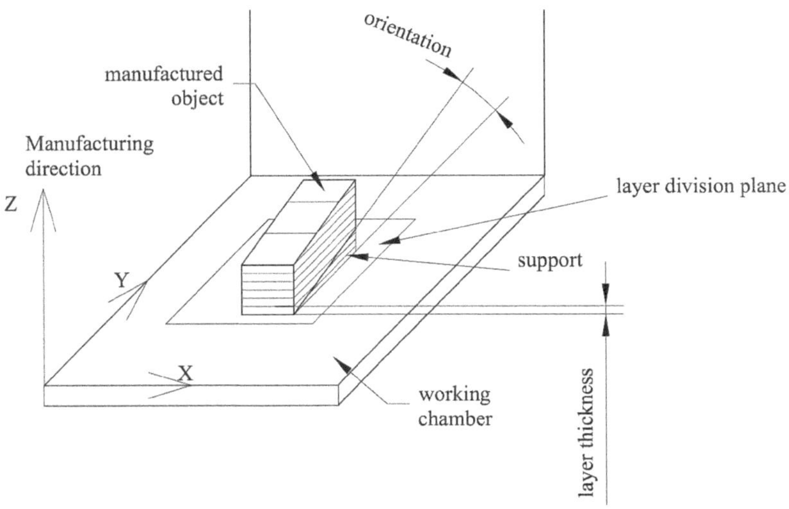

Fig. 2. Illustration of the orientation concept [3]

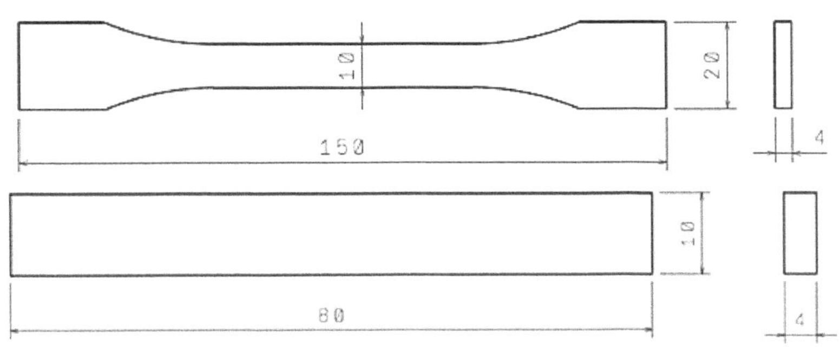

Fig. 3. Shape of samples for the experimental tests, top – tensile test, bottom – bending and Charpy impact test

Table 1. Summary of samples for the experimental tests

Lp.	Sample ID	Tensile sample prod. time [min]	Tensile sample support material [cm^3]	Bending / impact sample prod. time [min]	Bending / impact sample support material [cm^3]
1.	Flat-0	22	1.96	11	0.86
2.	Flat-15	109	20.68	n/a	n/a
3.	Flat-30	181	34.43	74	8.82
4.	Flat-45	218	26.68	95	10.87
5.	Flat-60	264	47.38	22	0.15
6.	Flat-75	246	31.99	n/a	n/a
7.	Vert	222	11.69	21	0.08
8.	Side-0	46	2.28	14	0.54
9.	Side-15	102	10.07	n/a	n/a
10.	Side-30	157	17.26	70	5.67
11.	Side-45	192	16.71	89	7.46
12.	Side-60	247	27.8	25	0.27
13.	Side-75	236	19.39	n/a	n/a

The samples were manufactured using the Dimension BST 1200 machine by Stratasys company, out of ABS material, using solid (monolithic) internal filling with 45° raster, the layer thickness was 0.254 mm. Five samples were manufactured for each sample type for the bending and the impact strength, three samples of each type were manufactured for the tensile tests. For the tensile and bending tests, a Zwick Roell Z020 machine was used (with varying equipment). For the impact test, a Charpy impact hammer was used. All the tests were carried out according to the aforementioned standards. In the bending test, the test was conducted up to deflection equal to 1.5 of sample thickness, which is 6 mm, or until the sample failure. The stress, calculated on the basis of recorded forces and deflection, is interpreted as the real bending strength σ_{fM} or the conventional (substitute) bending strength σ_{fC}, depending on whether the sample failed during the tests. The other tests always finished with the sample failure and the calculated results were interpreted as the tensile strength σ_m or the impact strength KC. For the reference and comparison purposes, the tensile tests and impact strength tests were also performed on samples of the same shape, manufactured out of the same ABS material, using the injection molding process.

RESULTS AND DISCUSSION

Results of all the strength tests are presented in Table 2, 3 and 4 (tensile, bending and impact strength, respectively). The two main groups of characteristics are presented: strength, interpreted as the maximal stress recorded in a sample and strain, presented as a percentage (elongation) at the maximal force. Impact strength is presented separately.

Fig. 4. Experimental tests on Zwick Roell Z020 machine – tensile test and bending test

Table 2. Results of the tensile strength tests

Sample ID	σ_m [MPa]	ε_b [%]	E [GPa]
Flat-0	19.0	4.6	1.6
Flat-15	18.1	2.0	1.6
Flat-30	13.8	1.3	1.5
Flat-45	12.9	1.5	1.5
Flat-60	13.7	1.2	1.6
Flat-75	11.1	1.0	1.5
Side-0	22.9	7.0	1.7
Side-15	21.6	3.1	1.7
Side-30	17.3	1.5	1.6
Side-45	14.9	1.2	1.6
Side-60	8.7	0.8	1.5
Side-75	10.9	1.0	1.5
Vert	11.0	1.0	1.5
Ref. (inj. molding)	30.1	27.2	1.7

Where: σ_m – tensile strength, ε_b – maximal strain, E – Young modulus.

The data is also presented in diagrams, in the Figures 5 (strength characteristics: tensile and bending strengths) and 6 (plasticity characteristics: elongation, impact strength).

In Figure 6, a critical range of orientations is shown. It is a range of orientation values, in which the transition between two types of material behavior – with a yield point and brittle – occurs. Inside this range, a dramatic decrease can be observed in all ductility (plasticity) characteristics. The curves were approximated and smoothed, as there is not enough data from all types of measurements – the variability of orientation angle was 15° in tensile strength and even higher in other tests. The critical orientation range was selected by the authors in between 5° and 30° in the Y orientation. The above orientation value of 30° in Y axis, all the samples in all the tests present the brittle behavior (no yield point, failure via layer disjoint, low elongation). This behavior was also presented by certain, random samples of 15° orientation. As no samples of orientation between 0° and 15° were tested, the 5° value was selected arbitrarily as the beginning of the critical orientation range – in further studies it may be verified.

Figure 7 shows an example of two different mechanisms of failure of the ABS samples made using the Fused Deposition Modelling process with various orientations. The course of the strength-deformation diagram is much different in both cases. In samples with the manufacturing orientation higher than a certain value, no yield point is present – they act as made out of a brittle material, the elongation at break is equal to elongation at maximal force and the measured elongation is very small (around 1%). Regarding the tensile test, the Y orientation value is a value of

Table 3. Results of the bending strength tests

Sample ID	dL [mm]	σ_{fM}, σ_{fC} [MPa]	ε_f [%]	Broken samples [%]
Flat-0	6.0	35.1	3.5	0
Flat-30	4.3	25.2	2.5	100
Flat-45	3.5	23.9	2.1	100
Flat-60	3.8	22.0	2.2	80
Side-0	6.0	41.1	3.5	0
Side-30	5.7	34.0	3.3	20
Side-45	3.8	27.0	2.2	100
Side-60	4.0	24.5	2.3	80
Vert	2.7	18.1	1.6	100

Where: dL – maximal deflection, σ_{fM} or σ_{fC} – bending strength, ε_f – maximal strain.

Table 4. Results of the impact strength tests

Estimator	Reference (inj. mold.)		Side-0		Flat-0		Side-45		Flat-45*		Vert*	
	[J]	KC [kJ/m²]	[J]	KC [kJ/m²]	[J]	KC [kJ/m²]	[J]	KC [kJ/m²]	[J]	KC [kJ/m²]	[J]	KC [kJ/m²]
average	2.53	62.86	1.20	28.80	0.74	17.71	0.03	0.60	0	0.00	0	0.00
median	2.40	59.70	1.20	28.73	0.71	16.99	0.03	0.60	0	0.00	0	0.00
max	3.23	80.35	1.42	34.13	1.02	24.58	0.05	1.20	0	0.00	0	0.00
min	2.16	53.73	1.00	24.04	0.60	14.46	0.01	0.24	0	0.00	0	0.00

* Values lower than sensitivity of the measuring equipment.

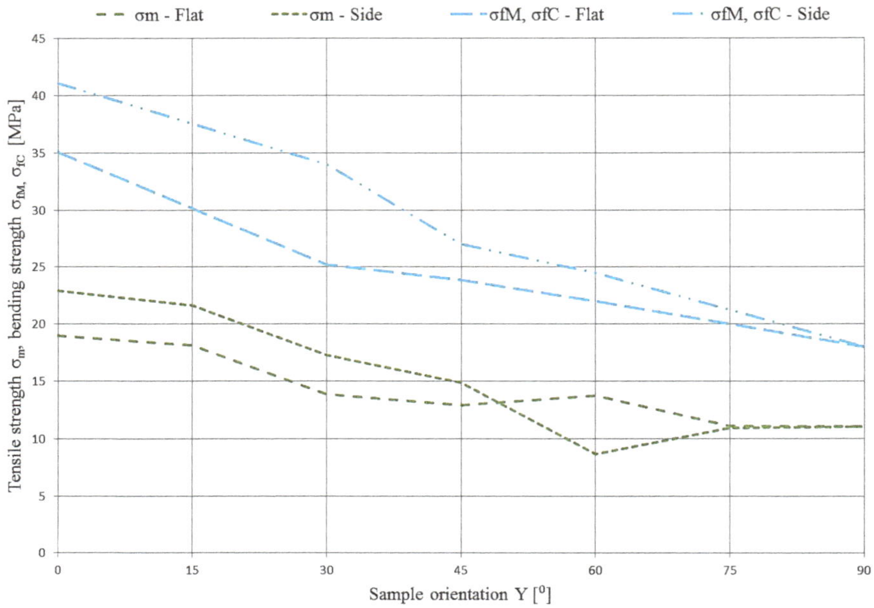

Fig. 5. Strength test results from the tensile and bending strength tests

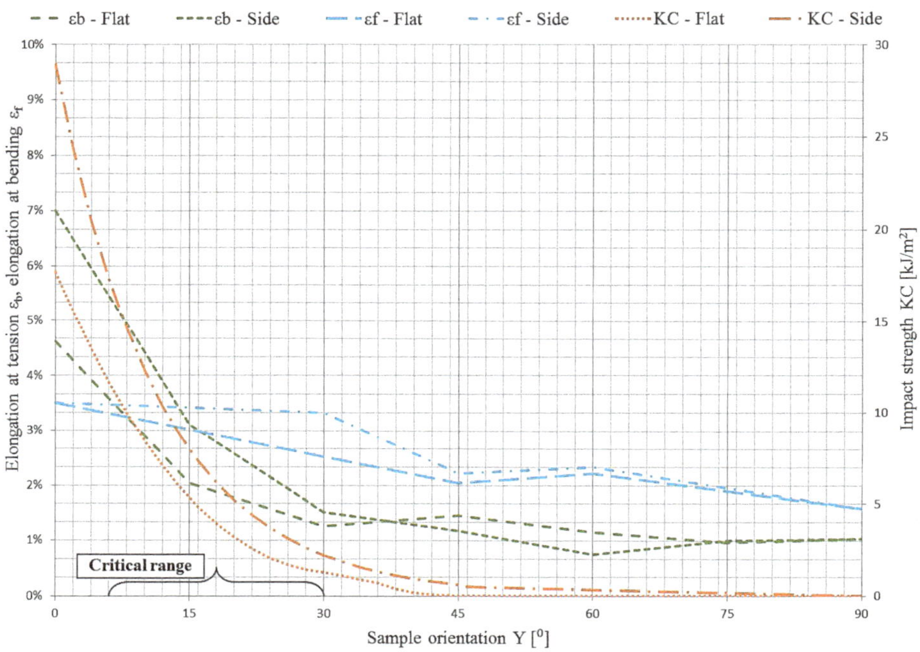

Fig. 6. Plasticity characteristics from the tensile, bending and impact strength tests

the angle between the loading force and the layer slicing plane, so it can be said that the higher the angle, the lower the strength. If this angle is higher than zero, the applied force is carried not only by the material threads – a certain portion of the load is applied to layer bonds, which are very weak due to lack of material fusion between adjacent layers. Therefore, after a certain angle is reached, the layer bond maximal strength is reached faster than strength of the material itself. It causes disjoint of the layers, macroscopically observed as a brittle fracture. The same mechanism applies to samples subjected to the bending tests. In the impact strength tests, the division is clearly visible, as the results differ drastically for various orientations – parts with the orientation of 45° and higher have no impact strength at all, the recorded values were randomly oscillating around zero, so it was assumed that they were lower than sensitivity of the measuring equipment.

When comparing the values of strength and plasticity characteristics achieved by the samples made using the FDM method and the injection molding out of the same ABS material, it is clear

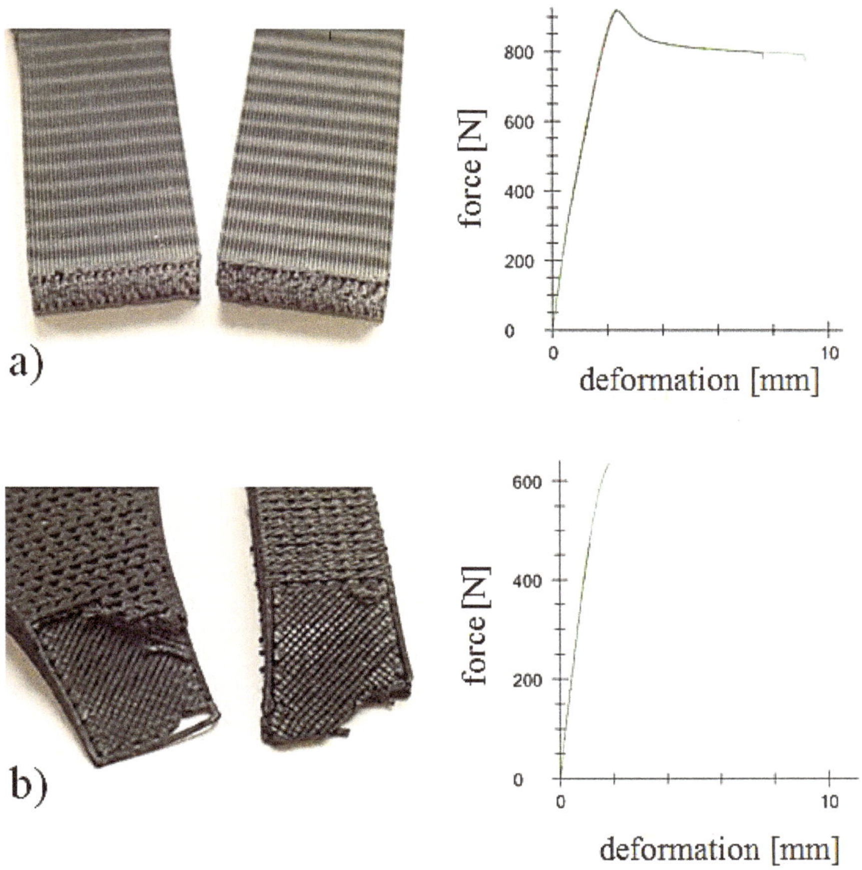

Fig. 7. Failure mechanisms in tensile samples: a) Side-0 – with a yield point, b) Flat-30 – brittle

that the injection molded parts are much stronger and more ductile at the same time. The difference is most visible in case of the impact strength tests, or plasticity characteristics in general. The raw values of tensile strengths for the best FDM samples do not exceed 80% values of the injection molded samples.

CONCLUSIONS

The samples made out of ABS using the additive FDM technology do not present satisfying mechanical properties. Moreover, the obtained properties can be much different if only one parameter of the process – the part orientation – is changed. There are two main reasons for this – the first one is the weak bond between layers in the FDM parts, which causes the strength characteristics to decrease, even dozens of times, in comparison to the injection molded samples. The second reason is the volume error, which is always present at the FDM samples – they are not monolithic. The volume error is the main reason why even the best parts made by the FDM process will never be as strong as the injection molded parts out of the same material – even if the weak interlayer bond effect is reduced to a minimum, the volume errors still make the effective loaded area smaller than in case of the fully monolithic part, which increases the overall stress and decreases strength.

This difference is the most visible for the impact strength tests. FDM products have many material discontinuities inside, which can act as notches during impact tests, this is why even the best FDM sample achieved only 47% of the impact strength of the monolithic sample produced by injection molding.

The studies presented in the paper allowed a discovery of a transition between one type of material behavior to another, with the only changing factor being a method of slicing the part into layers for the additive manufacturing by ABS material thread deposition in the FDM method. The two methods of failure of the additively manufactured samples are by thread failure or by layer disjoint. Macroscopically, they are perceived as a material with a yield point and a brittle material. The transition happens in a specific range of orientation values for the studied samples, this range was indicated in Figure 6. Most likely, for a certain load (tension or bending), more specific value or smaller range of values can be determined, in which the aforementioned transition occurs. This value or value range is defined as a critical orientation by the authors and it will be more deeply explored in further studies.

The future studies will consist in producing more ABS samples, within the preliminarily defined critical orientation range, to find out how much it is possible to narrow it down. The planned studies should allow for an experimental answer to the question: at which orientation value the layer bond is loaded up to its maximal strength faster than the material itself. The authors also plan to manufacture and test samples of other materials and on different manufacturing systems, including the low-cost 3D printers.

The results of the presented studies should be much of a help for engineers who plan to use additive manufacturing by the Fused Deposition Modelling method for parts which can be subjected to certain loads, as it is important to know what such a part is able to sustain and to which degree it could be comparable with parts manufactured using other, better known and explored manufacturing technologies. In the future, a set of guidelines will be probably formulated for the additive manufacturing process engineers on how to select a proper orientation knowing the purpose and probable load of the manufactured part. So far, it can be concluded that selection of the additive manufacturing process parameters is a complex problem. A general guideline is to focus on the economical effectiveness coefficients such as manufacturing time and costs, as the orientations which result in shorter manufacturing time usually result in smaller number of layers, which often makes for better strength. Still, this is true only for simple geometries – more studies are required to enable full control over effects of the additive manufacturing processes.

Acknowledgements

Work realized partially as a research grant in scope of statutory activities financed by Polish Ministry of Science for young scientists (decision no. 8710/E-362/M/2014). Part of the studies financed by the Polish National Science Centre, agreement number: 2011/01/N/ST8/07603.

REFERENCES

1. Chua C.K., Leong K.F., Lim C.S. Rapid prototyping: Principles and applications. World Scientific Publishing, Singapore 2003, p. 420.
2. Pająk E., Górski F., Wichniarek R., Dudziak A.: Techniki przyrostowe i wirtualna rzeczywistość w procesach przygotowania produkcji. Promocja 21, Poznań 2011.
3. Górski F., Wichniarek R., Kuczko W.: Influence of process parameters on dimensional accuracy of parts manufactured using Fused Deposition Modeling technology. Advances in Science & Technology Research Journal, 7(19), 2013, 27–35.
4. Bellini A., Guceri S., Mechanical characterization of parts fabricated using Fused Deposition Modeling. Rapid Prototyping Journal, 9, 2003, 252–264.
5. Ahn S.H. et al.: Anisotropic tensile failure model of rapid prototyping parts – Fused Deposition Modeling (FDM). International Journal of Modern Physics B, 17(8-9), 2003.
6. Han W.B, Jafari M.A, Seyed K. Process speeding up via deposition planning in fused deposition-based layered manufacturing processes. Rapid Prototyping Journal, 9, 2007, 212–218.
7. Masood S.H, Rattanawong W, Iovenitti P. Part build orientation based on volumetric error in fused deposition modeling. International Journal of Advanced Manufacturing Technology 16, 2000, 162–168.
8. Perez L.C.J. Analysis of surface roughness and dimensional accuracy capability of fused deposition modelling processes. International Journal of Production Research, 40, 2002, 2865–2881.
9. Rodrigues J.F, Thomas J.P, Renaud J.E. Mechanical behavior of acrylonitrile butadiene styrene fused deposition materials modeling. Rapid Prototyping Journal, 9, 2003, 219–230.
10. Ghorpade A, Karunakaran K.P, Wiwari M.K. Selection of optimal part orientation in fused deposition modeling using swarm intelligence. Journal of Engeenering Manufacture 221, 2007, 1209–1219.
11. Thrimurthulu K, Pandey P.M, Venkata R.N. Optimum part deposition orientation in fused deposition modeling. International Journal of Machine Tools &Manufacture, 44, 2004, 585–594.
12. Bagsik A., Schöppner V., Mechanical properties of Fused Deposition Modeling parts manufactured with ULTEM*9085. Proceedings of ANTEC 2011, Boston 2011.
13. Górski F., Kuczko W., Wichniarek R., Impact strength of ABS parts manufactured using Fused Deposition Modeling technology. Archives of Mechanical Technology and Automation, 31(1), 2014, 3–12.
14. Górski F., Wichniarek R., and Andrzejewski J., Influence of part orientation on strength of ABS models manufactured using Fused Deposition Modeling technology. Polymer Processing, 9, 2012, 428–435.

14

SELECTED PARAMETERS OF THE WORK OF SPEED LIMITER IN LINE STRAINING SYSTEM IN A FRICTIONAL LIFT

Paweł Lonkwic[1], Kamil Szydło[1]

[1] Lublin Factory of Personal Lifts LIFT SERVICE S.A., Roztocze 6, 20-722 Lublin, Poland, e-mail: pawel.lonkwic@lift.pl

ABSTRACT

The article presents the analysis of selected work parameters of speed limiter in line straining system. We analyzed the effect of changing the geometrical conditions of the new solution for the speed limiter in line straining system upon the working conditions in frictional lift braking system. Within the conducted simulations of the work of the system, which is responsible for lift braking with a tension with spring, a test bed was prepared, which simulated the work of tension-rope-limiter system. The tests were performed in the conditions reflecting the work of a lifting appliance. Analyzing the results obtained through empirical calculations, we can conclude that there is a possibility of applying the spring to eliminate the weight.

Key words: frictional lift, speed limiter, weight.

INTRODUCTION

According to the standard [7], the frictional lift car must be immobilized in case of exceeding its velocity by 0,3 m/s of the nominal velocity or breaking all the carrying strings. The main appliances protecting from uncontrolled increase of car velocity are: catchers and a speed limiter [6].

To examine if the braking system works correctly before it is admitted for operation the tests of braking system are performed by enforcing velocity increase by means of controlling system, and thus it is checked if the braking system works correctly. Such a test is performed with the nominal lifting capacity with the load increased by 125% of the lifting capacity. The braking system consists of a speed limiter, brakes (here called: catchers), tension and steel rope clamping all the above–mentioned sub-assemblies. The speed limiter is placed in the highest position of the lift shaft or in the engine room, placed over the lift shaft and it is an appliance controlling the speed of lift car movement. It consists of a cast iron wheel with an indented rope groove, the geometry of which is within the standard [7].

The line trims the speed limiter wheel by the angle of 180 degrees. Inside its structure the wheel has a system of latches, released by the centrifugal force. When the speed is exceeded, the speed limiter wheel revolves faster than the nominal speed, causing the latches to spread outside. The latches pulled apart like that wedge by the line wheel hollows, causing its blocking. The shape of groove and coefficient of friction between the steel rope and cast iron wheel stop the rope. A thorough kinematical analysis of steel limiter work was contained in [4]. It is the tension that is responsible for achieving appropriate friction of the rope on seed limiter wheel. It is supposed to generate the voltage in the rope of the value set by manufacturer of the brakes. The value of rope tension force is closely connected to the friction forces generated on the speed limiter wheel. To secure the speed limiter line tension and not to interfere with the remaining movable elements of the lift, the tension is placed in the lowest point of the lift shaft, which is shown in Figure 1.

In Figures 2 and 3 typical speed limiter (Figure 2) and tension (Figure 3) were shown.

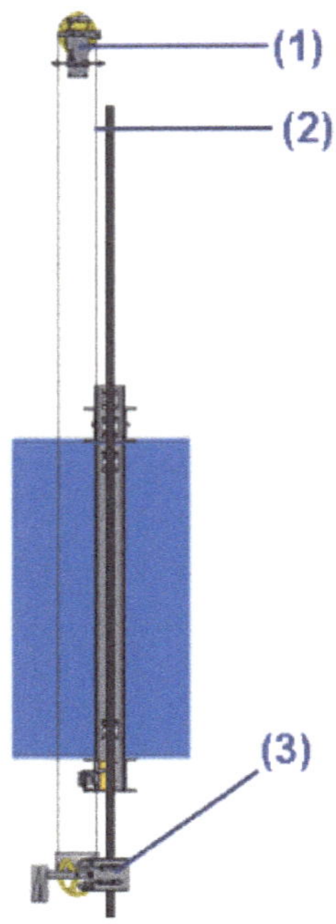

Fig. 1. Position of braking system of a frictional lift: 1 – speed limiter, 2 – speed limiter rope, 3 – tension [1]

Fig. 2. Typical frictional lift speed limiter [2, 3]

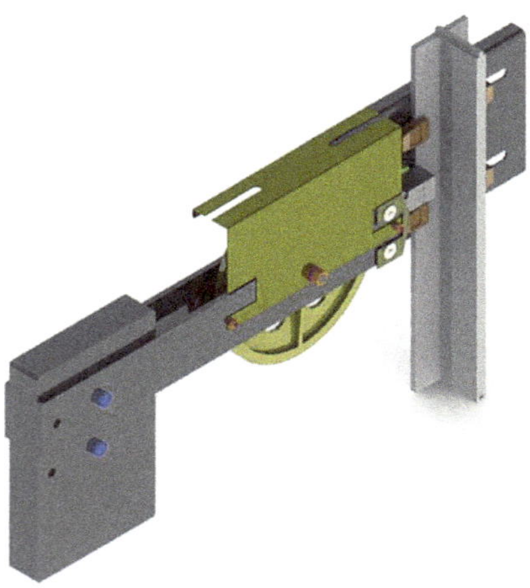

Fig. 3. Typical frictional lift tension [2]

NEW CONCEPT OF ROPE STRAINING METHOD

The main purpose of developing a new solution of tension was a maximum decrease of space available for tension fittings, which is connected with the obligatory trends in construction industry. Another purpose was to minimize the costs of manufacturing the product and improve the ergonomics of the assembly, which was connected with decreasing weight of the appliance by ca. 20 kg.

View of the tension with straining string was shown in Figure 4.

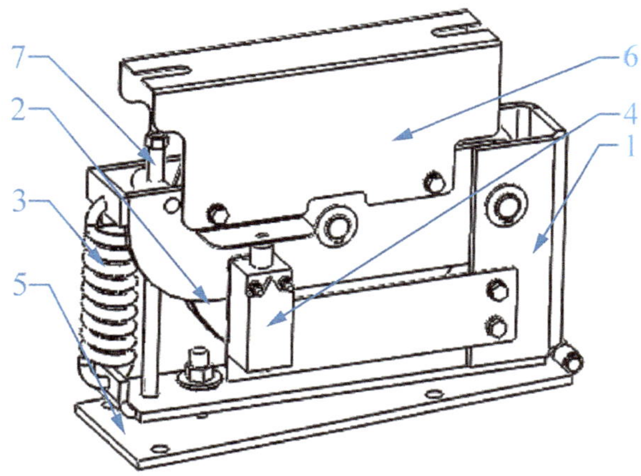

Fig. 4..New solution of frictional lift tension

The appliance consists of body 1, rope pulley 2, spring 3, tension screw 7 and electric connector 4. It is fastened to a base by means of the mount 5. The rope pulley 2 has got a shield 6, protecting from foreign matters permeating between the speed limiter rope and the rope pulley during work. The electric connector 4 is an element of safety circuit of the whole lifting appliance. The spring characterization in the suggested solution was selected so as the rope tension during the normal work of the appliance caused friction force between the speed limiter wheel and appropriate rope for catchers.

To eliminate all the tension structure defects, in the first phase of prototyping model #D was made, with simultaneous analysis of structure kinematical relationships. Structure optimization was performed on the basis of 3D model digitization and then MES simulation. The final stage of designing required calculations checking the correctness on the basis of standard records [3].

ALGORITHM OF SELECTING THE APPROPRIATE WAY OF ROPE STRAINING

The correctly selected braking system should be able to stop the car in motion, loaded with 125% of the nominal weighing capacity. To make it happen, the lift catchers must be selected proportionally to the loading and the tension and the speed limiter must meet the following conditions [7]:

- The tension must strain the limiter rope so as the force needed to make the brakes work do not cause rope friction loss. On the blocked speed limiter wheel. The minimum rope tension force is given by the manufacturer of speed limiter.
- The maximum forces generated in the speed limiter rope must be at least 8 times lower than the force that breaks the rope (specified by manufacturer of the rope).

In the classic solution, with the tension, the rope straining force G_n is the resultant of the force coming from tension G_o, working on arm R. In the solution with the application of straining spring straining force G_n is the resultant of force G_s coming from the stretched spring working on arm R. For that purpose we prepared the characterization of the spring, which is shown in Figure 5.

Considering the application of the spring in structure of the tension, it is justifiable to check what value of stretching force affects the rope. Using the above characterization and geometrical relationships of the braking system shown in Figure 6, we juxtaposed the characteristic working parameters for both types of tensions in Table 1.

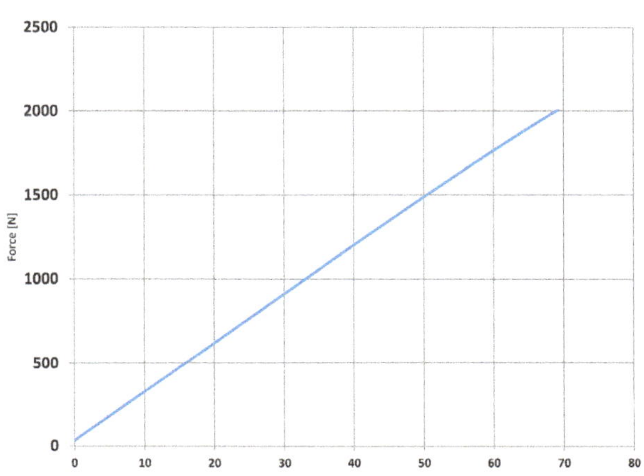

Fig. 5. Characterization of tension spring applied in the tension, drawn up on the basis of data obtained from the testing machine

Weight of the rope resulting from lifting height can be calculated from the relationship (1):

$$Q_l = g \cdot q_l \cdot H \text{ [N]} \quad (1)$$

where:

g – gravity force [m/s²],

q_l – rope unit mass [kg/m],

H – distance between limiter wheel and tension pulley (wheel) calculated from formula [m]:

$$H = H_p + H_n + H_d \text{ [m]} \quad (2)$$

where:

H_p – lifting height [m],

H_n – shaft top height [m],

H_d – shaft bottom height [m],

The stretching force in the rope for the tension with a weight can be expressed by means of an empirical relationship (3), and for the tension with a spring-by means of a relationship (4)

$$Gn = g \cdot [(\frac{R}{r} \cdot G_o) + ((0{,}5 \cdot \frac{R}{r}) \cdot G_r) + G_k] \text{ [N]} \quad (3)$$

$$Gns = g \cdot [(\frac{R}{r} \cdot Gs) + ((0{,}5 \cdot \frac{R}{r}) \cdot G_r) + G_k] \text{ [N]} \quad (4)$$

where:

G_r – weight of the tension arm [kg],

G_k – weight of the tension wheel [kg].

On the basis of relationships (3) and (4) speed limiter rope safety coefficient should be calculated, initially calculating the biggest force occur-

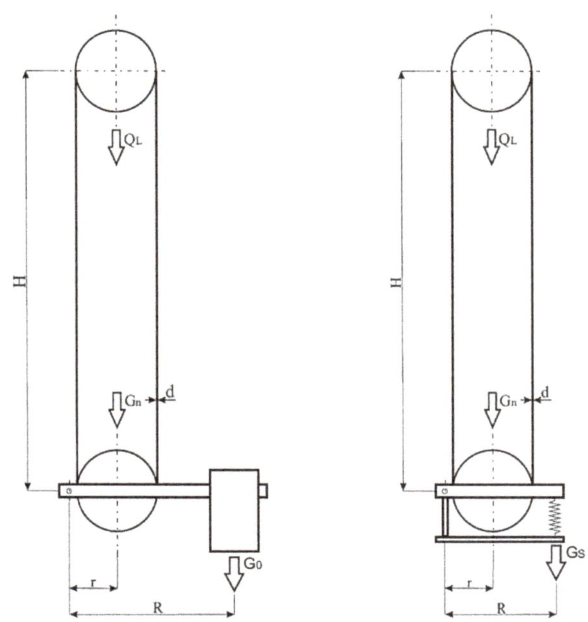

Fig. 6. Geometry of braking system in a frictional lift: R – length of the arm of the effect of force exerted by tension Go (spring Gs) [m], r – length of the arm of the effect of straining force Gn [m],Go – weight of the tension [kg], d – diameter of limiter rope [m], Q_l – weight of the rope resulting from lifting height [kg]

ring in the rope for the considered system, which can be described in relationship (5):

$$S = \left(\frac{G_n}{2} + Q_l\right) \cdot e^{\frac{\mu_0}{\sin\frac{\gamma_0}{2}}} \tag{5}$$

where:

F_z – minimum breaking force in the rope (rope manufacturer's data) [N],

On the basis of mathematical relationships (3), (4) and (5), the values of speed limiter rope safety coefficients must be calculated for both types of the tensions, which is expressed by formula (6):

$$k_{bo} = \frac{F_z}{S} \geq 8 \tag{6}$$

where:

S – the highest force in the rope.

Using the above empirical relationships and catalog charts of the manufacturers of catchers and steel ropes, in Table 1 we have collected the data for a model of frictional lift for the most unfavorable configuration, which is with lifting height H_p of 70 m. In order to perform comparative calculations we have used the characterization chart of speed limiter LK250 manufactured by P.F.B. S.r.l., Italy, for which the nominal rope straining force is G_n = 1020 [N]. The rope was selected on the basis of data contained in the technical charts by BRUGG. For experimental purposes we selected a rope of the structure 8.0 6x19S.

On the basis of relationships described above, in Table 2 we collected values calculated on the basis of data presented in Table 1.

Table 1. Geometrical parameters of the tensions juxtaposed on the basis of the performed project

Factor		Tension with weight	Tension with spring
G_o [kg]	G_s [kg]	22	48
R [m]		0.480	0.265
r [m]		0.114	0.115
G_k [kg]		0.5	
G_r [kg]		6	2
d [m]		0.08	
q_1 [kg/m]		0.25	
F_z [kN]		37.4	
μ_o		0.2	
γ_o [°]		40	
H_p [m]		70	
H_n [m]		3.6	
H_d [m]		1.4	

Table 2. Values of tension parameters obtained by means of empirical calculations

Specification	Tension with weight	Tension with spring
Rope weight resulting from the considered lifting height [N], relationship (1)	184	184
Strain force in rope [N], relationship (3) and (4)	1050	1130
Maximum force in rope [N], relationship (5)	4451	4789
Rope safety coefficient [-], relationship (6)	8.4	8

While the system of the tension with weight is simple to estimate the straining force value in the rope, however, the system with spring requires additional characterization analysis. To estimate the straining force value in the rope larger than that recommended by the manufacturer of speed limiter (in the discussed case: 1020 N), the force in spring should be selected so as the scope of minimum and maximum forces was on one hand larger than the recommended value of 1020 N, and on the other the value of safety coefficient was not higher than 8.

On the basis of the above records, as well as the conditions given in the introduction to point 3 of this study, in the Figure 7 we presented a fragment of characterization of a selected spring, taking into consideration the increased spring elongation, which shows the range of spring work where two conditions described above are met at the same time.

Within the conducted simulations of the work of the system, which is responsible for lift braking with a tension with spring, a test bed was prepared, which simulated the work of tension-

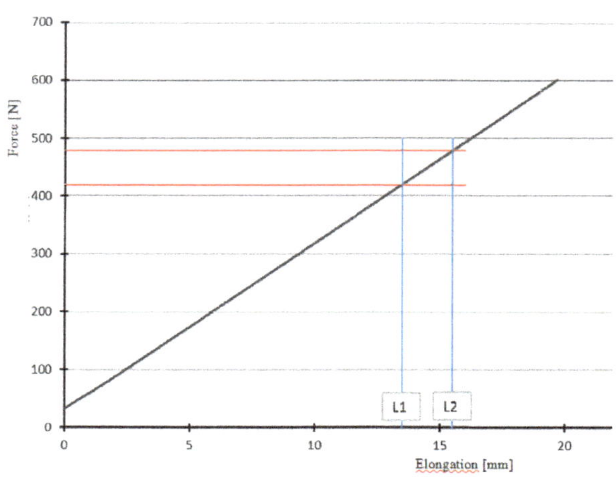

Fig. 7. Fragment of the characterization of spring for the case under discussion

rope-limiter system. The tests were performed in the conditions reflecting the work of a lifting appliance. The tests were performed within 1 year, what was presented in Figure 8.

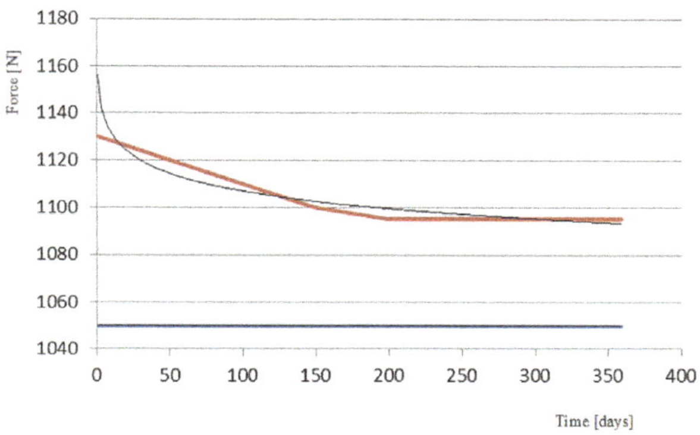

Fig. 8. Value of straining force Gn depending on time for the test performed within the period of 356 days: red line – tension with spring, blue line – tension with weight

Analyzing the above graph, we can notice that during testing in the whole period the spring characterization did not manifest losses of its properties, and at the same time, the value of straining force for the new tension solution did not fall below the value of 1020 N. in the initial testing period there was a visible decrease of straining force value from 1130 N to 1090 N, what is presented in drawing 8. That is caused by elongation of the new rope. This situation, stabilizing after about 200 days (for the conditions of experiment), and the rope tension reaches the value of 1090 N, which remained constant until the end of the experiment.

CONCLUSIONS

The conducted experimental studies and empirical calculations allow for the formulation of the following final conclusions:

- The presented initial studies on the application of tension with spring do not exhaust the subject matter. Subsequently, the dynamic studies should be performed to would take into account sudden impulses of force resulting from the work of lift system as a result of sudden braking.
- Analyzing the results obtained through empirical calculations, we can conclude that there is a possibility of applying the spring to eliminate the weight.
- A certain difficulty in selecting the correct straining force value is caused by the necessity of using the characterization of spring. Thus, it is important to create an appropriate algorithm of selecting the value of the force stretching the spring through changing its elongation.
- The presented concept of speed limiter rope straining appliance has a coherent structure securing appropriate work in the braking system. On the basis of the prepared tension model it is estimated that the manufacturing costs will be lower than these of a traditional solution by about 30%.
- Analyzing the development of construction market, including lift industry, it was therefore justifiable to undertake the subjects contained in the presented study.
- The conducted empirical analysis of spring selection and working conditions for the presented lift model does not exhaust the subject completely. Thus, it is justifiable to perform experimental studies, assembling a tension with a string on the lifting appliance to verify the stated project assumptions.

REFERENCES

1. Autivox brakes assembling instructions.
2. Instruction of operation and maintenance of the tension OB – 200–250 manufactured by Lift Service.
3. Instruction of DYNATECH speed limiter operation and maintenance.
4. Kaczmarczyk S., Andrew J.P.: Systems Engineering of Elevators. Published by Elevator World INC, Mobile, Alabama 2011.
5. Kwaśniewski J: Passenger and Freight Lifts – Structure and Operation. AGH Publishing House, Kraków 2006.
6. Lonkwic P.: Protecting Lift Systems against Uncontrolled Fall on the Example of a Passenger Lift. (in Polish) Narzędziowiec, 4, 2009.
7. Polish Standard PN EN 81.1+A3, Safety Regulations Concerning the Structure and Installation of Lifts, Part I. Electric Lifts.

15

WEAR TESTS OF SLIDING SURFACES IN THE ENVIRONMENT OF MIXTURES OF LUBRICATING OILS AND REFRIGERANTS

Kasper Górny[1], Przemysław Tyczewski[1], Wiesław Zwierzycki[1]

[1] Institute of Machines and Motor Vehicles, Poznan University of Technology, Piotrowo 3, 60-965 Poznań, Poland, e-mail: kasper.gorny@gmail.com; przemyslaw.tyczewski@put.poznan.pl; wieslaw.zwierzycki@put.poznan.pl

ABSTRACT

Elements of refrigeration compressors may be exposed to various types of wear processes, depending on the used oils and refrigerants. The presence of the refrigerant makes the lubricating properties and anti-oil mixture - factor much worse than in case of the lack of refrigerant oil. Current regulations on the use of ODS phase-out of working require refrigeration synthetic refrigerants HCFC (e.g. popular R22). The paper presents the effect of different mixtures of oils and refrigerants on friction coefficient and surface condition of cast iron and aluminium PA6. Tests were performed on a prototype machine using frictional node type block-on-ring, which is located inside a pressure chamber that simulates the refrigeration compressor. The results of the study confirm the possibility of using green refrigerant R290 (propane) as a substitute for R22, while keeping the existing mineral oil.

Keywords: refrigerant, refrigerating oil, the coefficient of friction.

INTRODUCTION

At the beginning of the XX century, compounds of chlorine, bromine, iodine and fluorine with carbon as refrigerants came into use. At the time, the substances, due to better thermodynamic properties, superseded such refrigerants as: ammonia, chloromethane, sulphur dioxide, propane and isobutane. Over 80 simple compounds which could be refrigerants were selected; among them were the following long time applied initiators R11, R12, R13 and R22. Only in 1974, the hypothesis by S. Rolanda and M. Moliny [10] on the emission harmfulness of the compounds destroying ozone in atmosphere, stopped their development. At present, tendencies concerning the application of the refrigerants are forced by respective regulations [11, 12]. On one side, they look for more ecological synthetic refrigerants and, on the other hand, there is a return to natural initiators (carbon dioxide, ammonia, hydrocarbons).

Changes of applied refrigerants usually require the design of refrigerating appliances to be adapted and lubricating oil to be changed. The refrigerant and lubricating oil create a mixture in friction nodes of refrigerating compressors. Complex mixture dependencies cause sooner compressors wear [1, 3].

The review of scientific literature on tribological investigations concerning the model friction nodes of refrigerating compressors showed that the majority of tests were performed on the prototype test stands [2, 9, 10, 11, 13]. No published work described a node type friction block-on-ring with the surface-contacting mating surfaces. According to the tests carried out by authors on real compressors wear, this node type is the most representative one for the operation of the radial slide bearings. The above mentioned test gap became an excuse to make the own elaboration, as well as the stand for wear tests of sliding friction nodes of refrigerating compressors in the environment of mixtures of lubricating oils and refrigerants. Using the elaborated method and test stand, tests are made to compar two kinds of lubricating oils in the mixture with natural refrigerant, such as propane R290. Literature gives little information on the influence of refrigerant R290 on the sliding surfaces of the refrigerating compressors. This paper presents also the tests carried out with the use of the refrigerant R22 being at present withdrawn from the refrigerating mechanisms.

TEST METHODS

In order to determine tribological properties of the mixtures of refrigerant and lubricating oil in the laboratory conditions, we made a prototype stand with the friction node type block-on-ring (Figure 1) situated in the high-pressure chamber (Figure 2).

The test stand, first of all, enables to present the operation conditions of the refrigerating compressors (compound temperature, refrigerant pressure, friction node loading, roller rotational speed), lubricating oil and refrigerant mixture production and the wear process occurrence in the conditions of the refrigerating compressors operation. The measuring position software enables to illustrate and register values such as the moment of friction, rotational speed, mixture temperature and refrigerant pressure in the chamber. The roller rotational speed is controlled with a frequency converter. The speed value can be set automatically or manually within the range 0 to 3000 rpm. The load value is set by the operator through weights within 10 to 150 N and changed of the minimum value 5 N.

The stand allows for the friction moment measurement within the range from 0 to 2 Nm. The operator can determine test duration and the frequency of the measurement registration. With the use of the thermostat connected with heaters located in the chamber, the temperature of the mixture of lubricating oil and refrigerant can be controlled.

The test stand with the friction node type block-on-ring is equipped with a high-pressure chamber with the sealing system preventing air entrance inside. The chamber enables obtaining conditions similar to vacuum ones as well as maximum refrigerant pressure of the value order 0.8 MPa. Sealing is very important due to high pressure occurring when the mixture is saturated with the refrigerant and the drive shaft output out-

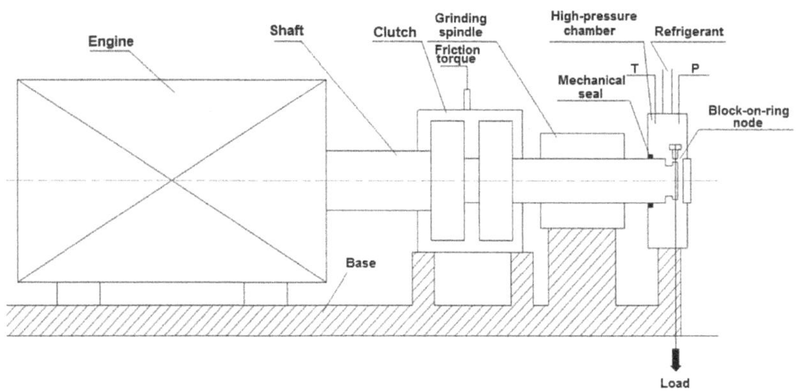

Fig. 1. Schematic tribometer test material consumption of refrigeration compressors. T – temperature in the chamber, P – measuring the pressure in the chamber, Mt – friction torque measurement

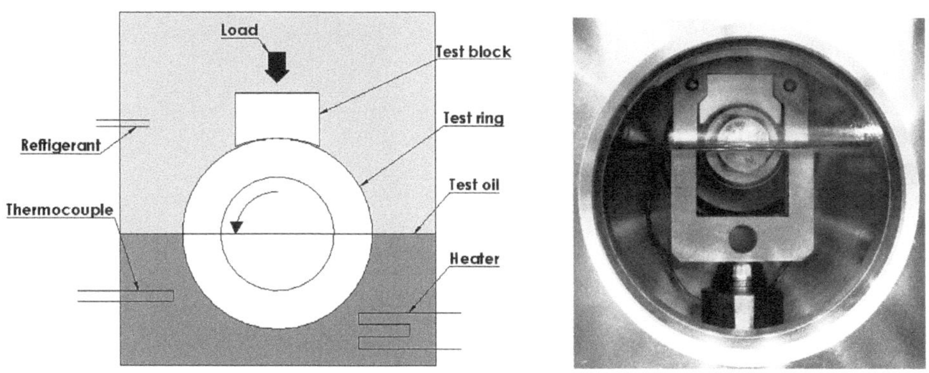

Fig. 2. Node type friction block-on-ring, a) schematic model of the chamber position, b) the chamber model with the actual node

side the chamber. All places susceptible to leakage are checked whenever the tested mixture is made.

Before starting the tests, all the elements contacting the latter mixture (the samples, handles and inside the high-pressure chamber) were thoroughly cleaned with naphtha extraction, and in case of samples an ultrasonic washer was used. After mounting the sample and counter-sample in the handles situated in the chamber and after closing it tightly, the vacuum ranging 10^{-5}MPa was made inside. Then, lubricated oil was poured into the chamber to the height covering half of the roller diameter (Figure 2). The mixture of lubricating oil with refrigerant was made according to the recommendations of Hong-Gyu [5] and Hoon Choa Sung [6], i.e. the high-pressure chamber was filled three times with refrigerant under pressure of 0.5 MPa and left there for 20 h for the right saturation of lubricating oil with refrigerant and in order to achieve a steady state. The tests were performed in the temperature of the beginning of the test at 298±1 K corresponding the starting conditions of the refrigerating compressors. During the test, pressure and temperature of the tested mixture were registered.

For the test, we used immobile concave block-shaped samples of aluminium PA6 (AlCu4MgSi) and rotating roller-shaped counter-samples of grey cast iron GJL-200. The tests were carried out at the load of 120 N thus achieving the block specific force on the roller being 1.155 MPa. The tests were performed for 3600 s at the roller rotational speed being 500 rpm. The friction path was 188.5 m.

The model friction node was tested in the environment of chosen mixtures of lubricating oils and refrigerants. They made tests with the use of the refrigerant R22, being now withdrawn, mixed with mineral oil and the refrigerant R290 in mixtures with mineral oil (MO) as well as with synthetic oil based on polyester (POE). The table 1 presents the most important properties of the tested refrigerants whereas the Table 2 gives the characteristics of the tested lubricating oils.

RESULTS AND DISCUSSION

Friction coefficient

The test stand computer system registered the friction moment values in the individual tests enabling the determination of the friction coefficient average values, of some mileages, in time. The Figure 3 presents average curves of the friction coefficient mileage for the tests of the mixture: lubricating oil RENISO KM 32 (MO) with the refrigerant R290, lubricating oil PLANETELF ACD 32 (POE) with refrigerant R290, and lubricating oil RENISO KM 32 (MO) with the refrigerant R22.

The friction coefficient had different values in the moment of the engine start and then after about 1500s it stabilized. In the tests carried out with the mixture R290/POE the friction coeffi-

Table 1. Basic properties of the tested refrigerants [4]

Property	Unit	Refrigerant	
		R22	R290
Chemical formula	–	CHF_2Cl	$CH_3CH_2CH_3$
Critical point	K	369.3	369.8
Critical pressure	MPa	4.99	4.28
Critical specific volume	m^3/kg	0.0019	0.004531
Ozone layer destruction potential ODP	–	0.055	0
Greenhouse effect creation potential GWP	–	1700	3

Table 2. Basic characteristics of lubricating oils studied [7, 8]

Parameter	Unit	Lubricating oil	
		Reniso KM 32 (MO)	Planetelf ACD 32 (POE)
Kinematic viscosity in the temp. 313 K in the temp. 373 K	m^2/s	$32 \cdot 10^{-6}$ $4.9 \cdot 10^{-6}$	$34.6 \cdot 10^{-6}$ $6.0 \cdot 10^{-6}$
Flow temperature	K	228.1	219.1
Flash-point	K	475.1	523.1
Density (in 288 K)	kg/m^3	881	981

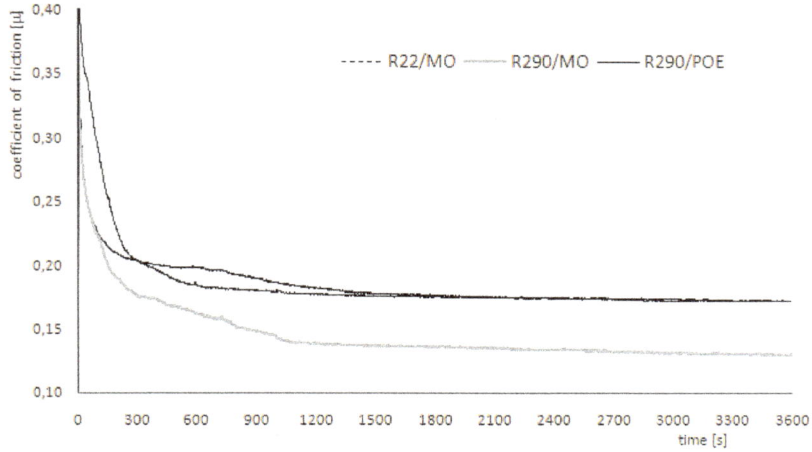

Fig. 3. Mileage coefficient of friction during the test for mixtures: lubricating oil RENISO KM 32 (MO) with refrigerants R290, lubricating oil RENISO KM 32 (MO) with refrigerants R22 and lubricating oil PLANETELF ACD 32 (POE) with refrigerants R290

cient was 0.137 whereas in case of the mixtures R290/MO and R22/MO it was 0.174. Such friction coefficient values suggest the grounds of substituting less ecological refrigerant R22 with R290 for the node friction wear criterion in the refrigerating compressors. In turn, one can state that when changing the refrigerant from R22 to R290, from the point of view of tribology, it is not necessary to change lubricating oil. However, at the occasion of changing refrigerant the change of lubricating oil should be taken into account individually when analysing the operation conditions of individual element of the refrigerating system.

Surface of friction

The friction nodes surfaces, i.e. roller and block surfaces were subjected to observation with the use of the metallographic microscope CARL ZEISS JENA. In case of block surface observation was made at enlargement ×63 and ×126. However, the roller observation was made at enlargement ×63, ×125, ×252 and ×504. The figure 4 presents the comparison of the new surface with the surfaces of the mating elements (grey cast iron / PA6) in the mixtures: R290/MO, R290/POE and R22/MO.

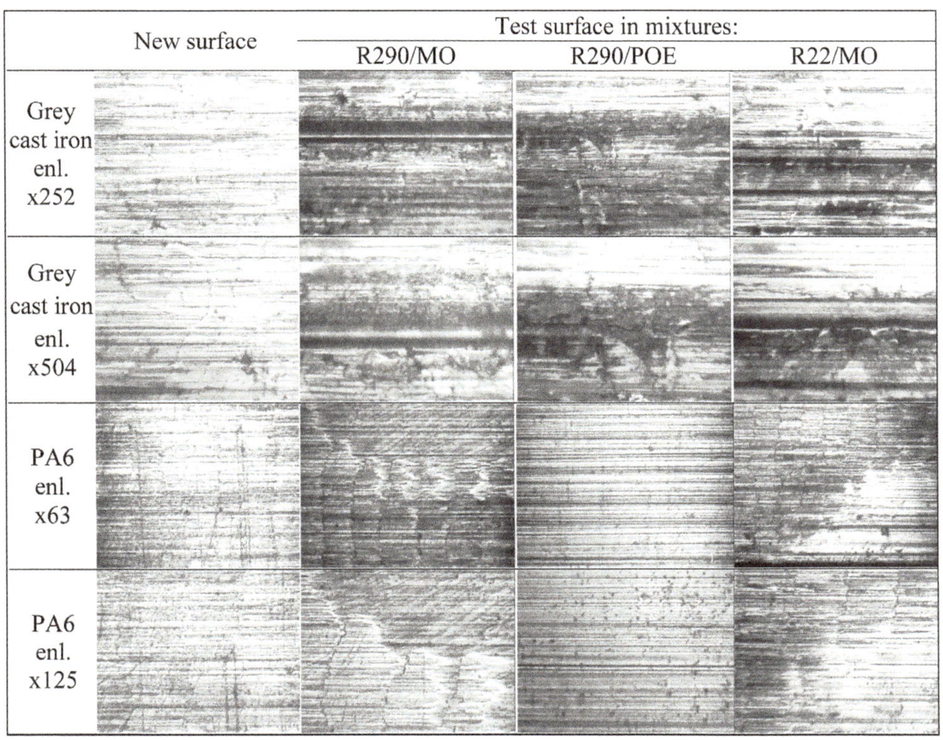

Fig. 4. Comparison of new surface after the test surfaces in a mixture R290/MO, R290/POE and R22/MO

Surfaces after tests with refrigerants are not equal. One can specially notice heterogeneity on all grey cast iron surfaces and on aluminium PA6 surface after tests with the use of mineral oil.

On the grey cast iron surfaces being in contact with mineral oil, researchers noticed traces of grooving in the presence of refrigerant R290 as well as R22. However, they were not noticed on the surfaces operating in the mixture of refrigerant R290 with oil POE.

Watching the surface after testing in the mixtures of refrigerant with oils for refrigerating compressors, the presence of deposits on the friction surfaces was found. The dark deposit layer occurred mainly on the surface of aluminium (PA6), except for the surface after testing in the mixture R290 with oil POE. This surface is relatively homogeneous. There are many shallow, uniform grooves suggesting homogeneous, slow surface wear without wear products (abrasive or deposit).

CONCLUSIONS

We noticed about 20% decrease of the friction coefficient between grey cast iron and aluminium in the mixture of propane with oil POE in comparison with mineral oil. However, the change of refrigerant only, R22 to R290 did not influence the change of friction coefficient. On the friction surfaces we observed the occurrence of deposits, except for aluminium surfaces contacting the mixture of refrigerant R290 with oil POE. Probably, the lack of deposits on the surface caused a significant decrease of friction coefficient.

At present, due to valid regulations concerning the application of substances impoverishing the ozone layer, there is the exchange of refrigerant R22 from the refrigerating systems to other alternative refrigerant. The performed tribological tests confirm that the interchangeable equivalent of refrigerant R22 can be the ecological one R290 (propane). Leaving mineral oil in the installation will not deteriorate tribological properties of the resultant mixture. However, additional exchange of mineral oil from the installation to oil POE will improve tribological properties of the new mixture, in comparison to the mixture of propane with mineral oil.

REFERENCE

1. Allison Y. Suh, Jayesh J. Patel, Andreas A. Polycarpou, Thomas F. Conry. Scuffing of cast iron and Al390-T6 materials used in compressor applications. Wear, 260(7-8), 2006, 735-744.
2. De Melloa J.D.B., Binderb R., Demasc N.G., Polycarpouc A.A. Effect of the actual environment present in hermetic compressors on the tribological behaviour of a Si-rich multifunctional DLC coating. Wear 267, 2009, 907-915.
3. Górny K., Tyczewski P., Zwierzycki W. Specification of lubricating oil operation in refrigeration compressors. Tribologia, 3, 2010, 63-73.
4. Grzebielac A., Pluta Z,. Ruciński A., Rusowicz A. Czynniki chłodnicze i nośniki energii. Oficyna Wydawnicza PW, Warszawa 2009.
5. Hong-Gyu Jeon, Se-Doo Oh, Young-Ze Lee. Friction and wear of the lubricated vane and roller materials in a carbon dioxide refrigerant. Wear, 267(5-8), 2009, 1252-1256.
6. Hoon Choa Sung. Tribological characteristics of various surface coatings for rotary compressor vane. Wear, 221(2), 1998, 77-85.
7. http://www.fuchs-europe.com/fileadmin/fuchs_upload/downloads/Englische_Prospekte/2013/RENISO _Refrigeration_Oils_2012-2013.pdf (20.06.2013).
8. http://www.yukselteknik.com/dokumanlar/Total/katalog/planetelf_ACD.pdf (20.06.2013).

ALTERNATIVE FUELS FOR DIESEL ENGINES

Jacek Caban[1], Agata Gniecka[2], Lukáš Holeša[3]

[1] Institute of Transport, Combustion Engines and Ecology, Faculty of Mechanical Engineering, Lublin University of Technology, 36 Nadbystrzycka Str., 20-618 Lublin, Poland, e-mail: j.caban@pollub.pl

[2] Maria Curie-Skłodowska University, Plac Marii Skłodowskiej-Curie 5. 20-031 Lublin, Poland

[3] Department of Road and Urban Transport, Faculty of Operation and Economics of Transport and Communications, University of Žilina, Univerzitná 1, SK-010 26 Žilina, Slovakia, e-mail: Lukas.Holesa@fpedas.uniza.sk

ABSTRACT

This paper presents the development and genesis of the use of alternative fuels in internal combustion ignition engines. Based on the analysis of the literature, this article shows various alternative fuels used in Poland and all over the world. Furthermore, this article describes the research directions for alternative fuels use in road transport powered by diesel engines.

Keywords: biofuel, diesel engine, eco-fuels.

INTRODUCTION

The increase in the emission of pollutants into the environment coming from road transport, depletion of natural resources, and economic considerations are the main reasons for the development of alternative fuels to power internal combustion engines. Fuel consumption by modern road vehicles is an interesting parameter for all users and producers. [18]. In order to meet the energy needs, there has been increased interest in alternative fuels, such as biodiesel, methanol, ethanol, biogas, hydrogen and gas to ensure adequate replacement of diesel fuel for internal combustion engines [2]. Vegetable oils were applied for the first time as power by Rudolf Diesel – in the engine he constructed [6]. Vegetable oils are a very promising alternative to diesel fuel because they are renewable and have similar properties [2]. Vegetable oils offer almost the same power with a slightly lower thermal efficiency when used in SI engines [21]. The interest in alternative fuels were noted again during the great oil crisis in the 1970's all over the world and this interest lasts until now, for economic and environmental reasons.

This paper presents the development of the use of alternative fuels in internal combustion engines. A careful analysis of the literature allowed us to present the main trends in the development of alternative fuels in the country and all over the world. This article describes the research directions in the use of alternative fuels in transport vehicles powered by diesel engines.

PHYSICO-CHEMICAL CHARACTERIZATION OF ALTERNATIVE FUELS

Research of alternative fuels based on biocomponents to power diesel engines focus mainly on identifying their impact on the environment [11]. Alternative fuels can be divided into the following categories:

- alcohols: methyl and ethyl,
- bioethanol,
- biogas,
- alkyl esters of fatty acids (FAME, FAEE),
- CNG gas (Compressed Natural Gas),
- vegetable oils (biodiesel),
- hydrogen.

In Europe and Poland, because of the climate and agronomic reasons, rapeseed oil methyl esters and their mixtures with diesel fuel [5] are most commonly used as liquid biofuels. Biofuel or motor fuel biocomponents will constitute 7.55% of the total fuel market in 2014 [8, 9]. Despite significant advantages, in the papers [1, 4] the drawbacks of biofuels and their negative impact on the sustainability and viability of the internal combustion engine, and the environment, are presented.

In Europe, the basic raw material for the production of biofuels is rapeseed. Basic biofuel used to power diesel engines are esters of fatty acids (FAME, FAEE). Esters of pure vegetable oils have good solubility in diesel fuel. This feature allows to create commercial mixtures of biofuels such as B10, B20. Due to high viscosity, low freezing point, water content and organic acids using pure rapeseed oil is not desirable. It may lead to seizure or damage of injection equipment and engine. The fuel in the supply system is also the only lubricating medium. The amount of hydrocarbons containing carboxyl group determines the lubricity of the fuel, what makes them chemisorbed on a clean metal surface [5]. Ethyl and methyl esters differ in products of incomplete combustion. In the process of incomplete combustion of the methyl esters, toxic substances, such as formaldehyde and radicals, are exuded, which do not occur in the case of incomplete combustion of the ethyl esters [23]. Exhaust gasses of diesel engines powered by RME contain three times more free methyl radicals than exhaust gasses of engines powered by REE [17]. The comparison of some properties of ON diesel oil, vegetable oil OR, and rapeseed oil methyl ester RME is presented in Table 1.

Subjecting the oil to esterification allows to break heavy molecular structure of tri-glycerides. The reduction of particle size is accompanied by the reduction in viscosity – which substantially simplifies and improves the fuel injection spray [8]. In this process, the reaction of esters neutralization occurs. The unquestionable advantage of esters of rapeseed oil is very low sulphur content translated into a reduction of particulate emissions [8]. The disadvantage is hygroscopy, which favors the proliferation of bacteria and fungi which translates into low chemical stability of the fuel. In Polish climatic conditions, the use of pure ester as a fuel without the additive depressant decreasing the pour point causes a lot of problems, especially in the conditions of autumn – winter [24]. Another problem associated with the esters are the adverse effects on the structure of the plurality of polymers (for example, rubber, some of the seals, coatings, etc.). Important from the operation point of view is the property of sediment dissolution resulting from diesel engines power. Diluted sludge can cause clogging of filters, and damage to the injectors. For this reason a number of engine manufacturers make restrictions on the use of biofuels and in case of their use they demand more frequent servicing. These prob-

Table 1. Comparison of selected physical and chemical properties of fuels ON, OR and RME [23]

No.	Parameter	Unit	ON Diesel	OR Rapeseed oil	RME Rapeseed oil methyl ester
1	Viscosity T_0 = 40 °C	mm^2/s	2–4.50	20	4.32
2	Mass calorific value	MJ/kg	42–43	36–38	36–38
3	Cetane number	–	Min 51	40–50	50–55
4	Theoretical air consumption	kg/kg	14.4–14.6	12.2–13	13.4–13.8
5	Density T_0 = 15 °C	kg/dm^3	0.8–0.845	0.90	0.88
6	Surface tension T_0 = 20 °C	N/m	24×10^{-3}	36×10^{-3}	–
7	The cloud point	°C	-12	18	-9
8	Cold filter plugging point	°C	-31 (IZ 40)	9	(-9) – (-11)
9	Freezing point	°C	-40	4	-15
10	The sulfur content	%	0.001	0.0002	0.0002
11	Carbon residue	%	0.01	0.17–0.5	0.5
12	Ignition point	°C	>55	200–300	130
13	Initial boiling point	°C	175	300	300
14	Average elemental composition	% m/m	86.4	77.6	76.8
	C	% m/m	13.4	11.7	12.1
	H	% m/m	–	10.5	11.0

lems, however, did not stop the development of biofuels. The authors in [8] indicate that the use of B100 fuel and CE (camelina esters) is accompanied by higher specific fuel consumption than in case of diesel. However, the authors of [9] confirm that the emission of nitrogen oxides is higher when the engine is fueled with biofuels than with diesel engine. Emission of hydrocarbons for fuel B100, CE and HE is on the same level, particulate emissions are highest for CE [9]. Reduction of CO and THC value during powering by esters of vegetable oils is related to the oxygen content chemically bonded in the molecules of esters. Motor power supply with B100 bioester very positively influenced the smoke opacity. According to the authors [11], at low and medium engine rotational speed of the engine crankshaft, the smoke opacity was more than three times lower on average, and at the high rotational speeds was lower by tens of percent, compared to smoke opacity of diesel-powered engine. Less clear is the effect of the application of bioester concerning the emissions of hydrocarbons and nitrogen oxides [11]. However, taking into account the fact that in real conditions the test engine works more time in the field of small and medium speeds than large ones [16], the impact of bioester on the emission of hydrocarbons is favorable and unfavorable in case of nitrogen oxides.

In [10], the authors draw attention to the mechanisms responsible for blocking by a mixture of biodiesel the fuel distributor or fuel filter of the vehicle. These mechanisms are complex and vary depending on the following factors [10]:

- FAME source and type of treatment used for their production,
- characteristics of the solvency of the base oil,
- the level of impurities and contaminants in the fuel,
- fuel storage and working conditions.

The authors [10] state that the increase in the tendency to block the filter of FAME blended mixtures is likely due to complex interactions of trace constituents present in various types of FAME and is also strongly dependent on the storage temperature, storage time, and the properties of the base oil. Filter blocking issues have been seen in a field trial using experimental B20 and B13 blends HVO7 fuel, where the FAME portions of the blends were produced using a 50:50 RME: SME mix. Residues from the blocked filters were confirmed to be a good match with sterile glucosides using IR analysis, and saturated monoglycerides were also found in some filter samples using GC analysis.

A further study attempting to assess the individual contribution of various trace components in FAME towards an increase in FBT, showed that there was no effect on FBT due to the addition of 0.3% by mass monoglyceride to B20, made from distilled RME. The presence of sterile glucosides and water in combination had the most significant detrimental effect to FBT in B20 RME blends. This effect was identified at a constant MG level of 0.3% by mass.

Ethanol has been considered an important alternative fuel for internal combustion engines for a long time, because it solves a number of problems of traditional crude based fuels such as emission of greenhouse gases and particulates. As a fuel for diesel engines, it suffers from a very low cetane number (around 5…8, while diesel is typically higher than 51), and consequently a high auto-ignition temperature (around 640 K while diesel is around 500 K). Ethanol has a heating value of about 60% of diesel. Igniting ethanol in a 4-stroke compression engine either requires use of a higher compression ratio, use of glow plugs or a catalyst [15]. Despite this, ethanol has successfully been used for example in Sweden as an alternative to diesel in bus fleets [20].

Diesel fuel-alcohol blends (known by a number of names – including E-diesel, M-diesel, Oxy-diesel and diesohol) also can be used as fuel in diesel engines [13, 19]. Generally, ethanol can be blended with diesel without engine modifications [12]. Because ethanol is highly hygroscopic, additives must be used in order to ensure solubility of anhydrous ethanol in diesel fuel under a wide range of conditions. Miscibility is limited, especially at lower temperatures. The addition of ethanol to diesel fuel can reduce lubricity and create potential wear problems in sensitive fuel pump and injectors designs. Ethanol possesses also a lower viscosity and calorific value, with the latter imposing minor changes to the fuel delivery system in order to allow injection of larger quantities of fuel, what is likely to be required for engine performance and for fuel injector/pump durability. Because ethanol has a very low cetane number it reduces the cetane level of the diesel-ethanol blend.

However, a major drawback in ethanol-diesel fuel blends is that ethanol is immiscible in diesel over a wide range of temperatures and water content due to their difference in chemical structure and characteristics [12]. These can result in fuel instability due to phase separation. Prevention of separation can be accomplished in two ways: by adding an emulsifier, which acts in order to suspend small droplets of ethanol within the diesel fuel, or by adding a co-solvent, which acts as a bridging agent through molecular compatibility and bonding to produce a homogenous blend [12]. Emulsification usually requires heating and blending steps to generate the final blend, whereas co-solvents allow fuels to be "splash blended", thus simplifying the blending process [7]. Blends of diesel and bioethanol can be used as fuel for compression ignition engines, but with limited scope, primarily due to the reduced mingle capability of ethanol with diesel, as well to the worse ignition properties of the mixture.

Promising results have been obtained for utilitarian bi-fuel (with initiating dose ON) diesel engines fueled, especially CNG [22]. Biogas is a renewable alternative fuel currently used to produce electricity and heat. Using it as fuel should strive to bring these properties, which , is unfortunately complicated by the widespread use of biogas. To meet the expectations associated with the ecological and economical operation of vehicle a number of modifications of standard engine is required. These include the change of shape of the piston crown and the engine head as well as a change of the fuel supply system. Additionally, the catalyst choice and special fuel tanks are required. Both manufacturers of new engines as well as users of many older vehicles will be able to take some advantage of natural gas application which in turn may lead to an increase in the number of new and adopted vehicles fuelled with this fuel.

Despite the many drawbacks, the technology of CNG fueling is becoming increasingly popular amongst the crowd of consumers, individual and fleet users, and public transport, as an economic and ecological fuel. The low price of CNG fuel and available technology of vehicle conversion play a dominant role.

CONCLUSIONS

Modern pro-ecological attitude associated with environmental protection against increased toxic exhaust emissions and over-exploitation of natural resources have forced to look for innovative liquid fuels. For efficient use of different fuels in existing engines, fuel or power systems of internal combustion engines modifications are applied. As demonstrated in the paper, the problem is to produce fuel similar to a classic one, whose properties can be used as best as possible, withdrawing from the available raw material.

Experience from the test stand and the operation show that the power of internal combustion engines with alternative fuels is a complex and long-term process and possible use of renewable natural resources are among the most environmentally friendly. It is stated that you cannot clearly determine which fuel is the best. In case of small motors it can be quite different fuels than for tractors' engines or trucks and buses. Climatic conditions of exploited vehicles are of great importance when using biofuel. Advantageous effects of RME on exhaust emissions from older diesel engines have been confirmed in many previous studies. However, in case of modern diesel engines, the influence of RME seems to be less recognized.

The main scientific problems relating to alternative fuels include: accurate knowledge of physical, chemical, thermodynamic, supplies and logistics processes. Issues mentioned in the article, the problems associated with the use and production of alternative fuels show that research on the development of alternative fuels is a complex and constantly current research field.

REFERENCES

1. Baczewski K., Kałdoński T. Paliwa do silników o zapłonie samoczynnym. WKiŁ, Warszawa 2008.
2. Banapurmath N.R., Tewari P.G., Hosmath R.S. Experimental investigations of a four-stroke single cylinder direct injection diesel engine operated on dual fuel mode with producer gas as inducted fuel and Honge oil and its methyl ester (HOME) as injected fuels. Renewable Energy, 2008, 33: 2007-2018.
3. Barabas I., Todorut I-A. Chassis Dynamometer and Road Test Performances of Biodiesel-Diesel Fuel-Bioethanol Blend. SAE Technical Paper, 2010-01-2139.
4. Gardyński L., Kiernicki Z. Wybrane właściwości smarne mieszanin oleju napędowego i rzepakowego. 5th International Scientific Conference on Combustion Engines KONSSPAL, Tadeusz Kościuszko Military Academy, Wrocław 2002.
5. Gil L., Ignaciuk P. Wpływ liczby kwasowej na smarność biopaliw. Postępy Nauki i Techniki, 2011, 11: 37-42.

6. Gradziuk P. Biopaliwa. Wydawnictwo „Wieś Jutra", Warszawa 2003.

7. Hansen A.C., Zhang Q., Lyne P.W.L. Ethanol diesel fuel blends - a review. Bioresource Technology, 2005, 96: 277-285.

8. Ignaciuk P., Gil L., Komsta H. Porównanie osiągów silnika o zapłonie samoczynnym zasilanego olejem napędowym i biopaliwami opartymi na estrach oleju rzepakowego i estrach oleju lnianki. Postępy Nauki i Techniki, 2012, 12: 40-45.

9. Ignaciuk P., Gil L., Liščak Š. Porównanie emisji związków toksycznych silnika ZS zasilanego olejem napędowym i biopaliwami opartymi na estrach oleju lnianki i estrach oleju rzepakowego. Postępy Nauki i Techniki, 2011, 11: 43-48.

10. Jolly L., Kitano K., Sakata I., Strojek W., Bunting W. A Study of Mixed-FAME and Trace Component Effects on the Filter Blocking Propensity of FAME and FAME Blends. SAE Technical Paper, 2010-01-2116.

11. Koszałka G., Hunicz J., Kordos P. Porównanie własności użytkowych i ekologicznych silnika o zapłonie samoczynnym zasilanego olejem napędowym i bioestrem. MOTROL, 2009, 11c: 86-94.

12. Kwanchareon P., Luengnaruemitchai A., Jai-In S. Solubility of a diesel-biodiesel-ethanol blend, its fuel properties, and its emission characteristics from diesel engine. Fuel, 2007, 86: 1053-1061.

13. Li D., Zhen H., Xingcai L., Wu-gao Z., Jian-guang Y. Physico-chemical properties of ethanol-diesel blend fuel and its effect on performance and emissions of diesel engines. Renewable Energy, 2005, 30: 967-976.

14. Masimalai S.K. Preparation, Characterization and Engine Test Analysis of Methyl Esters of Unrefined Palm Oil and D-Limonene Oil Mixture as CI Engine Fuel. SAE Technical Paper, 2010-01-2121.

15. Nguyen D., Honnery D. Combustion of bio-oil ethanol blends at elevated pressure. Fuel, 2008, 87: 232-243.

16. Niewczas A., Kordos P., Koszałka G. Laboratory method of reliability research of a diesel automotive engine. Transport and Telecommunication, 2006, 7(2): 312-317.

17. Park S.H., Cha J., Kim H.J., Lee C.S. Effect of early injection strategy on spray atomization and emission reduction characteristics in bioethanol blended diesel fueled engine. Energy, 2012, 39: 375-387.

18. Podstanickỳ I., Liščák Š., Droździel P. Modern method of fuel consumption measurement in vehicle transport. Maintenance and Reliability, 2005, 1(25): 16-20.

19. Rakopoulos D.C., Rakopoulos C.D., Kakaras E.C., Giakoumis E.G. Effects of ethanol-diesel fuel blends on the performance and exhaust emissions of heavy duty DI diesel engine. Energy Conversion and Management, 2008, 49: 3155-3162.

20. Simonsen H., Chomiak J. Testing and evaluation of ignition improvers for ethanol in a DI diesel engine. 1995, SAE 952512.

21. Srivastava A., Prasad R. Tryglicerides-based diesel fuels. Renewable and Sustainable Energy Reviews, 2000, 4(2): 111-133.

22. Stelmasiak Z. The Combustion Controlling in the Dual Fuel CI Engine by Pilot Dose Division. PTNSS-2011-SC-218. Combustion Engine, 2011, 3(146).

23. Struś M.S. Ocena wpływu biopaliw na wybrane właściwości eksploatacyjne silników o zapłonie samoczynnym. Oficyna Wydawnicza Politechniki Wrocławskiej, Wrocław 2012.

24. Tys J., Piekarski W., Jackowska I., Kaczor A., Zając G., Starobrat P. Technologiczne i ekonomiczne uwarunkowania produkcji biopaliwa z rzepaku. Acta Agrophysica, Lublin 2003.

17

ANALYSIS OF Z-AXIS KINEMATICS INFLUENCE ON MEASUREMENT RESULT

Maciej Gruza[1], Wiktor Harmatys[1], Piotr Gąska[1], Adam Gąska[1], Jerzy Sładek[1]

[1] Laboratory for Coordinate Metrology (M-10), Faculty of Mechanical Engineering, Cracow University of Technology, Jana Pawła II 37, 31-864 Kraków, Poland, e-mail: gruzam@interia.pl; wiktorharmatys@gmail.com; pjgaska@gmail.com; agaska@mech.pk.edu.pl; sladek@mech.pk.edu.pl

ABSTRACT

CMMs accuracy depend on a number of factors. In spite of using various methods for CMM correction, the differences can be observed in the metrological capability between the areas in the measuring volume of CMM. Authors attempted to estimate the impact of one of potential CMM accuracy influencing factors, a CMM Z-axis position during measurement, on measurement result. This article presents performed experiments and obtained results.

Keywords: CMM accuracy, CMM kinematics.

INTRODUCTION

The importance of the Coordinate Measuring Technique (CMT) in today's dimensional metrology has been steadily increasing. It is successfully used in many branches of industry around the world. Measurements using a Coordinate Measuring Machines become invaluable especially when measurement automation is required, as well as their high accuracy. Furthermore, a growing number of CMT users recognize the necessity of the measurement uncertainty estimation and the traceability assurance. Reputable research centres conduct studies aimed at better understanding of phenomena that occur during coordinate measurements. They try to disseminate knowledge about the components of the measurement uncertainty and implement the results of their experiences into industrial practice. Especially in recent years, it has brought tangible benefits to both parties. This way research laboratories can validate their research results, while cooperating companies often without bearing the costs, gain new technological solutions, which may improve accuracy of the measurement or reduce its duration. Currently, due to technical constraints and insufficient knowledge, many components of the measurement uncertainty cannot be precisely determined, however, their impact on the measurement accuracy may be at least estimated. The method which is based on experience and involves estimation of the impact of individual components on measurement accuracy is called the error budget. It is a method that requires extensive knowledge and long-standing practice, thus it is hard to apply on a larger scale, the application can be found mainly in calibration laboratories. The error budget can be utilized also for the so-called Virtual machine preparation. This solution allows multiple on-line simulation of measuring task. Results obtained in such a way can be then utilized during uncertainty estimation for example using methods of statistical analysis [1].

The most important part of the implementation of the virtual machine is the development phase which involves appropriate experiments aimed at identification of the main sources of errors that may occur during measurement. Virtual machine is usually based on several modules which model the major errors' contributors, such as machine kinematics errors and errors of the probe head. However, the machine should be also tested thoroughly for other factors and continuously developed. In this paper, the authors describe the experiment that was conducted in order to estimate the impact of the measured object position in the measuring volume of CMM on the measurement results.

THE COORDINATE MEASURING MACHINES AND ERRORS ASSOCIATED WITH THEIR FUNCTIONING

In CMT, data in a form of measured points coordinates is collected from the surface of the measured object. Sampled coordinates are then used for determination of dimensions of spatial shaped objects [2]. Measuring points can be gathered using different devices and utilizing various methods. The main tool of CMT is a Coordinate Measuring Machine (CMM). In conventional CMMs the kinematic pairs shift in mutually perpendicular directions, which define the axes X, Y and Z of the machine base Cartesian coordinate system. The machine movements in each axis are measured using linear scales. Essential element of each CMM is a probe head which provides a connection between measuring system and the measured object [3].

The machines may differ in construction and the methods used for probing. Methodology described in this paper refers to the CMMs with moving bridge and tactile probe head. The CMMs of such a type are commonly used in industry, moreover, they were also subject of a number of research programs. The main sources of errors are derived from machine kinematics and functioning of probe heads. The kinematics errors are associated with 21 components of geometrical errors which constitute so-called rigid body model of kinematic errors, further information on this subject can be found in [4]. The kinematic errors can be significantly reduced using different compensation methods for instance the Computer Aided Accuracy (CAA) matrix. In turn, probe head errors arise mainly due to phenomena related to the probe head work principle such as pretravel [5]. Additionally, probe heads are prone to errors connected with interactions between the measured surface and the tip ball [6].

The probe heads qualification process is used in order to minimize the above-mentioned influences. Other factors affecting the accuracy of the machines are the errors associated with the measurement dynamics [7]. During measurements performed on any CMM certain dynamic phenomena occur due to the movement of elements of a large masses. Between the measuring points the machine moves with high speed. Whereas the contact process itself is more complex and can be divided into four main stages: the approach of the tip ball with constant speed to the measuring surface, tip ball contact with measuring surface, identification of contact point position, retract from the measured surface. Different probing parameters, such as: probing speed, acceleration and approach distance have a substantial impact on the measurement results. The authors suspect that the location of the measured object in the measuring volume of the machine also contributes to the CMM errors. The experiment that is shown in this paper was conducted in order to estimate how changes in position of spherical standard would affect the measurement accuracy

EXPERIMENT AND RESULTS

In this study, the authors intended to investigate the errors associated with the z-axis kinematics. The experiment procedure involved measurements of the spherical standard which was positioned at selected locations in machine's measuring volume, in such a way that all possible measurement conditions would remain unchanged. An essential element of studies was to find a suitable holder, which would allow to change the position of a standard in the z-axis of CMM.

All measurements were performed on the ZEISS WMM 850S machine which is placed at Laboratory of Coordinate Metrology in Cracow University of Technology. During the measurements the machine was equipped with Renishaw PH10M articulated probe head with mounted TP20 STD probe. The ambient conditions were constantly monitored and during the whole experiment the temperature was at level of 20±0.5°C. The probing parameters were not changed between measurements. The spherical standard of 24.9946 mm which meets the requirements of form error according to [8] was used. As an attachment the holder for KOBA step gauge was chosen, due to its technical solutions and stability. It allows to fix a standard on eight different heights and adjust the tilt axis of the column. Three positions of the standard were used in experiment: the lowest and the highest possible height, and an intermediate position. The difference between the extreme heights was greater than half of the working range of the z axis. The three positions of the standard are shown on the Figure 1.

The standard sphere was measured at 82 points distributed evenly over the surface of the upper hemisphere of the standard. The location of measuring points was specified using two angles

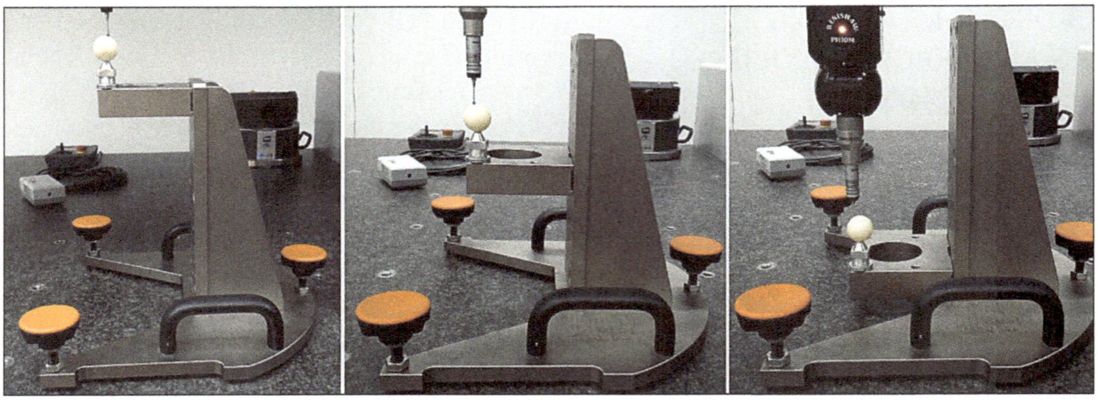

Fig. 1. The three positions of the standard during measurements

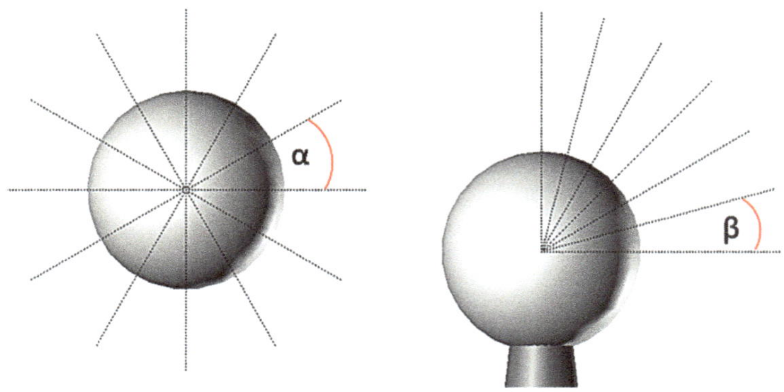

Fig. 2. The horizontal α and vertical beta β angles, which define measuring points

– horizontal and vertical ($\alpha = 40°$, $\beta = 10°$). The angles definition is presented in Figure 2.

The measurement sequence was repeated ten times in each of the standard positions. The measured sphere was evaluated in terms of diameter and sphericity deviation. Obtained results are presented in Table 1 and Table 2.

Additionally, in order to determine the distribution of errors the graphs that represents mean values of errors for each position was prepared (Figs. 3 – 5).

Table 1. The measured diameter for three positions of the standard

Position	Mean [mm]	Standard deviation [mm]
The lowest position	24.9963	0.00020
The intermediate position	24.9975	0.00046
The highest position	24.9966	0.00015

Table 2. The measured sphericity for three positions of the standard

Position	Mean [mm]	Standard deviation [mm]
The lowest position	0.0051	0.00046
The intermediate position	0.0060	0.00048
The highest position	0.0079	0.00023

CONCLUSIONS

On all the prepared graphs a characteristic distribution of errors with three positive and three negative extremes can be observed. Probably, it is caused by the construction of TP20 probe. In this solution the stylus is combined with three arms, which are oriented relative to each other by 120

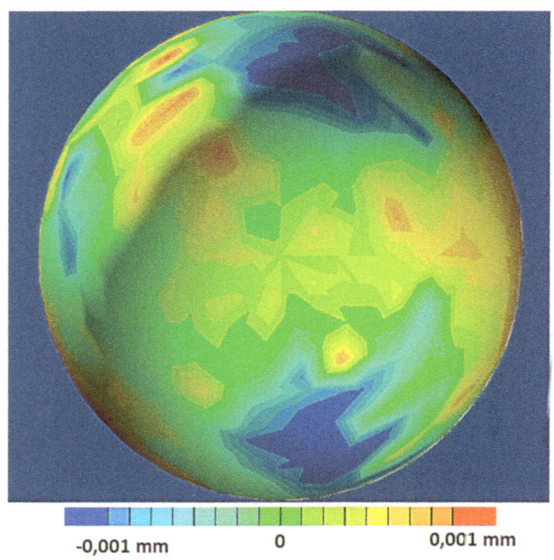

Fig. 3. The errors distribution for the lowest standard position

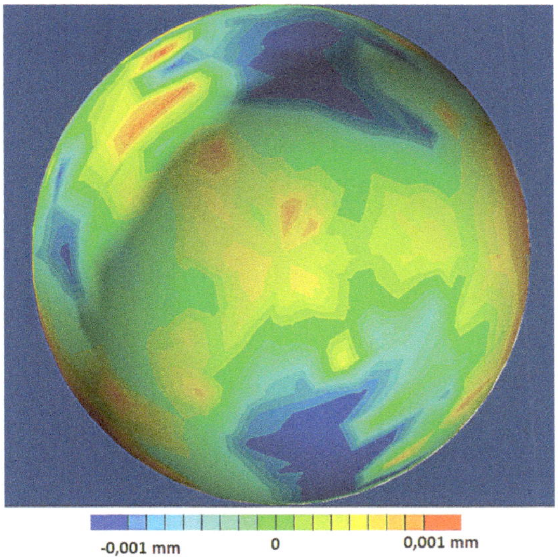

Fig. 4. The errors distribution for the intermediate standard position

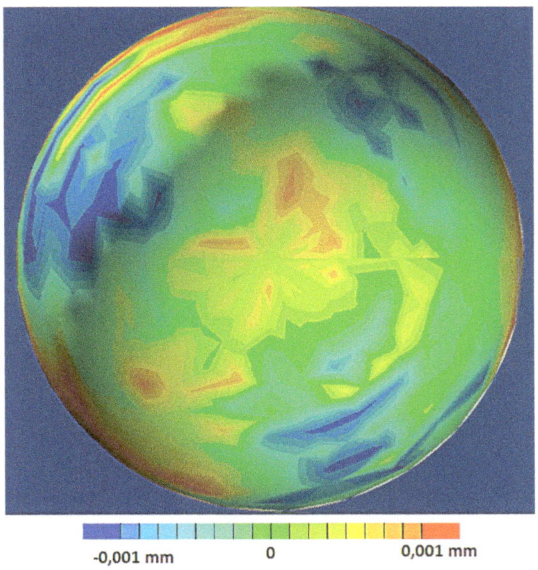

Fig. 5. The errors distribution for the highest standard position

degrees, and ended spherically. The spring presses arms to a prisms on which they rely. The contact impulse is generated when one of the arms lose contact with the prism and the electrical circuit is broken. The biggest errors were obtained at the equator of the standard.

The results presented in the Table 1 and Table 2 are not unequivocal. The diameters obtained for different positions did not differ significantly. The authors suspect that the results would be noticeably worse for the lowest location of the standard, because the z axis of the machine would be ejected the farthest. The results obtained for the sphericity suggest a completely the opposite conclusion. Perhaps the obtained results might be caused by the quill counterbalance system. On the other hand, the values of standard deviation for both parameters for this position are the smallest, what indicates that the machine works stably with low ejection of the quill.

The presented results are consistent with the experience of LCM, regarding the calibration of CMMs working in industrial conditions. Although, the obtained results are interesting, additional studies ought to be undertaken in order to draw more definitive conclusions.

Acknowledgements

Reported research was realized within confines of project financed by Polish National Center for Research and Development No: LIDER/06/117/L-3/11/NCBR/2012.

REFERENCES

1. Gąska A.: Modeling the accuracy of coordinate measurement with use of Monte Carlo method. Ph.D. Dissertation, Cracow University of Technology, 2011.
2. Sładek J. Accuracy of coordinate measurements. Cracow University of Technology, Cracow 2013.
3. Weckenmann A., EstlerT., PeggsG., McMurtryD.: Probing systems in dimensional metrology. CIRP Annals-Manufacturing Technology, 53, 2, 657–684.
4. Gąska A., Gruza M., Gąska P., Karpiuk M., Sładek J.: Identification and correction of coordinate measuring machine geometrical errors using Laser Tracer systems. Advances in Science and Technology Research Journal, 7, 20, 2013, 17–22.
5. Hocken R.J., Pereira P.H. (eds.): Coordinate measuring machines and systems. Second edition, CRC Press, 2012.
6. Sładek J.:Modelowanie i ocena dokładności maszyn oraz pomiarów współrzędnościowych. Praca habilitacyjna. Wydział Mechaniczny, Politechnika Krakowska, 2001.
7. Weekers W.G., Schellekens P.H.J.: Compensation for dynamic errors of coordinate measuring machines. Measurement 20, 3, 1997, 197–209.
8. ISO 10360 part 5, 2010. CMMs using single and multiple stylus contacting probing systems. Second edition.

18

AUTOMATING THE MODELLING PROCESS OF INVOLUTE SPUR GEARS WITH STRAIGHT TEETH

Tadeusz Markowski[1], Jacek Mucha[1], Waldemar Witkowski[1]

[1] Departament of Mechanical Engineering, The Faculty of Mechanical Engineering and Aeronautics, Rzeszow University of Technology, al. Powstańców Warszawy 8, 35-959 Rzeszów, Poland, e-mail: tmarkow@prz.edu.pl; j_mucha@prz.edu.pl; wwitkowski@prz.edu.pl

ABSTRACT

In the paper the possibility of using Delphi programming language in the design of the tooth profile was presented. A developed numerical application was used to generate a tooth profile of involute cylindrical gears with modification and without modification. The size of modifications of the tooth root and the addendum was compared with the parameters obtained from the analytical calculations and machining simulation of gears.

Keywords: involute cylindrical gears, Delphi programming language, gears modifications.

INTRODUCTION

In modern design the most commonly used model of gear is a virtual CAD model. It allows to carry out the analysis (including method FEM) without producing the prototype.

Cylindrical gears with straight teeth are suitable for the simplest mechanical gearboxes. To improve the strength parameters of gearing, various modifications of gear are made. Tooth profile can be modified, to ensure optimum dynamic parameters during meshing, by:

- tooth root modification (protuberance) – Fig. 1,
- addendum modification – Fig. 1.

Shape modification in the tooth line correction is a mirror of gear deflection (due to unequal distribution the load on the facewidth of toothed gears). This modification can be obtained by:

- crowning,
- modification of end relief,
- modification of direction of flank line.

It is often necessary, for operational reasons, it is necessary to remove some part of material on the side of the tooth, or to remove material on entire flank of the tooth. This modification is called a topological modification (three-dimensional). We can make this modification when the parameters of modification are greater than gear's dimensional deviation.

For involute spur gear mathematical formulas describing the following points on tooth profile were described [1–7]. The cutting edge of the tool (Fig. 2) is divided into sections, for which there are certain formulas defining the tooth profile [2, 7, 8].

To facilitate the calculation the computer application, developed in Delphi programming language was created. One of the application window is shown in Figure 3.

The developed application has a many options for determining the parameters. The most important of these are:

- entering parameters and characteristic points on cutting edge tools without modifying the tooth profile, the calculation of the profile points of the gear;
- entering tool parameters without modifying the tooth profile with the possibility of changing the thickness of the tool (tooth thickness of toll is given on the pitch diameter), the calculation of the profile points of the gear;

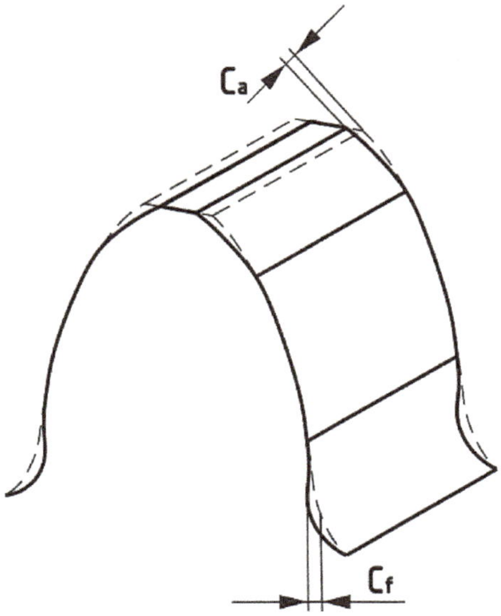

Fig. 1. Modification of tooth root and addendum of the gear (C_a – the depth modification of addendum, C_f – the depth modification of gear tooth root)

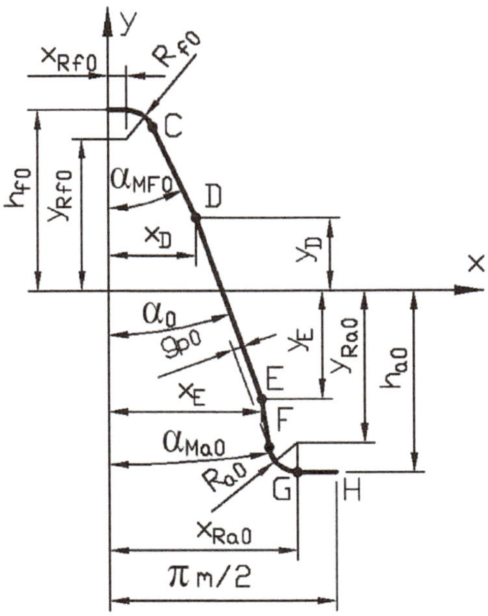

Fig. 2. Gear tool with addendum and tooth root modification

- entering parameters and characteristic points on cutting edge tools with tooth root and addendum modification, the calculation of the profile points of the gear;
- graphic representation of tooth profile for whole defined cutting tools;
- display coordinates of point on tooth profile, with possibility of writing them into .txt file.

Tooth profiles generated for example cutting tools are presented in Figure 4.

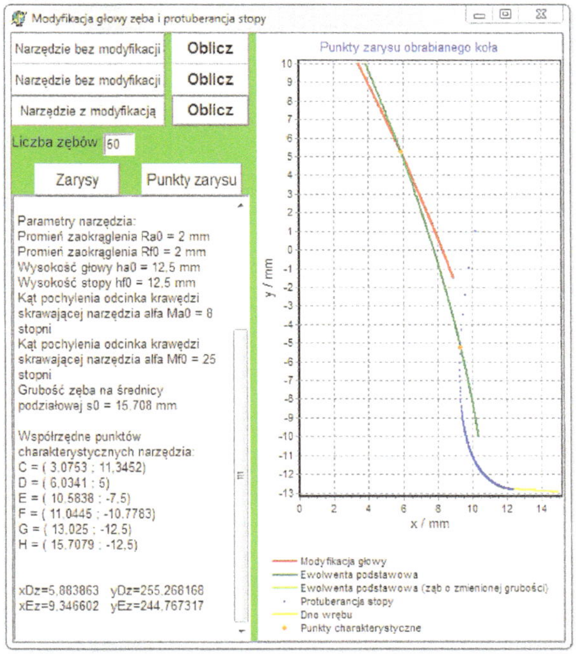

Fig. 3. Window of application for designing the tooth profile

Coordinates of point modification are based on the size of the cutting tool edge division and they are more accurate when the number of divisions is higher. For examples from Figure 4 the number of divisions on the cutting edge was 500.

While the division a greater number of profile points is increasing, more accurate tooth profile can be obtained from the outline of the side of the tooth. Unfortunately, in this case, the hardware requirements for smooth operation of the programme ware increased.

The result of calculations can be transferred to any CAD environment to automate the design process of tooth profile. From the program (Fig. 3) one side of the gear tooth is obtained, which can be used to prepare the whole model of the gear.

More labour-intense tooth profile modelling method is the simulation of machining in CAD environment. Virtual model (2D or 3D) is obtained by subtracting the mutual reeling and subtraction solid models of basic rack (as tool) and output gear.

To automate the process of virtual machining simulation an application (in Delphi programming language) was created. The program can generate a subsequent commands of simulation:

- copy the gear tool model,
- rotate and move the model of output gear,
- subtract the model of toothed tool from the output model.

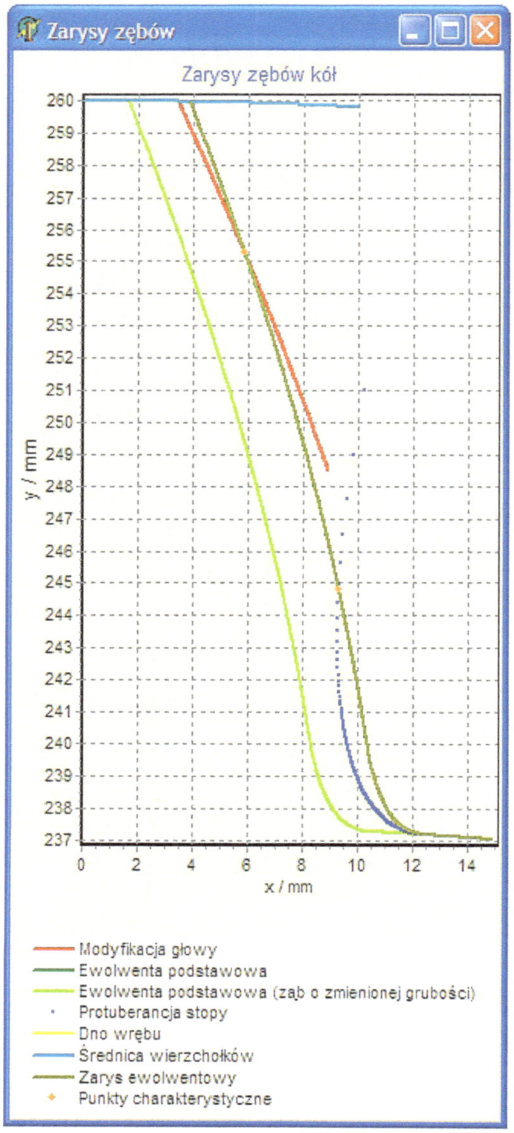

Fig. 4. Tooth profile comparison (gear parameters $m_n = 10$ mm, $z = 50$, $x_k = 0$, $\alpha = 20^{\circ}$, $c^* = 0.25$): with tooth root and addendum modification, without modification, with tooth width modification

Steps simulation are shown in Figure 5. Movements of models are presented by arrows.

File with simulation command saved in .scr file and can be open in AutoCAD environment. The simulation of the machining process does not have to run for cutting the whole gear. It can be run only for one tooth (one needs to pay attention and check if next step of simulation does not change our selected tooth space).

The sample values of gear modification: $m_n = 10$ mm, $z = 50$, $x_k = 0$, $\alpha = 20^{\circ}$, $c^* = 0.25$ were given in Table 1. Points of tooth profile were designed by using the application (Fig. 3) and by the virtual simulation of machining in CAD environment.

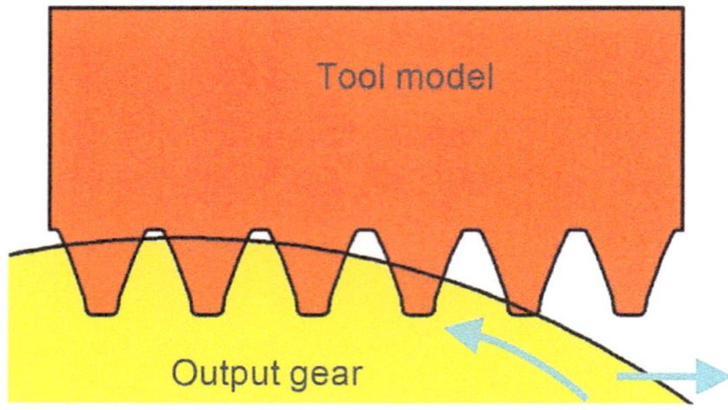

Fig. 5. Gear generating simulation

Reverse engineering can be used to reproduce an existing gear and its geometry. The real model is measured by appropriate tools, and so-called point cloud can be obtained. The next step is data processing in CAD system (creating a virtual model of gear). Too small number of measurements can lead to incorrect representation of the original gear. Example of surfaces generated by point cloud measured from cylindrical involute gear are shown in Figure 6. The next operation for that surfaces should be smoothing.

Table 1. Value of tooth root and addendum modification for sample gear

Descriptions	Analytic method [mm]	Cutting simulation [mm]
Value of addendum modification C_a	0.38407	0.39150
Radius of addendum modification	255.35947	255.28804
Value of tooth root modification C_f	0.72547	0.72604
Radius of tooth root modification	244.95065	244.97606

Fig. 6. Surfaces of cylindrical involute gear generated from point cloud

CONCLUSIONS

Computer Aided Design (CAD) allows to calculate a gear tooth form and change modification parameters without the need to manufacture the gear. Programming languages can be use to support the calculations of coordinate points of the designed tooth profile. Created applications allows to create many variants of gear with modification, which will be subjects to analysis (choose the most appropriate).

Parameters of tooth root and addendum modification were calculated with the analytical method (based on the analysis of counterpart gear). According to various studies, these parameters are different because various factors were included in the analysis (e.g. gearing stiffness, meshing ratio, etc.) [5, 7, 9]. The difference between the parameters obtained using the analytical models (depended on counterpart gear) and computer simulation of the machining fits within the numerical rounding errors.

REFERENCES

1. Bouzakis K.D., Lili E., Michailidis N., Friderikos O.: Manufacturing of cylindrical gears by generating cutting processes: A critical synthesis of analysis methods. CIRP Annals - Manufacturing Technology 2008.
2. Brauer J.: Analytical geometry of straight conical involute gears. Mechanism and Machine Theory, 2002.
3. Budzik G., Pisula J., Dziubek T., Sobolewski B., Zaborniak M.: Zastosowanie systemów CAD/RP/CMM w procesie prototypowania kół zębatych walcowych o zębach prostych. Mechanik, nr 12, 2011.
4. Kawalec A, Wiktor J.: Structural and technological features of involute spur gearings with undercut AT tooth root. Advances in Manufacturing Science and Technology, 2000.
5. Markowski T., Mijał M.: Contact stresses in cylindrical gear of the BBW meshing. Transportna Akademija Ukraini. Praci Zahidnogo Naukowogo Centru Logos Lwiw, no 7, 2000.
6. Michalski J., Skoczylas L.: Zastosowanie symulacji obróbki w diagnozowaniu procesu wytwarzania kół zębatych walcowych metodami frezowania obwiedniowego i dłutowania fellowsa. Projektuwannia, wyrobnyctwo ta ekspluatacja awtotransportnych zasobiw i poizdiw. Transportna Akademia Ukrainy, 10, 2003.
7. Wiktor J.: Wyznaczanie modyfikacji zarysu zęba w przekładni walcowej na podstawie sztywności zazębienia Zeszyty Naukowe Politechniki Rzeszowskiej nr 165, Mechanika z. 50, 1998.
8. Mijał M.: Synteza falowych przekładni zębatych. Oficyna Wydawnicza Politechniki Rzeszowskiej, Rzeszów 1999.
9. Műller L.: Przekładnie zębate projektowanie. Wydawnictwo Naukowo-Techniczne, Warszawa 1996.

19

EFFECTS OF RAIN ATTENUATION ON SATELLITE COMMUNICATION LINK

Ezeh G.N.[1], Chukwuneke N.S.[1], Ogujiofor N.C.[1], Diala U.H.[1]

[1] Electrical and Electronic Engineering, Federal University of Technology Owerri, P.M.B. 1526, Owerri, Imo State, Nigeria, e-mail: ugoezeh2002@yahoo.com; dialauchenna@yahoo.com

ABSTRACT

Rain attenuation is a major challenge to microwave satellite communication especially at frequencies above 10 GHz, causing unavailability of signals most of the time. Rain attenuation predictions have become one of the vital considerations while setting up a satellite communication link. In this study, rain attenuation models, cumulative distribution curves and other analytical tools for successful prediction of rain attenuation are presented. A three year Rain rate data was obtained from the Nigeria Meteorological Agency (NIMET) database in addition to experimental data. Of the three prediction models used in the study, Ajayi model gave the range of values closest to the experimental data. A correctional factor was determined as 1.0988 and used to modify the Ajayi model. This modification to Ajayi's model enabled its rain attenuation values conform more closely to the experimental result.

Keywords: correctional factor, cumulative distribution curve, outage rain rate, prediction model, rain attenuation, satellite communication.

INTRODUCTION

Rainfall intensity along terrestrial path is inhomogeneous in space and time and the raindrops have a non-spherical shape and thereby cause the attenuation on the horizontally polarized wave to be greater than the attenuation of the vertically polarized waves. The attenuation, due to these rain particles, is non-uniform and can only be statistically or experimentally determined from the rain rate measurements. The consequences of this rain attenuation are as follows:

- Loss of signal strength at the receiver.
- Wastage of transmission power in a bid to overcome this form of attenuation.
- Total loss of signal at the receiver in extreme cases.
- Unavailability of the satellite link for a great percentage of the time [1, 2].

Rain attenuation, because of the statistical variation in rainfall rates, is given as a probabilistic value which is often expressed as a rainfall rate exceeded for a certain percentage of the time. Therefore, in designing a satellite communication service system in a given area, it is often required that such experiment amidst other forms of analysis, should be carried out to ensure the least effect of rain attenuation and other forms of losses to that communication set-up.

This paper is aimed at studying rain attenuation, as it affects satellite communication links; to determine its nature and its possible effects on the propagation of satellite signals.

It will enable the designer to determine some technical parameters that are pertinent in the design process. The result of this investigation will also enable the communication system designer to determine the following:

- The optimum transmitting power in which rain attenuation will have the least effect.
- Whether it is economically feasible to Set-up such a system in the area in question.
- Determine the optimum microwave transmission frequency at which there will be the least effect of rain attenuation.

The experimental measurement data used in this study was carried out in Akwakuma, Owerri in Imo State, Nigeria. Owerri city lies in the south-eastern part of Nigeria. The area is located in the latitude 5.28° (5° 19'53") North of the equator and longitude 7.03°(7°4'30") East of the Greenwich Meridian.

GENERAL OVERVIEW OF RAIN ATTENUATION

Rainfall is a major cause of signal degradation for radio-communication systems operating at centimeter and millimeter wave bands, especially in the tropical region environment [5]. Determination of attenuation due to rainfall plays a significant role in the design of earth-satellite radio link at frequencies above 10 GHz [3, 4].

Rain attenuation of satellite signals do not simply affect the end users' resulting performance, it also affects the cost. Rain attenuation causes a greater power requirement from the transmitting units which hence leads to a higher cost per bit of transmission [5, 6].

In the determination of the statistical distribution of rain attenuation, many years of recording with sophisticated equipments like disdrometer, sensor rain gauges, power meters etc. are necessary to obtain the long-term rain attenuation statistics [5, 7]. Rain intensity measurements have been performed for many years in many countries hence their data is readily available for many places.

Consequently, in designing a satellite communication service system in a given area, it is often required that such an experiment, amidst other forms of analysis, should be carried out to ensure the least effect of rain attenuation and other forms of losses to the communication structure. However, from the economical point of view, it is not feasible to conduct measurement in one place every time a satellite receiver for communication services will be installed. It is usually very expensive and time consuming to conduct such an experiment and research. A minimum of one year data is needed to reveal the true effect of rain attenuation on satellite signals. Therefore, it is not only unfeasible but also impractical to carry out a one-year measurement and experiment, in advance, whenever a satellite telecommunication service is to be designed. With regards to this, many methods have been developed and employed to convert the available measured data in one geographical location to conform with the climatic and rainfall characteristics of the present location but this is not as accurate as using a locally measured data [8, 9, 10].

METHODOLOGY

An evaluation of the impact of rainfall on the propagation of earth-satellite signals operating in the Ku microwave frequency band is carried out. Considering the horizontal and vertical non-uniformity of rain structure, we therefore need a long period of measurement data to describe the seasonal variation of rainfall intensity. For a successful prediction of rain attenuation, the cumulative distribution of rainfall intensity obtained from long term measurements (at least 3-years data) is required. This research study is designed in such a way that raw data is retrieved from the Nigerian Meteorological Agency (NIMET). The data to be collected is rainfall rate data (mm/h) for a 3-year period (2010–2012). This data includes average monthly rainfall rate for the 3-year period, hourly rainfall rate for some selected months (highest rainfall intensity month, two average months and the least month) within the 3-year period. The data will then be processed using statistical means to find the probability of exceedance of each of the rainfall rate value in the data. A graph of rainfall rate against exceedance probability will be plotted. The value with the least probability of exceedance is determined and this value will be inputted into the available rain attenuation models to determine its corresponding rain attenuation value. The experiment will be carried out which will be used to determine a direct relationship between different values of rainfall rate and their corresponding values of rain attenuation. The results of the experiment will be used to compare with the result obtained using rain attenuation models. Then after proper analysis, a correction factor would be suggested to the model and hence a new model would emerge. The modified rain attenuation model would now be used as our working model. A 'Rain attenuation' – 'Rain rate' graph would be plotted using data obtained from the model, the experiment and the working model [11, 12, 13, 14].

The rainfall rate data retrieved from NIMET'S database shows hourly rainfall rate data. The rainfall rate data is measured in millimetre per hour (mm/h). The data covers a period of three years from 2010–2012. The long period data was to ensure an accurate prediction of rain attenuation. This is to make sure that the highest and the lowest possible rain rate expected in that region is not over-estimated or under-estimated. The 3-years data reveals the average rainfall rate for each month. The second data collected from NIMET reveals the hourly values of rainfall rate within some selected months. The second data would be used for comparison purposes. The exceedance probability of each value in the data is computed and is used to plot the cumulative distribution curve or Exceedance curve. The highest value of rain rate recorded in the experimental data and its corresponding value of rain attenuation is noted. This rainfall rate value is inputted in the three available rain attenuation models and the model that gives the closest value to the measured value will be selected as our active model. The different values of rainfall rate recorded during the experiment are loaded into the model and their corresponding model values of rain attenuation are also noted [15, 16]. A table is thus computed, showing the different rainfall intensity value, their experimental value for rain attenuation and their corresponding model value for rain attenuation. The two set of values are compared and used to suggest a correctional factor to the model. This correctional factor is used to modify our active model to yield a new model.

The working model

The working model is now used to generate the corresponding values of rain attenuation, by inputting the different values of rain rate into the working model. The different values of rain attenuation obtained are noted and then a graph of rainfall rate against rain attenuation is plotted for the different data sets. Finally, the cumulative distribution curve/exceedance curve for the rain attenuation values is plotted for the data.

RAIN ATTENUATION MODELS

The ITU-R model

The ITU-R method in the report 564-4 uses the concept of effective path length by means of a reduction factor. The method consists of the following procedure, which is proposed to calculate the long-term statistics of the slant path rain attenuation at a given location for frequency up to 30GHz (and provisionally for higher frequencies). This method involves several steps, which are presented below.

Step 1: The effective rain height h_e (km) is calculated from the latitude of the earth station, ϕ, as given by the following formulas.

$$h_R = 5.0 \text{ for } 0^\circ \leq \phi < 23^\circ \quad (1a)$$

$$h_R = 5.0 - 0.075(\phi - 23) \text{ for } \phi \leq 23^\circ \quad (1b)$$

Step 2: The slant path length L_s, below the freezing rain height is obtained, for antenna elevation angle $\theta > 5^\circ$, the slant path length, L_s (km) below the freezing rain height, h_R is given as,

$$Ls = \frac{(h_R - h_s)}{\sin\theta} \text{ for } \theta > 5 \quad (2)$$

where: h_s – the height of the earth station above the mean sea level, km.

For antenna elevation angle $\theta \leq 5^\circ$, a more accurate formula should be used which is given as:

$$Ls = \frac{2(h_R - h_s)}{\left(\sin^2\theta + \frac{2(h_R - h_s)}{R_e}\right)^{1/2} + \sin\theta} \text{ for } \theta \leq 5^\circ. \quad (3)$$

where: R_e – effective radius of the earth = 8500 km.

Step 3: The horizontal projection of the slant path length L_G (km) is found.

Step 4: Obtain the characteristic length of rain cell given by

$$L_o = 35e^{(-0.015R)} \quad (4)$$

where: R – rainfall rate value.

Step 5: Obtain the reduction factor, r given by,

$$r = \frac{1}{1 + \frac{L_G}{L_O}} \quad (5)$$

Step 6: Obtain the specific attenuation Y(dB/km) using the rain rate measurement data. This is contained in more details in Rec. ITU-R P.838-3.

Step 7: Finally, the rain attenuation is calculated using the expression below,

$$A = \gamma L_s r \quad (6)$$

Ajayi method

This method was proposed by Ajayi as described in [17, 18, 19]. The path reduction factor in the ITU-R prediction method converts a physical path length to an equivalent length along which the rain rate can be assumed constant. Since it turned out that the conversion process does not necessarily always lead to a reduction in path length, Ajayi modified the ITU-R method by introducing two reduction factors, the horizontal reduction factor and the vertical reduction factor.

The Ajayi method is broken down into steps for simplification.

Step 1: The height of the freezing level during rain,

$$hf_R = 5.0 \text{ for } 0^\circ \leq \theta < 23^\circ \quad (7a)$$

$$hf_R = 5.0 - 0.075(\theta - 23) \text{ for } \theta \geq 23^\circ \quad (7b)$$

The two reduction factors used here are the horizontal reduction factors, r_h and vertical reduction factor, r_v.

Step 2, 3, 4 and 5 remain unchanged, i.e. still the same procedure as that of the ITU-R method.

Step 6: The horizontal reduction factor, r_h can be calculated using,

$$r_h = \frac{1}{1+L_G^{0.002}(R)^{1.01}} \quad (8)$$

Step 7: The vertical reduction factor r_v can be similarly calculated using,

$$r_v = \frac{1}{1+\frac{h_{FR}}{5-0.4Q^{1.5}}} \quad (9)$$

Step 8: The effective path length, L_e through the rain can be evaluated from the triangle above by applying sine rule.

$$L_e = \frac{r_h \times L_G}{\cos \varepsilon} \quad (10)$$

where: $\varepsilon = \tan^{-1}\left(\frac{h_{FR} r_v}{L_G r_h}\right)$ (11)

Step 9: The value of specific attenuation (dB/km) is evaluated

$$\gamma = KR^a$$

Step 10: The value of rain attenuation is calculated, using the equation below.

$$A = \gamma L_e r \quad (12)$$

Dissanayake-Allnutt model

This method was developed by Dissanayake and Allnutt in their work [19]. The conversion of the physical path length to the effective path length as implemented in the ITU-R method does not always lead to an accurate reduction in path length. Therefore in the Dissanayake Allnutt method, the conversion factor is called the Adjustment Factor instead of reduction factor.

Dissanayake and Allnutt included two adjustment factors, in their proposed method, namely horizontal path adjustment factor and vertical adjustment factor. The horizontal path adjustment factor takes into account the inhomogeneity of rain along the propagation path horizontally, while the vertical adjustment factor take into consideration the vertical inhomogeneity of the rain. The two adjustment factors were devised by using available slant path attenuation measurement data. The derivation of these adjustment factor is beyond the scope of this paper, thus only the end results of the derivation will be included.

Step 1: The freezing height h_{fR} during rainy condition is given by,

$$h_{fR} = 5.0 \text{ for } 0 \leq \phi < 23^\circ \quad (13a)$$

$$h_{fR} = 5.0 - 0.075(\phi - 23) \text{ for } \phi \geq 23^\circ \quad (13b)$$

Step 2: The slant path length for elevator angle greater than 5° is

$$\sin\theta = \frac{h_{FR} - h_s}{L_s} \quad (14)$$

$$L_s = \frac{(h_{FR} - h_s)}{\sin\theta} \quad (15)$$

Step 3: Find the specific attenuation γ, using the ITU-R method,

$$\gamma = KR^a$$

Step 4: The horizontal path adjustment factor, r_h is calculated,

$$r_h = \frac{1}{0.628 + 0.194\sqrt{\gamma L_s}} \quad (16)$$

Step 5: Similarly, the vertical adjustment factor r_v is calculated,

$$r_v = \frac{1}{1+\sin\theta\,(1.282\sqrt{L_s} - 2.871} \text{ for } \theta > 23^0 \quad (17)$$

$$r_v = 1\text{-}0.5 \sin\theta \text{ for } \theta = 0^\circ \quad (18)$$

For latitudes between 0° and 23°, a linear interpolation between equation (13a) and (13b) is suggested.

Step 6: Calculate the horizontal projection L_G of the slant path as in the ITU-R method.

$$L_G = L_s \cos\theta$$

Step 7: Calculate the effective horizontal projection of the effective slant path length.

$$L_{Ge} = r_h L_G \quad (19)$$

Step 8: Calculate the effective slant path length, L_e

$$\sin\varsigma = \frac{h_{FR}}{L_e} \quad (20)$$

$$\text{or } L_e = \frac{L_{Ge}}{\cos\varsigma} \quad (21)$$

where: $\varsigma = \tan^{-1}\left(\frac{h_{FR}}{L_{Ge}}\right)$ (22)

Step 9: Calculate the rain attenuation

$$A = \gamma L_e r_v \quad (23)$$

EXPERIMENT

The result of the experiment and data collected from the experiment will serve as sources of primary data which will enable us to adjust the existing rain attenuation models to suit the present environment. The aim of this experiment is to physically measure the magnitudes of rain-induced attenuation and also measure the magnitude of point rainfall rate at that instant and possibly establish some form of relationship between the two parameters by adjusting the existing rain attenuation models. The experiment will also enable us to determine the outage rain attenuation and the outage rainfall rate.

The experiment is designed in such a way as to measure the average signal strength under normal propagation condition and note the signal strength. Then the signal strength is also measured during rainy conditions while taking note of the rain rate value of that rainfall. Then rain attenuation is computed as,

$$\text{Rain Attenuation} = \begin{pmatrix}\text{signal strength under}\\ \text{normal propagation}\\ \text{condition}\end{pmatrix} - \begin{pmatrix}\text{signal strength}\\ \text{under}\\ \text{rainy condition}\end{pmatrix}$$

This procedure was repeated for different rain events while taking note of their different measured rain rate value. The different rain rate values and their corresponding rain attenuation value were tabulated. The receiving antenna used is parabolic reflector antenna and the instrument used for measuring the signal strength is Digital Radio Frequency Power Meter. Rainfall rate was measured using a rain gauge and the timing was achieved using a digital stopwatch incorporated in a mobile phone. The tabulated values of rain rate and their corresponding rain attenuation value will thus be used for further analysis.

Instrumentation

Instruments used to carry out this experiment include the following:

- Parabolic reflector antenna having a specific known gain.
- Compass.
- Radio frequency power meter.
- Coaxial cable port connector.
- Connecting cable (coaxial cable).
- Rain gauge.
- Stopwatch timer.

Parameters pertinent to this experiment also include (Table 1).

RESULTS AND DISCUSSION

Experimental Results

The result of the experiment conducted is presented in the Table 2. The rain attenuation values are measured and recorded in db and the rainfall

Table 1. Table of value for experimental parameters

S/N	Parameter	Value
1	The satellite in space	Astra 2B (Multi TV Satellite) at 28.2° East
2	Frequency of the down link signal	12.527 GHz
3	Polarigation of the signal	Horizontal
4	Antenna elevation angle	42°
5	Latitude of site	5.28° North
6	Longitude of site	7.03° East
7	Height of antenna	2.9 m
8	Altitude of site	91.44 m
9	Antenna gain	35 dB
10	Antenna diameter	0.6 m

Table 2. Experimental result for rain rate and rain attenuation

S/N	Rain rate	Rain attenuation value (Q–P)	Measured signal power during rainfall (P)
1	154	15.62	36.88
2	150	15.68	36.82
3	142	15.60	36.90
4	136	15.64	36.86
5	128	15.42	37.08
6	120	15.12	37.38
7	112	14.54	37.96
8	104	14.07	38.43
9	92	13.10	39.40
10	86	12.39	40.11
11	72	10.73	41.77
12	64	9.62	42.88

rate is measured in mm/h. The signal power, Q measured under normal condition (absence of rainfall) was recorded as 52.5 dB.

Observation

Rain attenuation increases as rainfall rate value increases. At a rain rate value of 112 mm/h and above, there is total outage of reception (i.e. there is useful output at the destination of the receiver's output). The rain attenuation corresponding to this rain rate value as seen from the above table is 14.54 dB. This rain rate value and rain attenuation will be called Outage rain rate value and Outage rain attenuation value.

Cumulative distribution curve (Exceedance Curve) of rain attenuation was plotted for the four different months studied earlier to determine the probability of exceedance value and to estimate the chances/probability that there will be an outage that month. Since rainfall rate varies in a highly unpredictable manner, very long period of data is usually required for this estimation to be able to account for long-term variability in the rain rate within the years and thereby making the prediction process more accurate. However, for the purpose of this research, the moderate period of data used will suffice.

The highest rain rate value measured is tested to be 154 mm/h on the different models to see which model gives the closest value of rain attenuation to the measured value attenuation.

Evaluation of rain attenuation values using the available rain attenuation models

ITU-R model

Using ITU-R model, g = 9.884 dB/Km

$A = gL_s r = 9.884 \times 6.1015 \times 0.4338 = 26.16$

Rain attenuation = 26.16 dB

Ajayi method

g = 9.884 dB/Km

$A = gL_e r = 9.884 \times 3.3115 \times 0.4338$

A = 14.2 dB

Rain attenuation = 14.2 dB

Alnutt – Dissanayake model

$A = gL_e = 9.884 \times 5.4323 \times 0.70485 = 37.84$ dB

A = 37.84 dB

Comparison of rain attenuation values obtained from the different models

From the MATLAB program written, the rain attenuation corresponding to different rain rate values has been computed for different rain attenuation models, what is shown in Table 3. A comparison is made between different models and rain attenuation values obtained from the experimentation (Table 4).

Since Ajayi model has its rain attention values closest to the experimental results, the Ajayi model will be chosen as our working model.

The values marked by asterisk are correctional factors whose values are similar, i.e. whose values are more consistent than the others. These values would be averaged to find the mean correctional factor which would be used to modify the Ajayi model.

Mean value of $f =$

$$\frac{1.0983 + 1.9954 + 1.0998 + 1.1012 + 1.0978 + 1.0982 + 1.0997 + .0961}{8}$$

Table 3. Rain attenuation comparison table for different rain attenuation models

S/N	Rainrate	ITU-R	Ajayi	Disanayake	Measured
1	154	26.1637	14.2222	37.8431	15.62
2	150	26.2361	14.2615	36.6966	15.68
3	142	26.2442	14.2659	34.4192	15.60
4	136	26.1253	14.2013	32.7252	15.54
5	128	25.7936	14.0210	30.4862	15.42
6	120	25.2591	13.7305	28.2708	15.12
7	112	24.5202	13.3288	26.0802	14.54
8	104	23.5787	12.8170	23.9159	14.07
9	92	21.8000	11.8501	20.7226	13.10
10	86	20.7545	11.2818	19.1516	12.39
11	72	17.9499	9.7573	15.5596	10.73
12	64	16.1452	8.7763	13.5581	9.62

Table 4. Comparison of Ajayi, model with experimental result

S/N	RAIN RATE	AJAYI model	Measured	Suggested correctional factor (f)
1	154	14.2222	15.62	1.0983*
2	150	14.2615	15.68	1.0995*
3	142	14.2659	15.60	1.0935
4	136	14.2013	15.54	1.0942
5	128	14.0210	15.42	1.0998*
6	120	13.7305	15.12	1.1012*
7	112	13.3288	14.54	1.0978
8	104	12.8170	14.07	1.0909*
9	92	11.8501	13.10	1.1055
10	86	11.2818	12.39	1.0982*
11	72	9.7573	10.73	1.0997*
12	64	8.7763	9.62	1.0961*

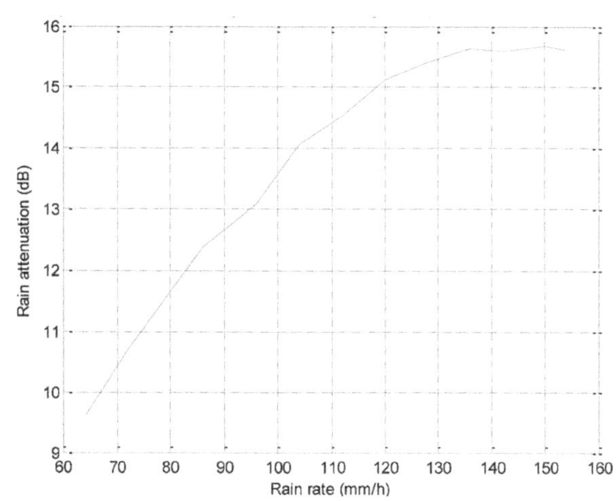

Fig. 1. Graph of rain attenuation against rain rate using the experimental measured values

$$\frac{8.7906}{8} = 1.0988$$

The correction factor to be used = 1.0988

For Ajayi; $A = \frac{3.1807R^{1.169}e^{-0.015R}}{35e^{-0.015R}+4.5343}$

Modified Ajayi model = $\frac{1.0988 x 3.1807 x R R^{1.169}e^{-0.015R}}{35e^{-0.015R}+4.5343}$

Modified Ajayi model = $\frac{3.495R^{1.169}e^{-0.015R}}{35e^{-0.015R}+4.5343}$

Table 5. Rain attenuation values using modified Ajayi model

Rain rate	Measured	AJAYI model	Modified AJAYI model
154	15.62	14.2222	15.6275
150	15.68	14.2615	15.6708
142	15.60	14.2659	15.6756
136	15.64	14.2013	15.6046
128	15.42	14.0210	15.4065
120	15.12	13.7305	15.0873
112	1.54	13.3288	14.6459
104	14.07	12.8170	14.0836
96	13.10	11.8501	13.4036
86	12.39	11.2818	12.3966
72	10.73	9.7573	10.7215
64	9.62	8.7763	9.6435

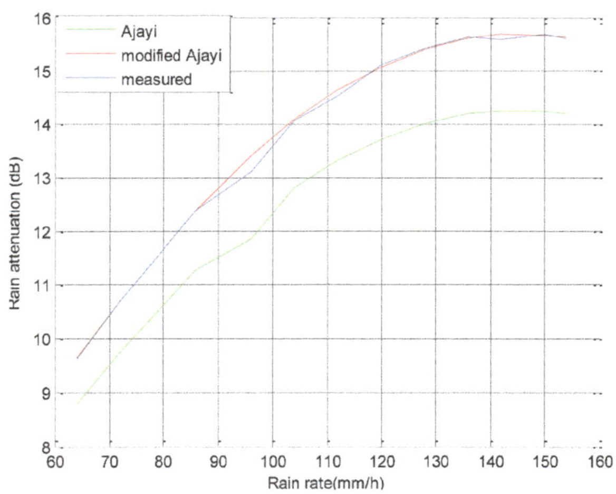

Fig. 2. Graph of rain attenuation against rain rate using, Ajayi, modified Ajayi and measured

Plotting of cumulative distribution curve (exceedance curve) of rain attenuation for different months using the modified Ajayi model

The MATLAB program for computing rain attenuation values is used to compute rain attenuation corresponding to different hourly rainfall

Table 6. Rain attenuation values of hourly rainfall rates in the month of August 2010

Rank	Rain rate	Frequency	Exceedance probability	Rain attenuation
1	154	2	0.0027	15.6275
2	152	4	0.0081	15.6525
3	150	1	0.0094	15.6708
4	149	7	0.019	15.6774
5	147	6	0.027	15.6856
6	146	7	0.036	15.6871
7	145	9	0.048	15.6869
8	143	5	0.055	15.6812
9	140	2	0.058	15.6592
10	138	9	0.070	15.6356
11	137	4	0.075	15.6210
12	135	6	0.083	15.5864
13	134	8	0.094	15.5663
14	133	7	0.103	15.5444
15	131	8	0.114	15.4949
16	130	10	0.128	15.4673
17	129	12	0.144	15.4378
18	128	7	0.153	15.4065
19	125	13	0.170	15.3011
20	124	6	0.179	15.2621
21	122	8	0.189	15.1785
22	121	2	0.192	15.1785
23	118	6	0.200	14.9884
24	117	3	0.204	14.9360
25	115	5	0.211	14.8257
26	114	7	0.220	14.7677
27	110	5	0.227	14.5166
28	109	4	0.232	14.4491
29	108	6	0.240	14.3797
30	106	3	0.244	14.2353
31	105	1	0.246	14.1604
32	104	4	0.251	14.0836
33	101	2	0.254	13.8421
34	98	3	0.258	13.5843
35	96	7	0.267	15.4036
36	93	2	0.270	13.1193
37	88	4	0.275	12.6114
38	84	1	0.277	12.1754
39	80	5	0.283	11.7143
40	75	5	0.290	11 1044
41	68	4	0.295	10.1924
42	65	3	0.300	9 7825
43	58	3	0.303	8.7884
44	55	2	0.306	8.3442
45	51	1	0.307	7.7421

rates of each month. The exceedance probability corresponding to the rain attenuation is also evaluated. A graph of rain attenuation against exceedance probability gives a cumulative distribution curve of rain attenuation. The shape of these curves will be used to estimate the likeliness or the probability or the chances that there will be an outage in a particular month.

In Figure 3, the rain attenuation value is the outage rain attenuation value which is 14.64 dB. This value was judged using two criteria; the exceedance probability of this rain attenuation value relative to other exceedance probabilities. The region of the graph in which this rain attenuation value occurs i.e. whether it occurs in region of high probability of occurrence or it occurs in region of low probability of occurrence. The region of high probability of occurrence is the region (point on the curve) having small gradient value relative to other regions, while the region of low probability of occurrence are the regions having high gradient value (more steep) relative to other regions. Then a decision is made using the points below:

- If the rain attenuation value has high probability of exceedance as well as lies in the regions of high probability of occurrence, this signifies a very high probability of outage that month.
- If the rain attenuation value has a high probability of exceedance but lies in the region of low probability of occurrence, this signifies a high probability of outage that month
- If the rain attenuation value has a low probability of exceedance but lies in the region of high probability of occurrence, this signifies a low probability of outage that month

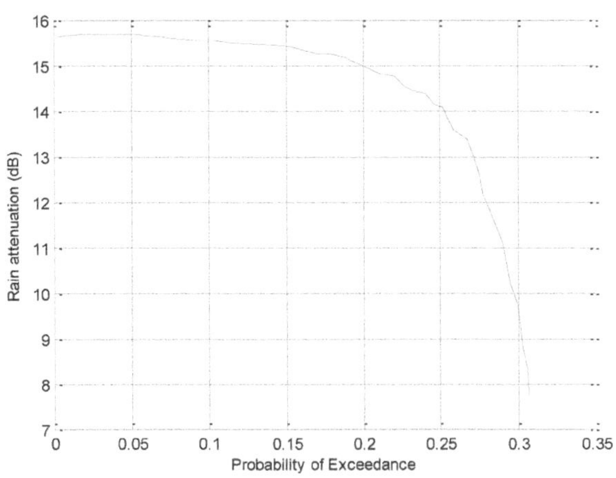

Fig. 3. Cumulative distribution curve (exceedance curve) of rain attenuation for the month of August, 2010

- If the rain attenuation value has a low probability of exceedance and also lies in the region of low probability of occurrence, this signifies a very low probability of outage that month.

In Figure 3, it is observed that the rain attenuation value of interest, 14.64 dB has a high probability of exceedance lies in the region of high probability of occurrence hence there is a very high probability of outage in the month of August, 2010.

From Figure 4, the Outage rain attenuation value 14.64 dB has a high probability of exceedance relative to other excedance probability but however lay in the region of low probability of occurrence, hence there is high possibility/probability of outage that month.

Looking at figure 5, it is seen that the outage rain attenuation value, 14.64 dB has a low probability of exceedance relative to the exceedance probabilities but lay in the region of high probability of occurrence, hence there is a low probability of outage this month.

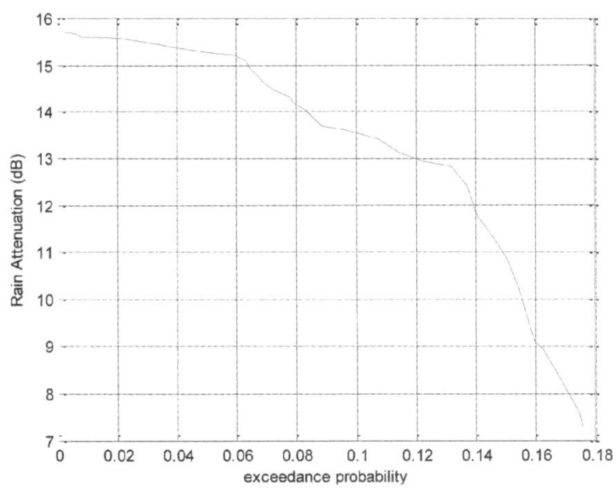

Fig. 4. Cumulative distribution curve of rain attenuation for the month of May 2012

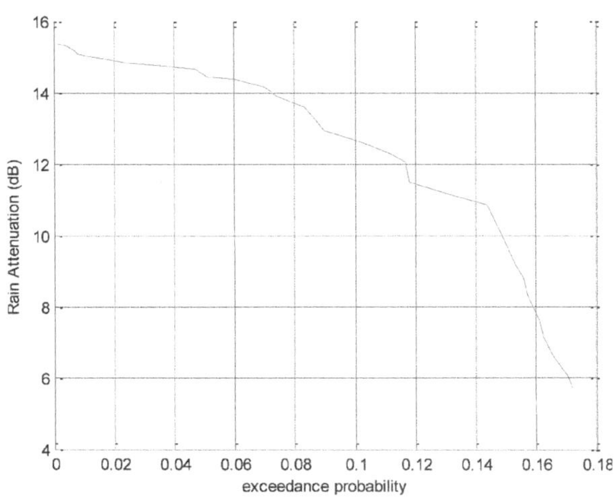

Fig. 5. Cumulative Distribution Curve of rain attenuation for the month of October 2010

Table 7. Rain attenuation values of hourly rainfall rates in the month of May, 2012

Rank	Rain rate	Frequency	Exceedance probability	Rain attenuation
1	144	2	0.0027	15.6849
2	142	2	0.0054	15.6756
3	138	1	0.0067	15.6356
4	136	1	0.0081	15.6046
5	135	4	0.0134	15.5864
6	134	6	0.0215	15.5663
7	132	4	0.0268	15.5206
8	130	4	0.0322	15.4673
9	129	2	0.0349	15.4378
10	128	1	0.0362	15.4065
11	125	8	0.0470	15.3011
12	124	5	0.0537	15.2621
13	123	4	0.0591	15.2213
14	120	3	0.0631	15.0873
15	117	1	0.0644	14.9360
16	114	2	0.0671	14.7677
17	112	1	0.0685	14.6459
18	109	3	0.0725	14.4491
19	107	4	0.0779	14.3084
20	105	1	0.0792	14.1604
21	103	3	0.0832	14.0049
22	99	4	0.0886	13.6721
23	98	6	0.0966	13.5843
24	96	8	0.1074	13.4036
25	93	5	0.1141	13.1193
26	91	6	0.1221	12.9212
27	90	7	0.1315	12.8196
28	86	4	0.1369	12.3966
29	81	2	0.1400	11.8319
30	80	1	0.1409	11.7143
31	78	2	0.1436	11.4747
32	75	3	0.1477	11.1044
33	73	2	0.1503	10.8505
34	68	3	0.1544	10.1924
35	62	3	0.1584	9.3621
36	60	1	0.1597	9.0763
37	59	2	0.1624	8.93319
38	56	3	0.1664	8.4925
39	52	4	0.1718	7.8939
40	50	2	0.1745	7.5896
41	48	1	0.1758	7.2822

Table 8. Rain attenuation values of hourly rain fall rates in the month of October 2010

Rank	Rain rate	Frequency	Exceedance probability	Rain attenuation
1	126	1	0.0013	15.3381
2	125	2	0.0040	15.3011
3	122	2	0.0067	15.1785
4	120	1	0.0081	15.0873
5	118	3	0.0121	14.9884
6	117	4	0.0174	14.9360
7	115	5	0.0242	14.8257
8	114	7	0.0349	14.7677
9	112	8	0.0470	14.7677
10	109	7	0.0564	14.4491
11	108	3	0.0604	14.3797
12	107	2	0.0631	14.3084
13	105	5	0.0700	14.1604
14	103	2	0.0725	14.0049
15	102	1	0.0738	13.9244
16	100	3	0.0779	13.7580
17	98	4	0.0832	13.5840
18	95	2	0.0859	13.3106
19	91	3	0.0899	12.9212
20	88	9	0.1020	12.6114
21	85	7	0.1114	12.2868
22	83	4	0.1168	12.0225
23	78	1	0.1181	11.4747
24	76	7	0.1275	11.2292
25	75	4	0.1329	11.1044
26	73	8	0.1436	10.8505
27	66	4	0.1490	9.9203
28	61	3	0.1530	9.2197
29	58	2	0.1557	8.7864
30	55	1	0.1570	8.3442
31	50	3	0.1611	7.5896
32	47	1	0.1624	7.1275
33	44	2	0.1651	6.6594
34	40	4	0.1705	6.0277
35	38	1	0.1718	5.7091

In Figure 6, it is clearly seen that the rain attenuation value, 14.64 dB has a low probability of exceedance relative to other exceedance probabilities and lies in the region of high probability of occurrence, hence there is a very low probability of outage this month.

Table 9. Rain attenuation values of Hourly rainfall rates in the month of November, 2011

Rank	Rain rate	Frequency	Exceedance probability	Rain Alttenuation
1	132	1	0.0013	15.5206
2	127	1	0.0027	15.3733
3	122	1	0.0040	15.1785
4	117	1	0.0054	14.9360
5	114	1	0.0067	14.7677
6	109	1	0.0081	14.4491
7	105	2	0.0107	14.1604
8	104	1	0.0121	14.0836
9	102	2	0.0148	13.9244
10	101	1	0.0161	13.8421
11	97	1	0.0174	13.4948
12	91	1	0.0188	12.9212
13	83	1	0.0201	12.0625
14	81	3	0.0242	11.8319
15	79	1	0.0255	11.5952
16	77	1	0.0268	11.3527
17	74	1	0.0282	10.9781
18	60	1	0.0295	9.0763
19	53	1	0.0309	8.0449
20	46	1	0.0322	6.9721
21	45	1	0.0336	6.8161
22	41	3	0.0376	6.1863

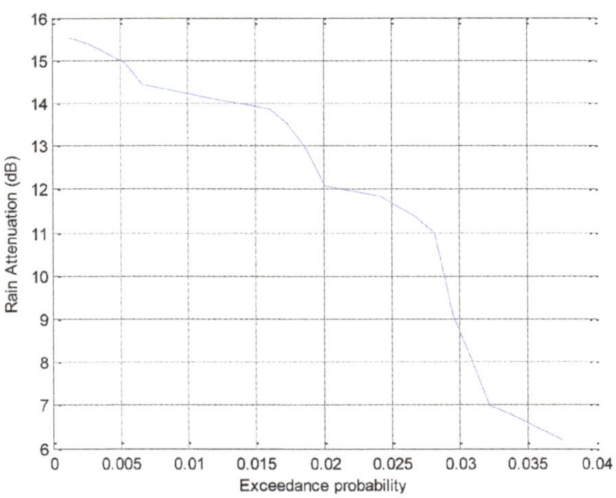

Fig. 6. Cumulative Distribution Curve of rain attenuation for the month of November 2011

Estimation of availability of the satellite link for the different months

The estimation of availability of the satellite link using the available data it is not very accurate because of lack of data showing the period for each rain rate value. Therefore, because of this constraint, the estimation of this availability is not very accurate but only serves to give us a value close to the actual value.

$$Availability = \frac{\text{Total period of availability}}{\text{Total period of time}} \times 100\%$$

Table 10 shows four different months studied and their estimated availability measured in percentage of the total period of time.

CONCLUSION

Rain attenuation is caused as a result of absorption of part or all of the signal's radiation power by the raindrop. This absorption is as a result of scattering effect (diffraction and refraction) of the rain drop to the signal. Rain attenuation is prevalent at frequencies above 10 GHz and increases with frequency. This is because as frequency increases, the signal wavelength decreases and approaches the size of a rain drop and hence giving the rain drop more scattering and absorption capabilities to the signal. At a particular frequency, rain attenuation increases exponentially with increasing rain rate values because, the higher the rainfall rate, the higher the rain drop size (rain drop diameter). Hence, it is more its scattering effect on the signal, because as the rain drop size increases, it tends to approach the wavelength of the signal.

The month of August, 2010 is characterized by very high intensity rainfall rates with a very high probability of outage. This means that outage is very likely to occur this month and hence has a very low availability of 78%. The month of May, 2012 characterized by moderately high intensity rainfall rates, has a high probability of outage and a low availability of 93%. For October 2010,

Table 10. Table of availability for the different months studied

S/n	Month, Year	Outage rain Events (hr)	Non outage rainevents (hr)	Non-rain Events (hr)	Total period of availability (hr)	Total period of time (hr)	Availability
1	August, 2010	164	65	515	580	744	77.965%
2	May, 2012	51	80	613	693	744	93.15%
3	October, 2010	33	93	618	711	744	95.56%
4	November, 2011	5	23	692	715	720	99.31%

it is characterized by low intensity rainfall rates, has low probability of outage and hence has a high availability of 96%. The month of November, 2011 characterized by very low intensity rainfall rates has very low probability of outage, hence the outage is not very likely to occur in this month and so consequently has a very high availability of 99%. Therefore, the higher the probability of outage in a month, the lower is its expected availability.

The outage rain rate was found to be 112 mm/h and also the corresponding outage rain attenuation was also found to be 14.54 dB. Since the outage rain attenuation value is 14.54 dB and the highest possible rain attenuation (i.e. highest recorded rain attenuation) is 15.0756 (as predicted by the model) therefore, in link budget analysis, an extra power of (15.6756 – 14.54 = 1.1356) dB is needed for the downlink signal so as to compensate for rain attenuation losses. By doing this, the probability of outage of each of the month will enormously decrease and their corresponding availabilities will also increase. By doing this we can achieve a satellite link having the availability of 99.9% in an average year.

High gain antennas (having high efficiency and high directivity) should be used in area characterized by high intensity rainfall and having heavy rain attenuation. Parabolic reflector antennas having gain of 35–40 dB is recommended for use so as to compensate for the negative effects of rain attenuation in satellite communication links. More so, antenna having large effective area should be used.

Though reasonable amount of work was done on this topic, there is still a lot of room for improvement. There is a need for long period of data showing the period of each rain rate value so as to enable us to accurately estimate the availability of the satellite communication link.

There is serious need for a model or equation which will enable us calculate the specific numerical value of outage probability of a given rain attenuation value by simply inputting it's exceedance probability and the gradient of the point or region of the curve where that value occurs instead of using relative and comparative values.

REFERENCES

1. Stallins W., Data and Computer Communications. Upper Saddle River, New Jersey, United States, Pearson prentice hall, 2007.
2. Laplante P.A., IEEE Electrical Engineering dictionary, N-W Boca Raton, CRC press ILC, 2000.
3. Rain Fade. Retrieved from http://en.wikipedia.org/wiki/Rain_fade
4. Bryant H., Adinula I., Riva C. and Brussand G., Rain Attenuation Statistic from Rain Cell Diameters, and Heights. Int. J. Satell. Commun., 19, 2001, 263-283.
5. Mc Cormick K.S., Rain Attenuation Measurements in South East Asia. Moscow, Dompars, June 1994.
6. Ajayi G.O., Rain intensity and Randrop Size Measurement in Nigeria. Seminar on Radioware propagation in Tropical Regions, Trieste (Italy), 1993.
7. Propagation in non Lonized media. CCIR, Vol. V, Geneva 1990.
8. Barclay P.A., Benett J.A. and Boston R.C., Properties of Rain for Microwave Propagation Studies. Report from the Electrical Engineering Department, School of Engineering, Monash University, Melbourne (Australia), 1978.
9. Akeyama A., Morita K., 11 and 18 GHz Radiowave Attenuation due to Precipitation on a Slant Path. IEEE Transaction Antennas and Propagation, Vol. AP-28(4), 1980, 580-585.
10. Moupfouma F.L, Martin N., Rain Structure on Attenuation due to Precipitation based on Measurement in Tropical Zone. 3-7 December, 1990.
11. Moupfouma F., Rainfall Rate Cumulative Distribution Modelling for the Determination of Rain Attenuation Statistics in Terrestrial and Satellite Radio Links in Tropical areas. Seminar on Radiowave Propagation in Tropical Regions, Triete (Italy), 1993.
12. Facts about Rain Fade and Satellites. http://www.phonetvinternet.com
13. Laws J.O. and Parsons D.A., The Relation of Raindrop sizes to Intensity. Trans Amer. 138 Geophys. Union, Vol. 24, 1943, 432-460.
14. Marshall J.S. and Palmer W.M.K. The Distribution of Raindrops with size. Journal of Meteorology, Vol. 5, 1948, 165-166.
15. Joss J., Thams J.C. and Waldvogel A., The Variation of Raindrop Size Distribution at Lorcano. Switzerland, Proc. Int. Conf, on Cloud Physics Toronto (Canada), 1968, 369-373.
16. Ajewole M.O., Scattering and Attenuation of Centimeter and Millimeter Radio Signals by Tropical Rainfall. Ph.D thesis, Federal University of Technology Akure Nigeria (unpublished), 1997.
17. Availability. Retrieved from http://en.wikipedia.org/wiki/Availability
18. Ojo J.S. and Ajewole M.O., Rain Rate and Rain Attenuation Prediction for Satellite Communication in Ku and Ka bands over Nigeria. Progress in Electromagnetic Research B, Vol. 5, 2008, 207–223.
19. Ajayi G.O., Slant Path Rain Attenuation measurement in Africa. Seminar on Radiowave Propagation in Tropical Regions, Trieste (Italy), 1993.
20. Alnutt J.E. and Rogers D.V., The INTELSAT Slant Path Radiowave Propagation Experiment in the Tropical Regions. Int. J. Satel. Commun., 8(3), 1990.

20

APPLICATION OF ADDITIVELY MANUFACTURED POLYMER COMPOSITE PROTOTYPES IN FOUNDRY

Wiesław Kuczko[1], Radosław Wichniarek[1], Filip Górski[1], Paweł Buń[1], Przemysław Zawadzki[1]

[1] Chair of Management and Production Engineering, Poznań University of Technology, Piotrowo 3, 60-965 Poznań, Poland, e-mail: wieslaw.kuczko@put.poznan.pl; radoslaw.wichniarek@put.poznan.pl; filip.gorski@put.poznan.pl; pawel.k.bun@doctorate.put.poznan.pl; przemyslaw.zawadzki@put.poznan.pl

ABSTRACT

The paper presents a method, developed by the authors, for manufacturing polymer composites with the matrix manufactured in a layered manner (via 3D printing – Fused Deposition Modeling) out of a thermoplastic material. As an example of practical application of this method, functional prototypes are presented, which were used as elements of foundry tooling – patterns for sand molding. In case of manufacturing prototype castings or short series of products, foundries usually cooperate with modeling studios, which produce patterns by conventional, subtractive manufacturing technologies. If patterns have complex shapes, this results in high manufacturing costs and significantly longer time of tooling preparation. The method proposed by the authors allows manufacturing functional prototypes in a short time thanks to utilizing capabilities of additive manufacturing (3D printing) technology. Thanks to using two types of materials simultaneously (ABS combined with chemically hardened resins), the produced prototypes are capable of carrying increased loads. Moreover, the method developed by the authors is characterized by manufacturing costs lower than in the basic technology of Fused Deposition Modeling. During the presented studies, the pattern was produced as a polymer composite and it was used to prepare a mold and a set of metal castings.

Keywords: additive manufacturing, Fused Deposition Modelling, foundry tooling.

INTRODUCTION

The Additive Manufacturing Technologies (AMTs), widely known under the name of "3D printing" are more and more widespread in various branches of industry. They become a crucial tool, allowing the companies to significantly reduce time of preparation of a product for manufacturing, which results in a significant reduction of costs of implementing a new product into market. Among many advantages of the AMTs, the most important are: lack of need for manufacturing of special tooling, manufacturing directly from the digital CAD model, wide range of available devices and applied materials (ceramics, plastics, waxes, metals). Thanks to building a product layer by layer, it is possible to drop practically all the manufacturing limitations regarding shape complexity of a product, which are present in conventional manufacturing technologies (machining, foundry, plastic forming).

Among many AMT methods, the Fused Deposition Modeling is one of the most widespread and globally available technologies. The reason for this is a relatively low cost of buying and using a machine for production and non-complex and eco-friendly process (odorless and non-toxic material, minimal waste volume, low energy consumption). The mechanical properties of the FDM products are good enough to use them as visual or concept prototypes, mock-ups or design aids. In case of a production of short series of products or functional prototypes, the strength of the FDM-made products is often not satisfying [1].

Foundries are production plants which have one of the largest necessities for functional prototypes, especially if it comes to specialization in short series of individualized castings. Manufacturing of the special tooling (e.g. models for sand molds) through conventional machining is very costly and in many cases economically ineffective. An alternative is manual prototype production by experienced model-makers. Disadvantage of this method is an insufficient accuracy of the obtained models and time consumption, that is why foundries are constantly looking for new solutions. The research work focused on possibility of application of Rapid Prototyping or Rapid Tooling techniques (utilizing the AMTs) in foundry are conducted widely in the world [2, 3, 4]. However, applications of prototypes made by FDM technology are usually limited to serving as a pattern for making silicone molds, which are further used to cast wax models for investment casting [5, 6].

Requirements regarding prototypes formulated by the companies which use the sand mold casting are usually as following:

- appropriate mechanical properties allowing to use the prototype for making many molds,
- high dimensional accuracy,
- rapid time of manufacturing,
- possibly low cost of making the prototype.

The Fused Deposition Modeling technology allows making prototypes with sufficient accuracy. However, in order to ensure a proper strength, the prototype must be built with a solid, monolithic internal filling (deposing material inside the part in a form of strictly adhering paths). It is directly transferred to the amount of material used in the manufacturing process and time of the process, which significantly affects the final price of the prototype. On the basis of many years of experience, the authors can state that almost 80% of enquired, needed prototypes are not produced because the price is too high. The method proposed by the authors – described on the example of a certain case study in this paper – allows reduction of these costs by limiting the use of material and shortening the manufacturing time in the Fused Deposition Modeling technology. Thanks to possibility of controlling the parameter of internal filling degree, the volume of material used for its manufacturing can be reduced, depending on the model size and shape, from several up to several dozen percent. It is directly transferred to reduction of the manufacturing time. Such a prototype can be, however, used only as a visual model due to its low strength [7]. A solution for making partially filled, usable and strong models is the authors' patented technology of filling inside of a model with a liquid, chemically hardened resin. It allows using such models as elements of foundry tooling because they become strong, but the filling does not significantly increase their costs and allows keeping the manufacturing time acceptable.

THEORETICAL INFORMATION

The FDM method consists in linear deposition of plasticized thermoplastic material, extruded by a nozzle of a small diameter (for the BST 1200 device used in the studies described in this paper, the nozzle diameter is 0.4 mm) by a special head, which can move in two axes (X and Y), in parallel to the modeling table. After building a geometry inside a given layer, the table changes its position regarding the head (Z axis of the machine) by a defined value of the layer thickness (for the used BST 1200 device, 0.25 mm and 0.33 mm layer thicknesses are available). The extruded material solidifies in contact with the previously deposed layer, creating a stable connection. To ensure the support for geometries which protrude out of the contour of a previous layer significantly, it is necessary to build special support structures. They are also made out of thermoplastic material, but this material has different properties (namely, lower adhesion) – it is possible to mechanically separate them from the main part after the manufacturing process is finished [8].

The most frequently used FDM material is the acrylonitrile – butadiene – styrene (ABS), which ensures a relatively good strength and allows further processing of the manufactured objects (grinding, painting, gluing etc.) [9]. In the FDM method, it is possible to manufacture elements with the solid or sparse internal filling (Fig. 1a and 1b).

In the practical industrial applications, very often the elements manufactured additively are subjected to complex loads. An example of such loading is manufacturing of molds for casting in the mold sands. During the mechanical consolidation of the mold sand, the prototype (used as a pattern) is loaded by multiple forces, which leads to a complex stress state inside the part. The forces act in a dynamic way, which usually leads to failure of prototypes with the partial (sparse)

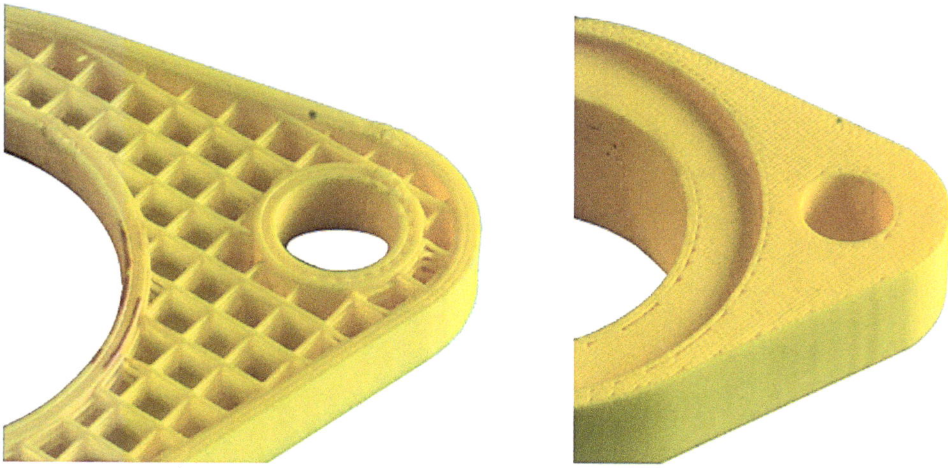

a) b)

Fig. 1. Prototypes made using the FDM method with filling: a) sparse (partial), b) solid (full)

filling. That is why the authors have proposed an alternative method of manufacturing of the sparse prototypes, in which the empty spaces are filled with a liquid resin, forming a fibrous composite. Two phases can be distinguished in such a composite: the matrix, made out of the ABS material using the FDM process and the reinforcement in a form of resin (Fig. 2).

Usually, in case of traditional composites, the internal fibers are a reinforcement, while the matrix holds the shape of the object and fixes the composite together. In case of a composite proposed by the authors, the contour produced using the FDM technology is the final product shape, while the resin ensures better mechanical properties [10].

Method of manufacturing the composite prototypes of matrix according to the FDM process and filled with a resin was patented by the authors in the Patent Office of Republic of Poland [11]. The developed method allows, depending on the used resin and the level of complexity of the matrix geometry, to conduct the filling process while the FDM process is still running or then it is finished and the support structures are removed.

STUDY DESCRIPTION

The model described in this paper is a casting of a faucet for spring for water tapping. Because of a characteristic shape of the faucet model, which was inspired by a shell, it was not possible to use a solution in a form of a ready product available in catalogues of the fittings producers. The model of the faucet was designed as stylistically coherent with the whole spring tap concept, so the prototype of a functional casting needed to be produced, for a fair event presentation. Figure 3 presents a final

Fig. 2. Fibrous composite with the FDM-made matrix, filled with resin

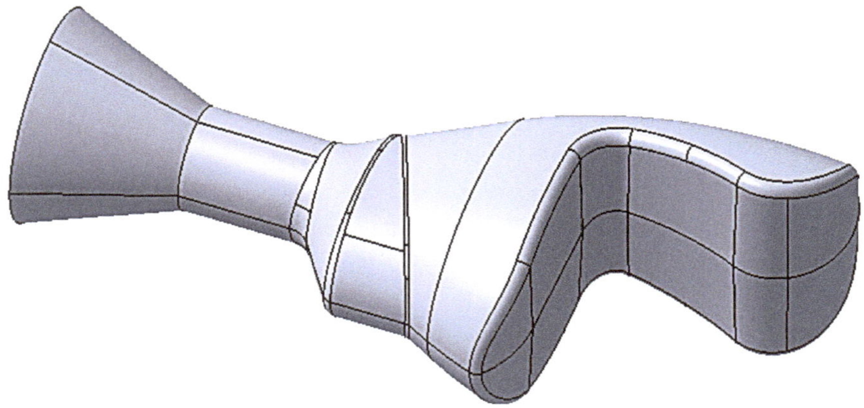

Fig. 3. Model of the faucet with the gating system

3D model of the faucet with the designed gating system, shaft element later used as a fixture, introduced drafts and allowances for shrinkage of brass, out of which the element was planned to be casted.

The model was divided into two halves. Each half was converted into a triangular mesh and saved to the STL format. After importing the model to the system controlling the FDM machine, it was sliced into layers of 0.254 mm thickness (103 layers total) and paths for the machine head deposing the model material were generated. For making the matrix of the ABS-resin composite, the Dimension BST 1200 machine from the Stratasys company was used. The material of the matrix was the ABS material of P400 type, colored in black. To shorten time of manufacturing the matrix, the partial internal filling was introduced with the maximum possible thread separation distance, equal to 12.7 mm. Because of the complex shape of the prototype, filling the matrix was realized gradually, during the FDM manufacturing process. To fill the whole matrix correctly, the first filling was performed after the layer no. 43 was deposed (Fig. 4a), while the second one – after the layer no. 90 was deposed (Fig. 4b).

To fill the matrix, polyurethane, chemically hardened resin Axson F19 was used. The filling was conducted gravitationally. The polyurethane resin was decided to be used as a reinforcement of the composite, because it makes good adhesive connection with the ABS material and has generally good mechanical properties [12]. The polyurethane resins also have good castability, which makes filling of the matrix internal space easier. Their lifetime is long enough to perform the whole process without perturbations. Other advantages of the polyurethane resins include low shrinkage during hardening and low chemical aggressiveness. After hardening process is

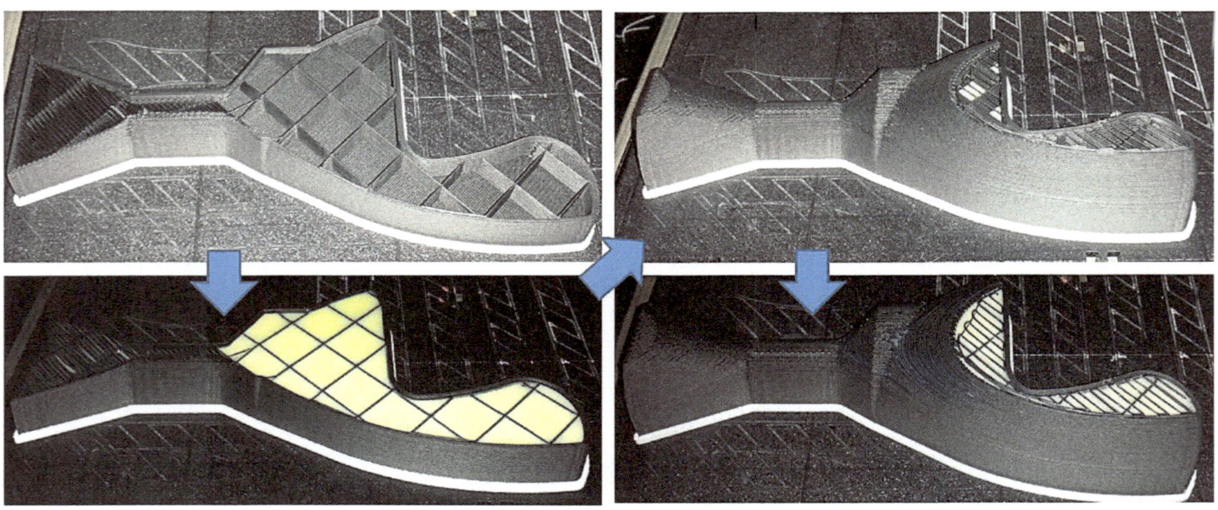

a) b)

Fig. 4. Stages of filling the matrix with the resin: a) filling after layer 40, b) filling after layer 90

finished, the resin is not a toxic waste and there is no problem with its utilization. Its relatively low price is also an important factor – it is less costly than the ABS material (used in commercial FDM systems), which allows to make the functional monolithic prototypes in a less expensive way.

After the FDM process was finished, the prototype was left until the resin was totally hardened, then the support material was removed (gray material visible in the Figure 4). The products manufactured additively are characterized with the staircase effect on their surface. It is related to characteristics of the FDM process itself – deposing the ABS material threads layer by layer. The layer thickness has direct influence on the size of the staircase effect. In the described case, the applied layer thickness was equal to 0.254 mm (the lowest possible value for the used BST 1200 machine) what was not sufficient to obtain a proper surface quality The prototype was subjected to the post-processing in form of grinding. In Figure 5, the prototype after removal of the support material is shown (staircase effect visible).

Despite low shrinkage of the F19 resin (not exceeding 1%), its influence on dimensional and shape accuracy of the obtained composite was also studied. The finished prototype and the matrix itself have been measured using a 3D scanner Atos Compact Scan 5M and then compared with each other, the results were presented in a form of a colored deviation map. The next stage was preparation of a pattern plate and a molding box. Both halves of the faucet were based to each other using fixing pins and assembled mechanically to the pattern plate made out of plywood of 10 mm thickness (Fig. 6a). Other elements of the molding box were also manufactured out of plywood. The assembled box is presented in Figure 6b. To make the mold, molding sand composed from quartz sand mixed with liquid sodium glass. After the box was filled, the sand was consolidated mechanically and hardened by blowing carbon dioxide through it. One half of the ready mold is presented in Figure 6c.

The prepared mold was filled with a liquid brass, then it was opened and the casting was cleaned. The Figure 7 presents the obtained raw

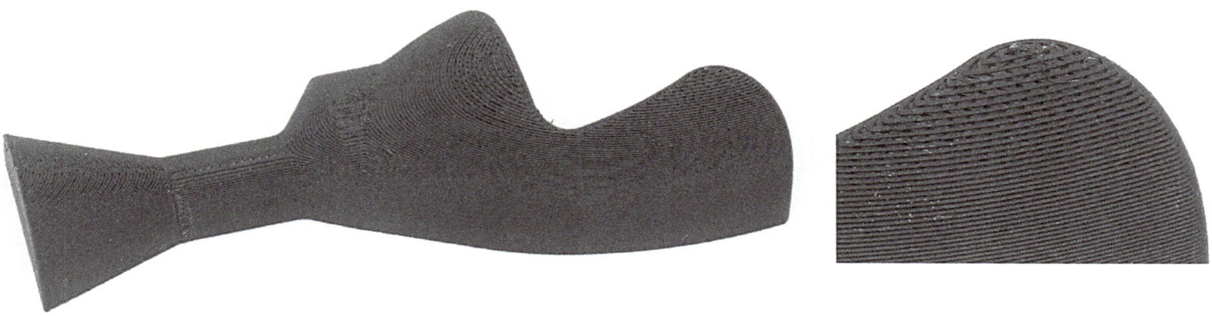

Fig. 5. The prototype after removal of the support material (staircase effect visible)

a) b) c)

Fig. 6. a) pattern plate with the composite prototype, b) ready molding box, c) the mold

casting. The casting was subjected to machining (milling of allowances, drilling holes, threading) and then galvanized, to put a chrome coating on its surface. A final product – the faucet – is presented in the Figure 8.

To study the economical aspect of the production of composite prototypes in relation to the production of prototypes with solid internal filling with the FDM method, the setup times were measured, as well as times of manufacturing prototypes of the same geometry with both methods. The costs of used materials were also taken into account.

RESULTS DISCUSSION

Use of composite products with matrix manufactured using the FDM method and filled with a polyurethane resin as a foundry tooling (a pattern) have proven an efficiency of the proposed methodology. As opposed to prototypes with the sparse internal filling, the produced composite prototypes were characterized with an appropriate strength and ability to carry loads related to the process of consolidation of the applied molding sand. The authors have also prepared a pattern plate with a prototype made solely out of ABS material with the sparse filling. During the process of mechanical consolidation of the molding sand, the prototype was destroyed – disjoint of threads of the ABS plastic occurred in the area where the staircase effect was most apparent (Fig. 5). Under pressure of the molding sand, the outer layer of the material collapsed to the inside, destroying the internal framework of the partial internal filling. It was therefore proven that the FDM process with the partial filling is not a viable option to prepare patterns for castings from the sand molds.

Total time of manufacturing of both halves of the prototype with the monolithic (solid) internal filling was 280 minutes, while the time of manufacturing the matrix itself with the internal thread distance equal to 12.7 mm was 133 minutes. The

Fig. 7. The raw casting

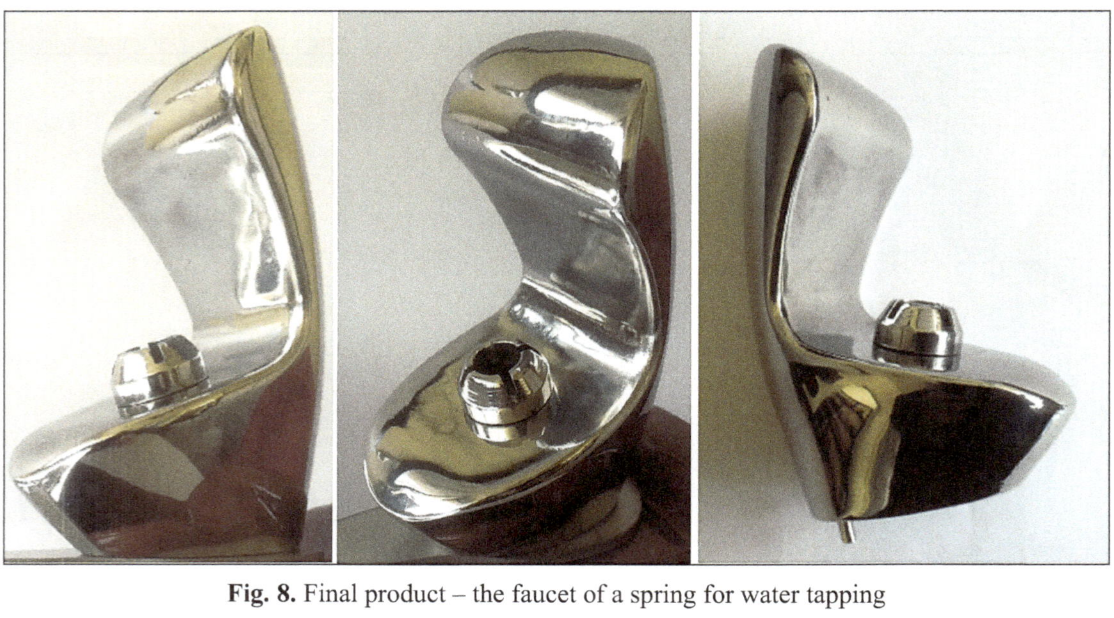

Fig. 8. Final product – the faucet of a spring for water tapping

total time of manufacturing the composite prototypes, considering the matrix production time, time of filling it with the resin and time needed for resin hardening was 250 minutes. For the prototypes of solid internal filling, 273 cm^3 of the P400 ABS material was used, while the composite prototypes required 51 cm^3 of the same ABS material and 260 cm^3 of the Axson F19 resin.

Table 1 presents a juxtaposition of costs and time of manufacturing of the prototype of the faucet pattern using both approaches (standard and proposed by the authors).

The Figure 9 presents results of comparison of the matrix manufactured using the FDM method with the partial filling and the composite prototype (the same matrix after filling it with the resin).

An average error of fitting both models was 0.06 mm. The greatest deviation is present in the end part of the faucet (red and navy blue color). The deviation in these areas is equal to 0.26 mm in the bottom part and –0.24 mm in the upper part, which indicates that a local deformation is present (buckling directed to the bottom part). In case of 85% of the other points, the deviations are inside accuracy and repeatability of the BST1200 machine, which is ±0.2 mm.

CONCLUSIONS

The studies conducted by the authors have proven, that the developed method of manufacturing the composite prototypes can be used for production of patterns for foundry. Thanks to the use of the FDM process to build the composite matrix (which is an external shell of the composite), it is possible to obtain an accurate representation of the desired shape, while the application of a polyurethane resin for filling the inside of the matrix makes it possible to increase the total strength of the product. Moreover, selection of resins of appropriate properties is possible (such as increased hardness, abrasion resistance or other properties), which will affect the final properties of the composite. The applied Axson F19 resin is much cheaper than the ABS P400 material used in the BST1200 machine, which allowed reduction of the material costs to 59% in the described case. The total manufacturing time was also shortened by 24%. It is noteworthy that the total composite manufacturing time presented in the paper includes also the time needed for solidification (hardening) of the resin (90 minutes for the described case). The hardening is realized outside the building chamber of the FDM machine, which is an additional advantage – time of operation of the device is only 50% of time needed for manufacturing a prototype of the same geometry, but with the solid internal filling. Still, the attention must be paid to the fact, that the presented times can be very different for different geometries.

The methodology of manufacturing composite prototypes developed and patented by the authors is currently used in the Laboratory of Rapid Manufacturing of Poznan University of Technology during cooperation with industrial companies for making functional prototypes. The authors are also continuing work aimed at determination of how the basic properties of the obtained composites change depending on the type of applied resin.

Table 1. Comparison of times and costs of materials for manufacturing of monolithic and composite prototypes

Manufacturing of composite prototypes	Total relative manufacturing time		Relative cost of the used material	
	[min]	[%]	[PLN]	[%]
FDM-made ABS prototype with solid filling	280	100	456.00	100
ABS-resin composite prototype	250	89.3	111.00	24.3

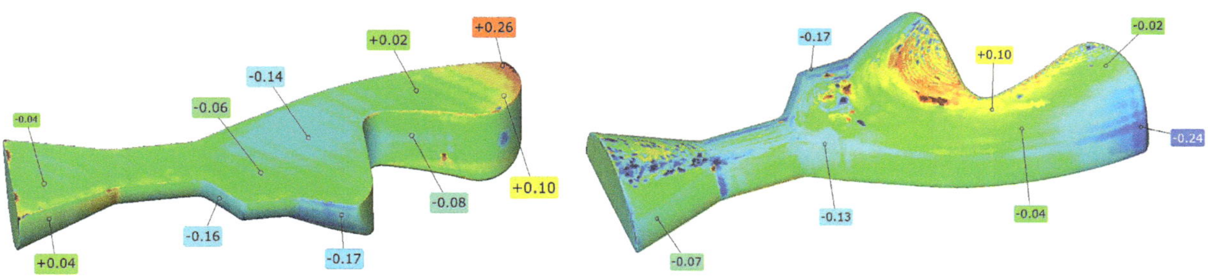

Fig. 9. Comparison of the matrix manufactured using the FDM method with the composite prototype (matrix filled with resin)

REFERENCES

1. Kuczko W., Wichniarek R., Górski F.: Odtwarzanie zużytych lub zniszczonych elementów maszyn za pomocą inzynierii odwrotnej oraz technik szybkiego wytwarzania. Inżynieria Maszyn (Journal of Machine Engineering), 18(1), 2013, 3–7.
2. Rosochowski A., Matuszak A.: Rapid tooling: the state of the art. Journal of Materials Processing Technology, 106, 2000, 191–198.
3. Wang W., Conley J.G., Stoll H.W.: Rapid tooling for sand casting using laminated object manufacturing process. Rapid Prototyping Journal 5, 1999, 134–140.
4. Cheah C.M., Chua C.K., Lee C.W., Feng C., Totong K.: Rapid prototyping and tooling techniques: a review of applications for rapid investment casting. The International Journal of Advanced Manufacturing Technology 25, 2005, 308–320.
5. Lee C.W., Chua C.K., CheahC.M., Tan L.H., Feng C.: Rapid investment casting: direct and indirect approachesvia fused deposition modelling The International Journal of Advanced Manufacturing Technology 23, 2004, 93–101.
6. Raju B.S., Chandrashekar U., Drakshayani D.N., Chockalingam K.: Internet based e-manufacturing of prototypes for one-off intricate castings by using rapid prototyping technology. International Journal of Applied Engineering Research, 5(20), 2010, 3357–3365.
7. Sood A.K., Ohdar R.K., Mahapatra S.S.: Parametric appraisal of mechanical property of fused deposition modeling processed parts. Materials and Design, 31, 2010, 287–295.
8. Kuczko W., Wichniarek R., Górski F.: Influence of process parameters on dimensional accuracy of parts manufactured using Fused Deposition Modelling technology. Advances in Science and Technology Research Journal, 7(19), 2013, 27–35.
9. Kuczko W., Wichniarek R., Górski F.: Influence of filling type on strength of parts manufacured by Fused Deposition Modelling. Journal of Machine Engineering, 14(3), 2014, 113–125.
10. Oczoś K.: Kompozyty włókniste – właściwości, zastosowanie, obróbka ubytkowa. Mechanik 7, 2008, p. 579.
11. Kuczko W., Wichniarek R.: Sposób wytwarzania kompozytowych wyrobów o osnowie z tworzywa termoplastycznego („Method of manufacturing of composite products with the matrix out of thermoplastic material"), patent no P.401988, date: 10.12.2012.
12. Arenas J., Alia C., Blaya F., Sanz A.: Multi-criteria selection of structural adhesives to bond ABS parts obtained by rapid prototyping. International Journal of Adhesion & Adhesives 33, 2012, 67–74.

21

PERIODIC SIGNAL DETECTION WITH USING DUFFING SYSTEM POINCARE MAP ANALYSIS

Valeriy Martynyuk[1], Mykola Fedula[2], Oleksii Balov[3]

[1] Depatrment of Radio Engineering and Communications, Faculty of Programming, Computer and Telecommunication Systems, Khmelnytskyi National University, Khmelnytskyi, Ukraine, e-mail: martynyuk.valeriy@gmail.com; mailfm2000@gmail.com; superbalov@gmail.com

ABSTRACT

In this article the periodic signal detection method on the base of Duffing system chaotic oscillations analysis is presented. This work is a development of the chaos-based signal detection technique. Generally, chaos-based signal detection is the detection of chaotic-to-periodic state transition under input periodic component influence. If the input periodic component reaches certain threshold value, the system transforms from chaotic state to periodic state. The Duffing-type chaotic systems are often used for such a signal detection purpose because of their ability to work in chaotic state for a long time and relatively simple realization. The main advantage of chaos-based signal detection methods is the utilization of chaotic system sensitivity to weak signals. But such methods are not used in practice because of the chaotic system state control problems. The method presented does not require an exact system state control. The Duffing system works continuously in chaotic state and the periodic signal detection process is based on the analysis of Duffing system Poincare map fractal structure. This structure does not depend on noise, and therefore the minimum input signal-to-noise ratio required for periodic signal detection is not limited by chaotic system state control tolerance.

Keywords: periodic signal detection, chaotic system, Duffing oscillator, Poincare section, SNR.

INTRODUCTION

Weak periodic signal detection is one of the most important problems in many fields of modern technology. The main of these fields is the diagnostics of industrial machinery and equipment, biomedicine signal processing, radar information detection and communications. The new chaos-based signal detection method was proposed in 1992 [1]. This method is based on the Duffing oscillator state transition under weak periodic signal influence. The modern works related to the above focus on the chaos identification algorithms and methods [2]. Also there are different modifications of the Duffing oscillator which have a higher sensitivity [3]. These works are based on the results of chaotic system modeling and numerical simulation. But they are not connected closely with the general signal processing theory.

It should be noted that such methods are not popular in practical applications because of the oscillator critical state control errors. So we propose a new detection method with using Duffing oscillator without the state transition.

CHAOS-BASED SIGNAL DETECTION

The Duffing oscillator model is given by Eq. (1):

$$\ddot{x} + k\dot{x} - x + x^3 = s(t), \qquad (1)$$

where: $s(t)$ – the driving signal,
x – the output signal,
k – the damping constant [2, 4, 5].

In this work we consider the damping constant value $k = 0.5$. The driving signal is defined by Eq. (2):

$$s(t) = A_0 \sin(\omega_0 t + \varphi_0) + A_s \sin(\omega_s t + \varphi_s) + N(t) \quad (2)$$

where: $A_0\sin(\omega_0 t)$ – the exciting signal;
$A_s\sin(\omega_s t)$ – the input signal to-be-detected;
$N(t)$ – the input noise.

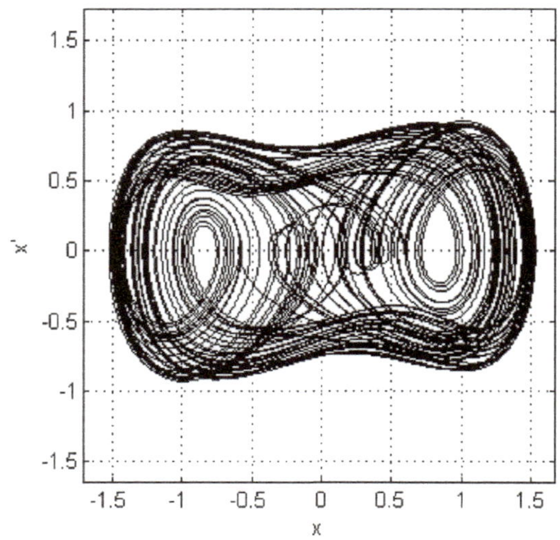

Fig. 1. Typical phase trajectories in chaotic state

The potential function [6, 7] of Duffing oscillator is defined by expression:

$$H(x) = \frac{1}{2}\cdot x^4 - x^2 \quad (4)$$

The state of Duffing oscillator depends on the average potential value. For the Duffing oscillator (1,2) the chaotic oscillations are possible in the interval:

$$0 < H(x) < 1 \quad (5)$$

This is shown in Figure 2.

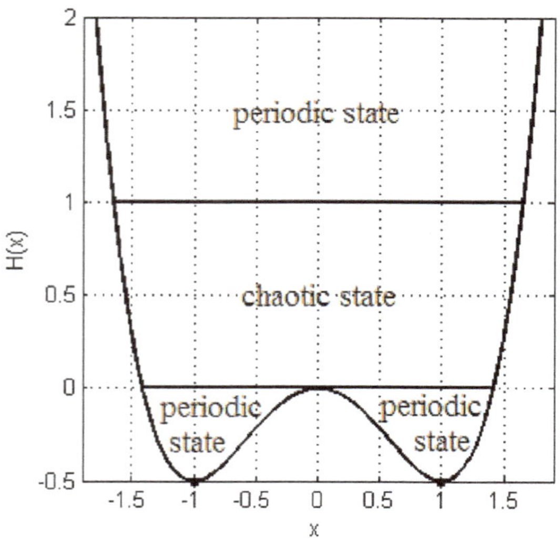

Fig. 2. Phase trajectories in the chaotic state

Generally, the dynamics of Duffing oscillator can be described by the phase trajectories which are projected on the Hamiltonian surface [8, 9, 10]. The Hamiltonian surface $H(x, x')$ of unforced Duffing oscillator is given by expression (3).

$$H(x,x') = (x')^2 + \frac{1}{2}x^4 - x^2 \quad (3)$$

The projection of Duffing oscillator phase trajectories on its Hamiltonian surface is shown in Figure 3.

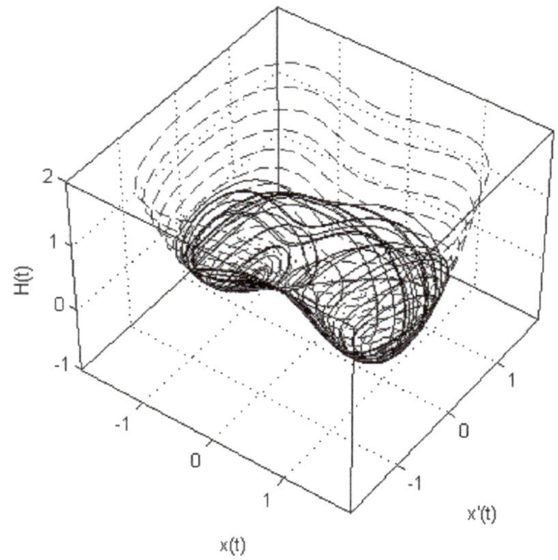

Fig. 3. Phase trajectories projection on the Hamiltonian surface in the case of chaotic state

For periodic signal detection the initial Duffing oscillator state is set to critical chaotic by the exciting periodic signal amplitude A_0 = 0.8245.

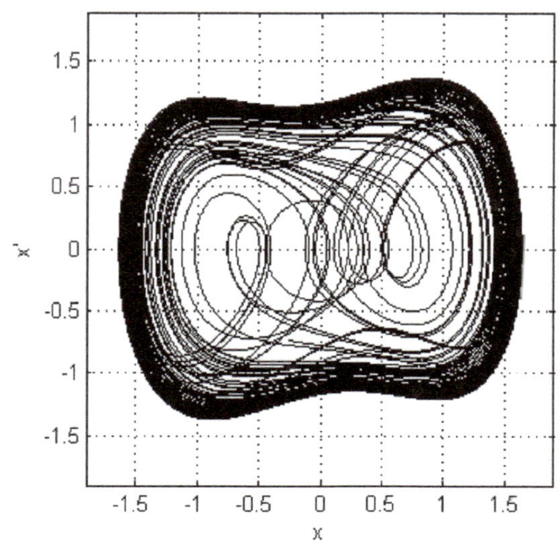

Fig. 4. Phase trajectories near the critical chaotic state

If $\omega_0 \approx \omega_s$ and the input signal periodic component A_s exceeds its threshold value, the system transforms from chaotic state to periodic state [2, 11, 12].

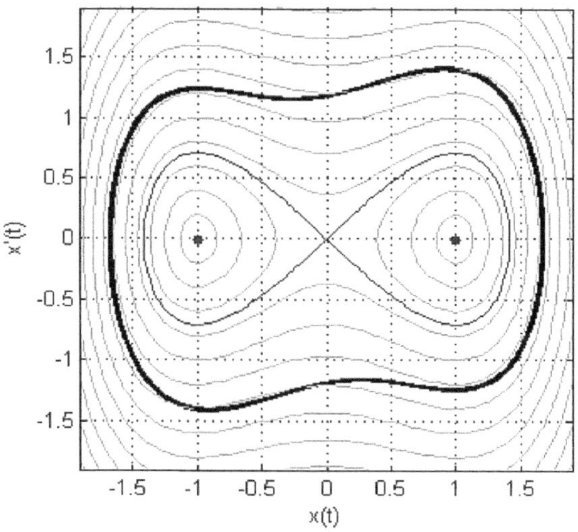

Fig. 5. Phase trajectories in the periodic state

The method described allows to detect weak periodic signals under strong background noise conditions. The main disadvantages of this method are related to the critical state control problem. In critical state the Duffing oscillator is sensitive to non-stationary input signals but in periodic state it is not sensitive to weak periodic signals. These factors may cause significant errors during the periodic signal detection process.

THE FRACTAL-BASED DETECTION METHOD

In this paper we propose the signal detection method which is based on Duffing oscillator Poincare map fractal structure analysis.

Let us consider the point $(x(kT,\varphi_0); x'(kT,\varphi_0))$ on the phase plane with respect to the Duffing oscillator Poincare map structure. This point moves along oscillator Poincare map structure with time.

If we increase the input signal periodic component, the point $(x(kT,\varphi_0); x'(kT,\varphi_0))$ shifts along the Poincare map structure, as shown in Figure 6.

The direction of this shift is noise-independent. But it varies with respect to the chaotic dynamics of the Duffing oscillator.

It is known that the Duffing oscillator Poincare map has a complex fractal structure. At any time the $(x(kT,\varphi_0); x'(kT,\varphi_0))$ point shift direction depends on its position in fractal structure of Duffing oscillator Poincare map.

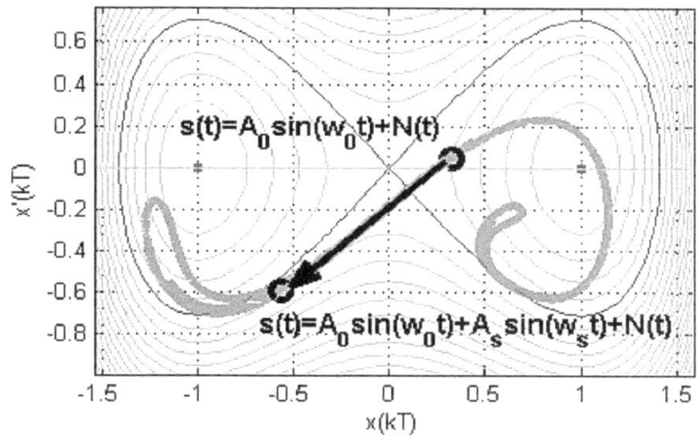

Fig. 6. The effect of periodic component increase

The main bifurcation point of Duffing oscillator has coordinates (0; 0). Near this point the phase trajectory divergence speed is much higher than in other places. When a segment of Poincare map structure passes this region, then it is stretched as shown in Figure 7.

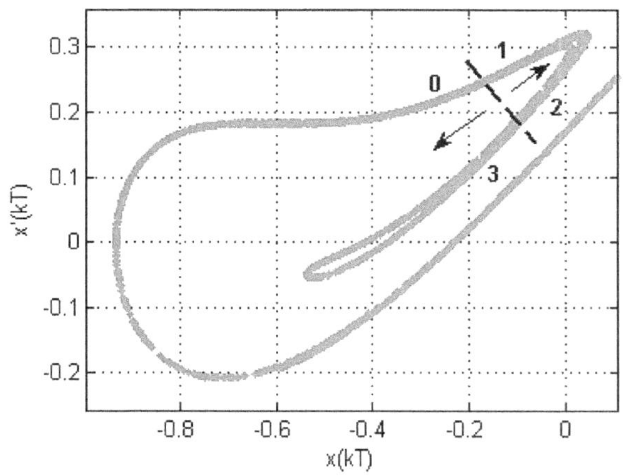

Fig. 7. The stretching of Poincare map fractal structure

If the fractal structure segment passes the stretch region, it can be divided into four parts with different dynamics. In Figure 7 the parts are numbered in accordance with the input periodic component amplitude increase direction. Parts 1 and 2 move from left center to right center. But the parts 0 and 3 stay at the orbit of the current center. After some time the parts 0, 1, 2, 3 stretch as well as the segment {0123}, and the fractal structure complexity increases.

Denote the length of the $(x(kT,\varphi_0); x'(kT,\varphi_0))$ point path along the fractal structure by $\lambda(t, A)$. If we consider the signal frequency $\omega_s = \omega_0$, $A = A_0 + A_s$ is the input periodic component amplitude.

Therefore, generally $\lambda(t, A)$ depends on the input periodic component amplitude monotonically. So the expressions $\lambda(t)\big|_{A=const}$ and $\lambda(A)\big|_{t=t_0}$

are monotonic increasing functions, where t_0 is any time value from the experiment time interval. Thus, it is advisable to estimate the path length $\lambda(t, A)$ with the tolerance up to one fractal structure element. Then the path length $\lambda(t, A)$ can be estimated as a normalized number $\Lambda(t, A)$ in the quaternary system value:

$$\Lambda(t, A) = L_M \cdot 4^0 + L_{M-1} \cdot 4^{-1} + ... + L_0 \cdot 4^{-M} \quad (6)$$

where: M – the number of digits,

L_0, ..., L_M – the weights of the corresponding digits.

The weight value can be taken from the array $\{0, 1, 2, 3\}$. The fractal structure corresponding to $\Lambda(t, A)$ is shown in Figure 8.

The fractal structure in Fig.8 shows that in the case of input noise absence we can define the input periodic signal amplitude $A = A_0 + A_s$ from Duffing oscillator output signal with very high accuracy. If we know the $\Lambda(t_k, A_0)$, then we can compare this value to $\Lambda(t_k, A_0 + A_s)$, and determine the presence or absence of periodic component with amplitude A_s at the input. The detection process is the decision between two hypotheses, H_0, signal absent, and H_1, signal present.

For convenience, we consider the discrete functions $V_{H0}(k)$ and $V_{H1}(k)$:

$$V_{H0}(k) = \Lambda(t_k, A_0), \quad (7)$$

$$V_{H1}(k) = \Lambda(t_k, A_0 + A_s). \quad (8)$$

The fractal-based periodic signal detection process was simulated in MATLAB/Simulink environment under the folowiwng conditions:

$$A_0 = 0.4V, \ A_s = 0.005V,$$
$$\omega_0 = \omega_s = 1\, rad / \sec.$$

The noise is white and the signal-to-noise ratio is 3dB in the band $[0.1\omega_0; 10\omega_0]$. The graphs of the functions $V_{H0}(k)$ and $V_{H1}(k)$are presented in Figure 9.

The Figure 9 shows that if the signal $A_s \cdot \sin(\omega_s t)$ is present, then $V(k)$ is much greater, than if it is absent. The case of input noise presence is shown in Figure 10.

The results in Figure 10 show that the input noise causes the increase of $V_{H0}(k)$ value, but $V_{H1}(k) > V_{H0}(k)$ as well.

The experiments show that if the input SNR is higher than 0dB by power, we can set a constant threshold value $V_{thr} \in (V_{H0\max}; V_{H1\min})$. Therefore, the detection process is robust and reliable.

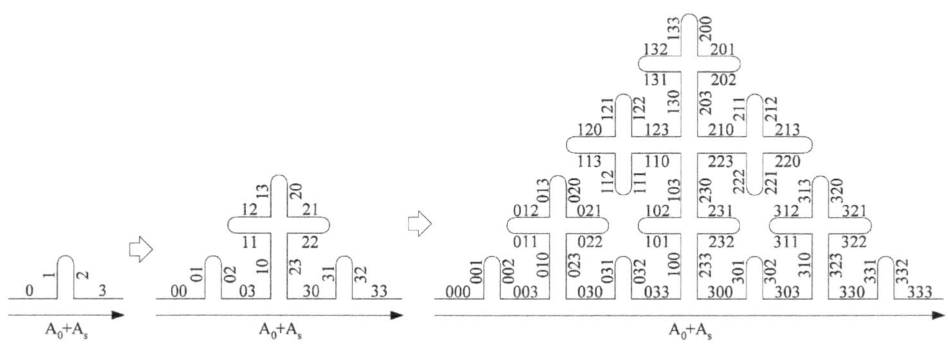

Fig. 8. The fractal structure corresponding to path length estimation $\Lambda(t, A)$

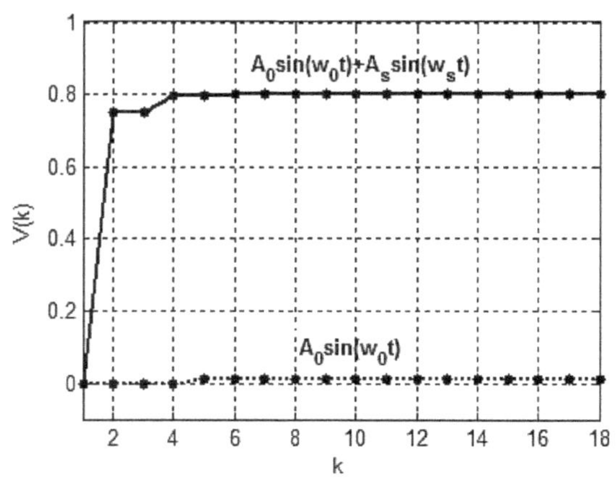

Fig. 9. Functions $V_{H0}(k)$, $V_{H1}(k)$ without input noise

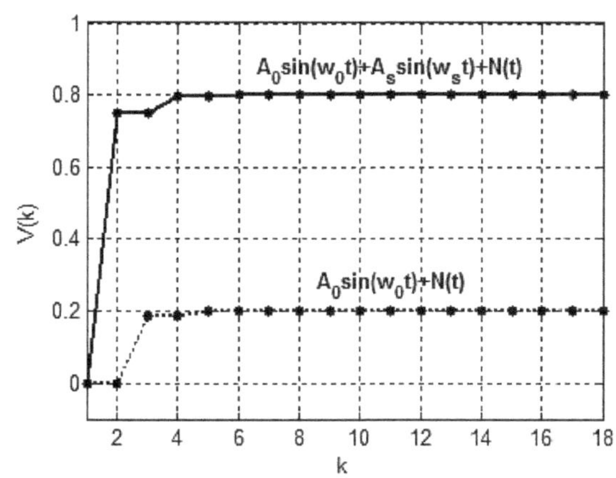

Fig. 10. Functions $V_{H0}(k)$, $V_{H1}(k)$ with input noise

If the input SNR is lower than 0 dB by power, then the first digits $L_M,...,L_{M-p}$ of $\Lambda(t_k,A_0)$ can be greater than the same digits of $\Lambda(t_k,A_0+A_s)$. So we cannot detect the signal $A_s\cdot\sin(\omega_s t)$ directly with the constant threshold value. In this case we must analyze the next digits $L_{M-p-1},...,L_1$ of $\Lambda(t_k,A)$ without respect to the first $L_M,...,L_{M-p}$. This is the direction of our future work.

CONCLUSION

The main advantages of the method proposed are the following features:

- the critical state setting and control are not required;
- Duffing oscillator works continuously in chaotic state and the chaos identification is not required;
- the estimation of input periodic component amplitude is robust and reliable;
- the minimum input SNR required for periodic signal detection is not limited by chaotic system state control tolerance.

This method is more resistant to non-stationary and pulse noise in comparison with the known method [2]. The direction of the future work is connected with the analysis of Duffing oscillator Poincare map fractal structure under SNR less than 0 dB.

REFERENCES

1. Birx D. I. Chaotic oscillators and CMFFNS for signal detection in noise environments. IEEE International Joint Conference on Neural Networks, Vol. 22, 1992, 881–888.
2. Jalilvand A., Fotoohabadi H. The Application of Duffing Oscillator in Weak Signal Detection. ECTI Transactions on Electrical Eng., Electronics, and Communications, 9(1), 2011, 1–6.
3. Li Yue, Yang Baojun. Chaotic system for the detection of periodic signals under the background of strong noise. Chinese Science Bulletin, 48(5), 2003, 508–510.
4. Dai Y. The Weak Signal Detection Method and Instrument. The National Defence Industrial Publishing House, Beijing 1994, 265–278.
5. Liyun S., Qian Y., Yuli Z., Jiaojun L. Noise Immunity of Duffing Oscillator and its Applications in Weak UWB Signal Detection. Journal of Networks, 7(3), 2012, 540–546.
6. Thomson J.M.T., Stewart H.B. Nonlinear Dynamics and Chaos. John Wiley & Sons, Ltd, Baffins Lane, Chichester, West Sussex, England 2002.
7. Ott E. Chaos in Dynamical Systems. Cambridge University Press, Cambridge 1993.
8. Moon F.C. Chaotic Vibrations. An Introduction for Applied Scientists and Engineers. John Wiley & Sons, Inc., Hoboken, New Jersey 2004.
9. McLachlan N.W. Ordinary Non-linear Differential Equations in Engineering and Physical Sciences, 2nd Ed. Clarendon, Oxford 1958.
10. Armitage J.V. and Eberlein W.F. Elliptic Functions. LMS 67, Cambridge 2006.
11. Ferrer S., Lara M., Families of Canonical Transformations by Hamilton-Jacobi-Poincare Equation. Application to Rotational and Orbital Motion, Journal of Geometric Mechanics, Vol. 2, 2010, 223-241.
12. Wang Guanyu, Chen Dajun, Lin Jianya, Chen Xing. The application of chaotic oscillators to weak signal detection. IEEE Transactions on industrial electronics, 46(20), 1999, 440-443.
13. Yang Jianhong, Zhang Rencheng, Fang Huaiying, et al. Application of Duffing Oscillator signal detection to arcing faults protection. Automation of Electric Power System, 30(24), 2006, 69-72.

MODERNIZATION OF TECHNOLOGICAL LINE FOR CELLULAR EXTRUSION PROCESS

Tomasz Garbacz[1], Ludmila Dulebova[2], Volodymyr Krasinskyi[3], Lenka Markovicova[4]

[1] Lublin University of Technology, 36 Nadbystrzycka St., 20-618 Lublin, Poland, e-mail: t.garbacz@pollub.pl

[2] Technical University in Košice, 74 Mäsiarska St., 040 01 Košice, Slovakia, e-mail: lidmila.dulebova@tuke.sk

[3] Lviv Polytechnic National University, 12 St. Bandery St., Lviv, Ukraine, e-mail: vkrasinsky82@gmail.com

[4] University of Zilina, 1 Univerzitna St., 010 26 Zilina, Slovakia, e-mail: lenka.markovicowa@fstroj.uniza.sk

ABSTRACT

In recent years, cellular extrusion of polymers has been one of the fastest growing processing methods. It is used especially to obtain sections, bars, pipes and cellular coatings free from strains on the product surface, displaying reduced density and minimum shrinkage, at the same time maintaining similar properties of products extruded in a conventional way. In order to obtain cellular structure, product properties are modified by application of proper plastics or adding blowing agents. The article provides a description of manufacturing processes of cellular product. It gives characteristics of blowing agents used in the extrusion process and of processing conditions. Also, it discusses selected results of examined properties of the obtained cellular products. As part of the modernization of the cellular extrusion technology the extrusion head was designed and made. During the designing and modeling of the head the Auto CAD programe was used. After the prototyping the extrusion head was tested. In the article specification of cellular extrusion process of thermoplastics was presented. In the research, the endothermal chemical blowing agents in amount 1,0% by mass were used. The quantity of used blowing agent has a direct influence on density and structure of the extruded product of modified polymers. However, these properties have further influence on porosity, impact strength, hardness, tensile strength and another.

Keywords: cellular extrusion, extrusion head, simultaneous construction, CAD, blowing agents, physical properties.

INTRODUCTION

Extrusion of cellular plastics differs from extrusion of solid plastics in that the product obtained as a result of processing procedure is formed into a diphase structure plastic material-gas, with possibly smallest and evenly distributed gas bubbles. Cellular structure is obtained due to the insertion of the blowing agent (porophor) in the form of an inert gas, low-boiling liquid or a solid body into the input plastic which, when in liquid or solid state, transforms into gas under the determined conditions of extrusion process [1, 2, 3, 4, 5, 19, 20]. When in the proper temperature the process of gas emission is started, numerous microspheres which are then generated, dissolve in the ambient plastic material due to the operation of pressure and surface development. The emerging cells may be filled with air or with any other gases, for instance CO_2 and N_2, yet later they are replaced by air as a result of diffusion [6, 7].

The blowing agent (porophor) is selected so as to match the type of plastic material, and its decomposition temperature has to be higher than the melting point of the plastic material, yet lower than the temperature during plastic extrusion. Cellular plastic in a liquid form is not yet a stable system because as a result of interfacial tension at the material–gas contact point, and as a result of diffusion, the number of cells decreases, whereas their dimensions enlarge, which constitutes an undesirable effect. The obtained cells keep enlarging until gas pressure and interfacial tension are balanced. Favourable structure of a small-cell plastic can be maintained in the product by its immediate cooling and solidification. [2, 8, 9].

Currently, the most significant group of porophors used in extrusion and injection moulding of plastics are organic porophors. They belong to a number of groups having different chemical structures. Of those, the greatest significance is displayed by an azo compound; azo-formic acid diamide (azodicarbanamide). Blowing agents used in cellular extrusion of plastics may have exothermic or endothermic decomposition characteristics [1, 2, 8, 10, 11, 12]. Porophors that has been used so far usually have exothermic decomposition characteristics. This may be the cause of local overheatings and generation of irregular cellular structure of the product. When blowing agents have endothermic decomposition characteristics, generation of gas during processing is violently stopped when the energy flow is cut off. The application of such blowing agents considerably shortens the cooling time. Examples belonging to this group are among others bicarbonates e.g. sodium bicarbonate, ammonium bicarbonate.

The studies on cellular extrusion process conducted by authors of the articles [6, 7, 9, 10, 13] are focused, among others, on the generation of products from cellular plastics. This results in the reduction of purchase costs of the plastic material, of process energy costs and of transport costs. Study on cellular extrusion of thermoplastics regards the process in which the modification of product properties is done by alternating the extrusion conditions and design features of technological line elements. The extruded product obtained may have a fully solid or cellular structure, both cellular core and topcoat, or cellular core but solid topcoat [2, 7, 14]. Extrusion process which involves the application of blowing agents results in obtaining new, modified physical and technological properties of the extruded products.

EXPERIMENTAL PROCEDURE AND RESULTS

As a part of the modernization of the cellular extrusion technology the extrusion head was designed, with the using Auto CAD program. Next, the extrusion head was made according to the prepared design solution. After the prototyping the extrusion head was testing in order to apply in technological line of cellular extrusion. Figure 1 shows the schemes of the designed extrusion head, while in Figure 2 and 3 the actual appearance of the designed extrusion head is presented.

The body (2) of extrusion head consists of two parts, one of which allows the connection of head to the extruder plasticizing system, the second part of the body enables the assembly of the extruder die (1). Between the two parts of body the distributive mandrel (4) is mounted. Extrusion die is a replaceable item, which allows the extrusion of profiles with different shapes and sizes, both symmetrical and asymmetrical.

The extrusion head has two heating zones and two corresponding annular heaters mounted on the body of the head and the die body, and two temperature sensors (thermocouples) installed in two heating zones of the head. Interchangeable dies allow the production of symmetrical shapes.

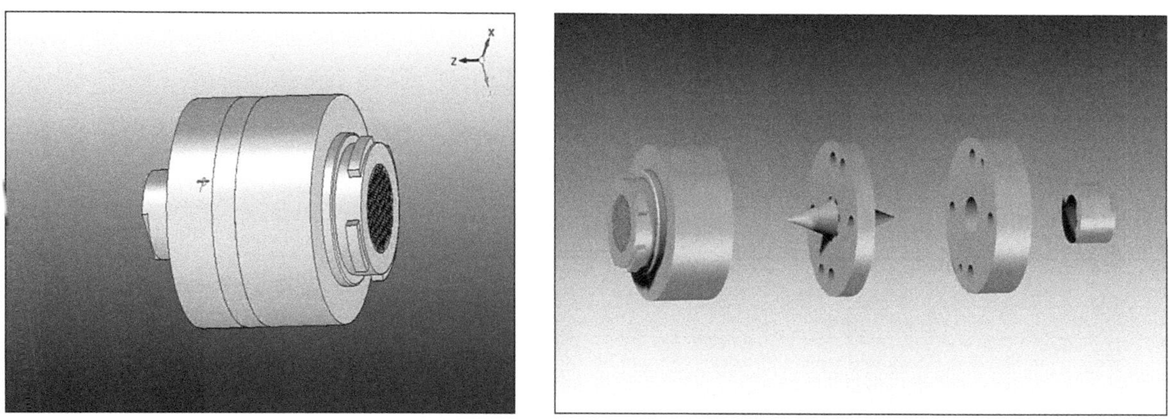

Fig. 1. Examples of the designed schemes of extrusion head for profiles

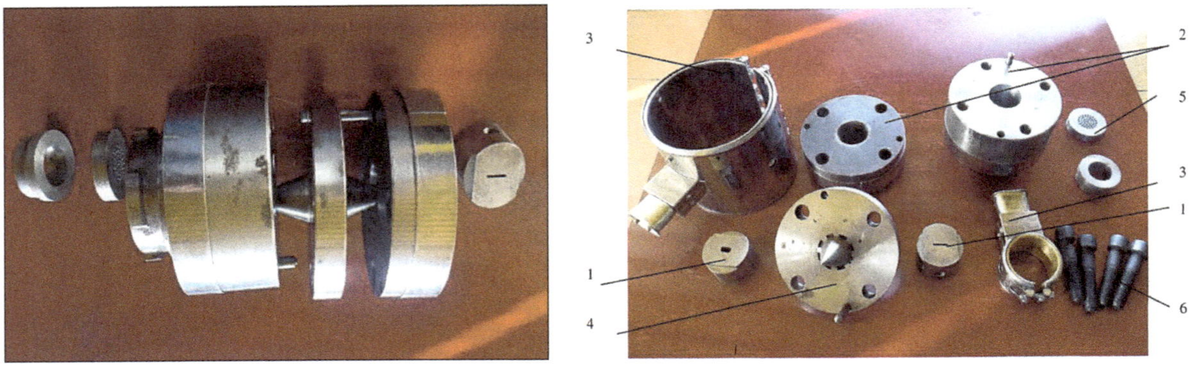

Fig. 2. View of the parts of extrusion head for profiles: 1 – die, 2 – body parts, 3 – annular heaters, 4 – distributive mandrel, 5 – filter of polymer

Fig. 3. View of the extrusion head for profiles

In the study presented in the article, the used blowing agent was Hydrocerol BIH 70. It has endothermic decomposition characteristics and it assumes the form of granules with diameter of 1.2 to 1.8 mm. Hydrocerol BIH 70 contains 70% of the blowing agent mass having the initial decomposition temperature of 140°C. The blowing agent is a mixture of sodium bicarbonate and 2-hydroxypropane- 1,2,3-tricarboxylic acid (citric acid). In the presented study, Hydrocerol BIH 70 was dispensed as 0.2% to 1.0% of the mass.

The investigation of the extrusion process was carried out in the processing line for extrusion .The extruder plasticizing unit was equipped with four heating zones. The screw had an L/D ratio of 24 and external diameter D of 45 mm (Figure 4). The line also contained the extruder head for moulding rods with external extrusion die diameter equal 8.00 mm, cooling device (cooling bath) and collection device. Extrusion was carried out under the following parameters: temperature of the heating zones 150, 160, 160 and 165°C respectively; extruder head temperature within two heating zones 155°C.

As a result of the carried out process of extruding modified polyethylene with blowing agents, the extruder product was obtained in a form of a rod of 8.00 ± 0.01 mm in diameter. Extrusion and cooling conditions were properly selected, which allowed to obtain a product having a solid topcoat and a cellular core.

The specimens of of the extruded product were then examined in structural properties in compliance with relevant recommendations and norms [15–18]. Their density and porosity were investigated. In addition to that, their

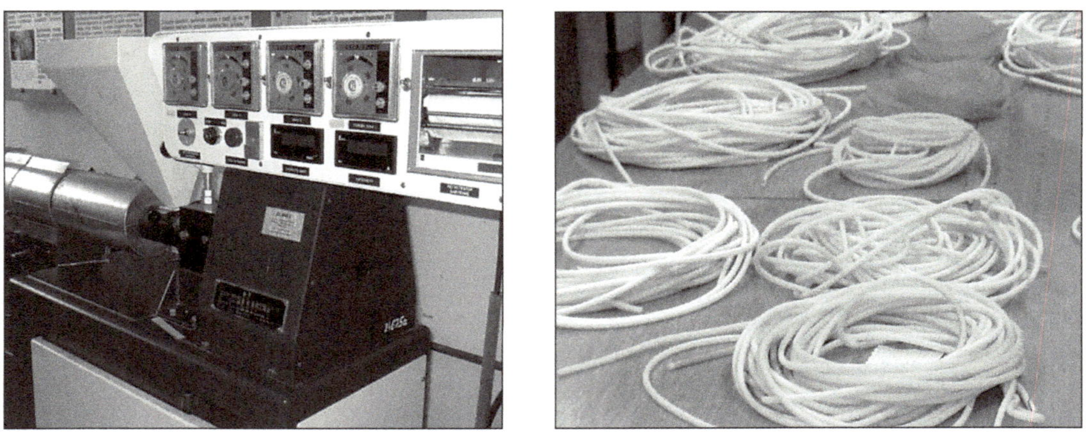

Fig. 4. View of the technological line section for extrusion process and extruded product

surface structure was analyzed based on digital images made.

The method of standard density determination of plastic materials was applied during studies on density of sample products from cellular plastics. Measurements were taken in accordance with recommendations of a relevant standard. Refined particles of sample, whose mass ranged from 1 to 5 g, were the object of the study.

The hardness tests by Shore's method were conducted using an Affri-manufactured hardness tester, type ART13 – Shore's method D. Such a choice was due to the method employed for determining solid polymers hardness.

Water and oil absorptivity examination was carried out in accordance with PN-EN ISO 62:2000: Plastic materials. Determination of water absorptivity standards and with PN-EN ISO 175: 2002: Determination methods of effects of immersion in liquid chemicals.

The investigation and analysis of the porous structure of the produced extruded parts were conducted using an confocal microscope, type Olympus FluoView FV1000 and copyright position of image analysis of porous structure and the author's stand for porous structure image analysis. Microscope FV1000 was equipped with 1.3 Mpix camera, enabling direct viewing of microscaning image on computer screen. Observation of specimens structure and its recording was made in reflected light with suitable magnification.

In line with recommendations and in accordance with proper standards, density, porosity, hardness, water and oil absorptivity in relation to the extruded product obtained from cellular polyethylene were examined. Exemplary results are demonstrated in Tables 1 and 2.

The macroscopic structure of the produced cellular products was examined at the stand for polymer cellular structure image analysis. The examples of the cellular structure of coatings produced are shown in Figures 5 and 6.

Extrusion products made of PE with 0.4% blowing agent content do not have sufficiently cellular structure (Figure 5). The cell dispersion patterns in the structure could be attributed to

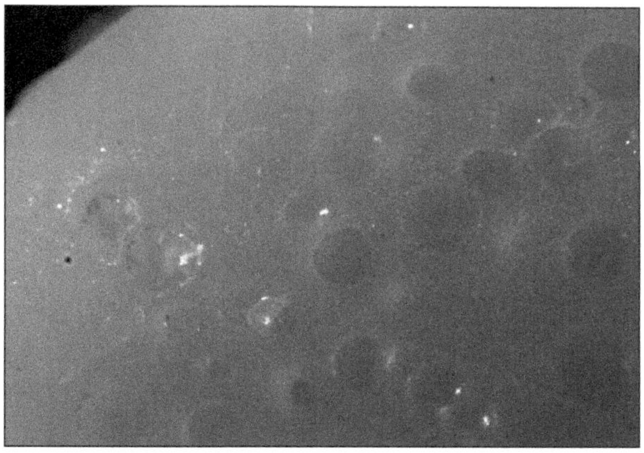

Fig. 5. Cross-sectional view of a cellular polyethylene product, with 0.4% content of the blowing agent Hydrocerol BIH 70

Table 1. Results of the studies of the apparent density, porosity and hardness of product modified by the blowing agent

Sample No.	Content of blowing agent, %	Porosity, %	Density, kg/m^3	Hardness, °Sh D
1	0	0	940	65
2	0,2	19	762	60
3	0,4	22	733	60
4	0,6	36	602	55
5	0,8	40	564	50
6	1,0	42	545	51

Table 2. Results of studies on water absorptivity and oil absorptivity of cellular polyethylene

Sample No.	Content of blowing agent, %	Water absorptivity, %	Change in mass following immersion in oil, %	Change in mass following desiccation, %
1	0	0,25	0,15	0,08
2	0,2	0,22	0,20	0,07
3	0,4	0,20	0,19	0,08
4	0,6	0,30	0,15	0,06
5	0,8	0,30	0,20	0,06
6	1,0	0,30	0,20	0,06

the non-uniform dosing of a very small blowing agent amount into the polymer and the characteristics of the applied blowing agent with the endothermic decomposition behavior.

Based on the analysis of the photographs taken, it can be seen that the coating with the 0.6 and 0.8% blowing agent contents has a visible solid surface and most uniform distribution of similarly sized pores (Figure 6).

The number of pores increases proportionately to the increase in the blowing agent content in the polymer being extruded. The number of pores also depends on their position and distance from the surface.

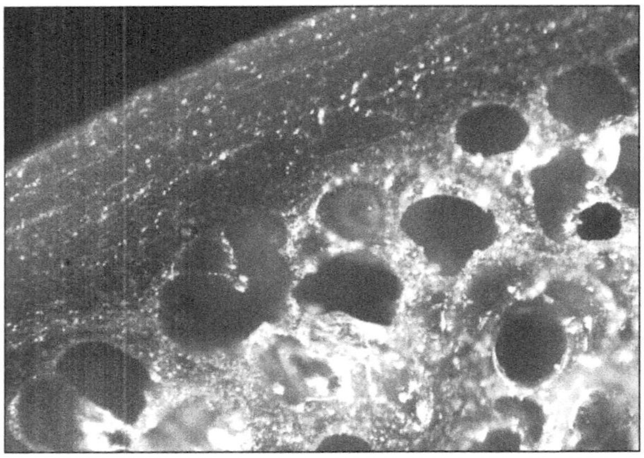

Fig. 6. Cross-sectional view of a cellular polyethylene product, with 0,8% content of the blowing agent

CONCLUSIONS

1. As part of the modernization of the cellular extrusion technology the extrusion head was designed, with the using Auto CAD program. Next, the extrusion head was made according to prepared design solution. This extrusion head was used in the presented studies of cellular extrusion.
2. Type and quantity of the blowing agent applied in the study was matched in such a way that under assumed conditions of the extrusion process an extrudedr product having a solid topcoat and cellular core was obtained.
3. Quality of the obtained product, including features such as external look, degree of porosity and density should be evaluated as very good. The obtained product was a rod having degree of porosity of 19–42%, density ranging from 762 to 545 kg/m^3 respectively and hardness ranging from 50 to 65 Sh D.
4. As a result of the conducted measurements, values for amounts of water absorbed by a sample ranging from 0.25 to 0.30% were obtained. Quantity of absorbed water by sample surface does not change significantly along with the increase in blowing agent content. Values for sample mass gain after immersion in oil ranging from 0.15 to 0.20% were obtained. Neither water absorptivity nor mass gain after immersion in oil do not change along with the increase in blowing agent content in the processed plastic material. This may show that cells generated in the cellular extrusion process are closed and prevent penetration of water and oil into the interior of the cellular extrusion product.

Acknowledgements

This paper is the result of the project implementation: Technological and design aspects of extrusion and injection molding of thermoplastic polymer composites and nanocomposites (PIRS-ES-GA-2010-269177) supported by The international project realized in range of Seventh Frame Programme of European Union (FP7), Marie Curie Actions, PEOPLE, International Research Staff Exchange Scheme (IRSES).

REFERENCES

1. Garbacz T.: Properties of triple-layered PVC coatings synthesized in the micropore coextrusion method. Polimery 56, 7-8, 2011, 129-134.
2. Dulebová Ľ., Greškovič F.: Influence of regrind on properties of plastics processing by injection moulding. Materials Engineering – Materiálové inžinierstvo 18, 2011, 44-48.
3. Samujło, B., Sikora, J.W.: The impact of selected granulometric properties of poly(vinyl chloride) on the effectiveness of the extrusion process. Journal of Polymer Engineering 33, 1, 2013, 77-85.
4. Ľ. Dulebová (et al.): The impact of the amount of filler on the properties of polymer materials. Kovárenství 44, 2012, 33-36.
5. Garbacz T., Krasinskyy V.: Title evaluate the effectivesness of the extrusion process. Progresivne Strojarske Technologie a Materialy. PRO-TECH-MA 2012, ÚVZ Herľany - TU Košice, Slovakia, 2012 (in CD).
6. Sikora J.W., Samujło B., Dzwonkowski J.: Influence of a feed-opening section on the output and selected mechanical properties of a poly(vinyl

chloride) extrudate. Advances in Polymer Technology 33, 1, 2014 (published online).

7. Bociąga E., Palutkiewicz P.: The influence of injection moulding parameters and blowing agent addition on selected properties, surface state and structure of HDPE parts. Polymer Engineering and Science 53, 4, 2013, 679-704.
8. Matuana L.M., Park C.B., Balatinecz J.J.: Structures and mechanical properties of microcellular foamed polyvinyl chloride. Cellular Polymers, 17, 1998, 1-16.
9. Tor-Świątek A., Samujło B.: Use of thermovision research to analyze the thermal stability of microcellular extrusion process of poly(vinyl chloride). Maintenance and Reliability 15, 2013, 58-61.
10. Garbacz T., Dulebova L.: Porophors during the extrusion process. Chemistry and Chemical Technology 7, 1, 2013, 113-118.
11. Tejeda E.H., Sahagún C.Z., González-Núñez R., Rodrigue D.: Morphology and mechanical properties of foamed polyethylene–polypropylene blends. Journal of Cellular Plastics 41, 2005, 417-435.
12. Dulebova L., Garbacz T.: Thermal analyzes of polymer composites. Transfer innovacii 26, 2013, 130-134.
13. Bociąga E., Palutkiewicz P.: The Impact of Mould Temperature and Blowing Agent Content on Structure and Properties of Injection Moulded Parts. Cellular Polymers, 2013, 32(5), 257-277.
14. Garbacz T., Dulebova L., Krasinsky V.: Effectiveness of cellular injection molding process. Advances in Science and Technology Research Journal 8(18), 2013, 74-80.
15. PN-EN ISO 1183-1 (2006): Methods for determining the density of plastics.
16. PN-EN ISO 868 (2003): Plastics and ebonite. Determination of indentation hardness by means of a durometer (Shore hardness).
17. PN-EN ISO 62 (2000): Plastics. Determination of water absorption.
18. PN-EN ISO 175 (2002): Plastics. Methods of test for the determination of the effects of immersion in liquid chemicals.

23

WIND TURBINE OPERATION PARAMETER CHARACTERISTICS AT A GIVEN WIND SPEED

Zdzisław Kamiński[1], Zbigniew Czyż[1], Mirosław Wendeker[1]

[1] Department of Thermodynamics, Fluid Mechanics and Aviation Propulsion Systems, Mechanical Engineering Faculty, Lublin University of Technology, 36 Nadbystrzycka Str., 20-618 Lublin, Poland, e-mail: z.kaminski@pollub.pl; z.czyz@pollub.pl; m.wendeker@pollub.pl

ABSTRACT
This paper discusses the results of the CFD simulation of the flow around Vertical Axis Wind Turbine rotor. The examined rotor was designed following patent application no. 402214. The turbine operation is characterised by parameters, such as opening angle of blades, power, torque, rotational velocity at a given wind velocity. Those parameters have an impact on the performance of entire assembly. The distribution of forces acting on the working surfaces in the turbine can change, depending on the angle of rotor rotation. Moreover, the resultant force derived from the force acting on the oncoming and leaving blades should be as high as possible. Accordingly, those parameters were individually simulated over time for each blade in three complete rotations. The attempts to improve the performance of the entire system resulted in a new research trend to improve the performance of working turbine rotor blades.

Keywords: wind turbine, renewable energy, computational fluid dynamics, numerical analysis.

INTRODUCTION

Numerical investigations unlike experimental ones enable the validation of design intents at the design stage. Accordingly, costly preparation of model or prototype for stand testing can be avoided. CFD methods can be applied to compare aerodynamic characteristics of the forces acting on the working rotor surfaces and determine the parameters of its operation, including torque or power generated by the turbine system. If various design changes are made, one can explore their impact on characteristics and improve the entire system. If the geometry of wind turbine blades is modified, numerous aerodynamic phenomena are triggered.

Turbine performance depends largely on the pressure distribution on its working surface. The challenge is to achieve the best possible recovery of kinetic energy derived from the wind by blade shaping so that the oncoming blade could show the smallest possible drag coefficient at the possible highest drag coefficient of the leaving blade as defined in the following relation [1, 4].

$$P = P_1 - P_2 \qquad (1)$$

$$P = c_{x1} \cdot \frac{A \cdot \rho}{2} \cdot (v - u)^2 - c_{x2} \cdot \frac{A \cdot \rho}{2} \cdot (v + u)^2 \qquad (2)$$

where: P – resultant force acting on the turbine rotor;
A – front surface of the turbine blade;
ρ – density of the agent flowing around the rotor;
c_{x1} – leaving blade drag coefficient;
c_{x2} – oncoming blade drag coefficient;
u – peripheral speed of the blade tip;
v – wind speed

The resultant force P as a difference of the component forces P_1 and P_2 is a driving force of the entire rotor. Actually, that force determines rotor torque. This study identifies the forces acting on the blade during its entire rotation range.

MODELS FOR THE SIMULATION STUDIES

The geometrical model of the rotor that is numerically simulated is a construction according to patent application no. 402214 (Figure 1). The rotor consists of blades that enable us to change an active surface, i.e. that can receive the kinetic energy of wind. Principally, its operation involves adjusting the angle α between the top and bottom part of the blades mounted on the assembly axis (Figure 2). Increasing the angle of opening α between the upper and lower part can increase the active surface which absorbs kinetic energy of wind.

This study is part of the large research and refers to the calculations for a single angle of blades, i.e. a = 90° only. Moreover, the relation in [1] was applied to modify the blade working surfaces to improve their performance. The geometric model of the turbine assembly was developed in CATIA v5. Its size was defined by the size of the wind tunnel used to verify the CFD results. The measurement chamber is 292×292 mm (cross-section). The geometry applied in the simulation reflects the wind tunnel conditions. The turbine under study was in the space of transverse dimensions corresponding to those in a wind tunnel.

The values of the parameters of the rotor and their indications as in Figure 4 are given in Table 1.

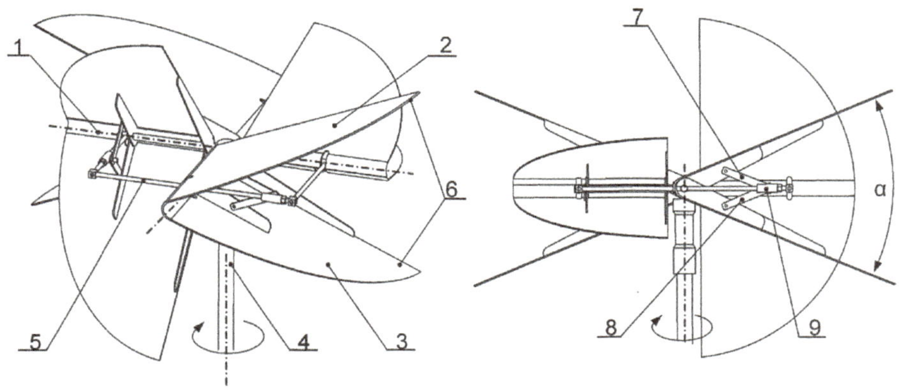

Fig. 1. Isometric view of the rotor with adjustable blade working position (left) and side view of the rotor with adjustable blade working position (right) [5]: 1 – assembly axis, 2 – upper part, 3 – lower part, 4 – bearing axis, 5 – leading axis, 6 – blade, 7, 8 – linkage connecting the sleeve with the upper and lower blade part, 9 – sleeve to adjust spacing blade elements, **α** – angle between the upper and lower blade part

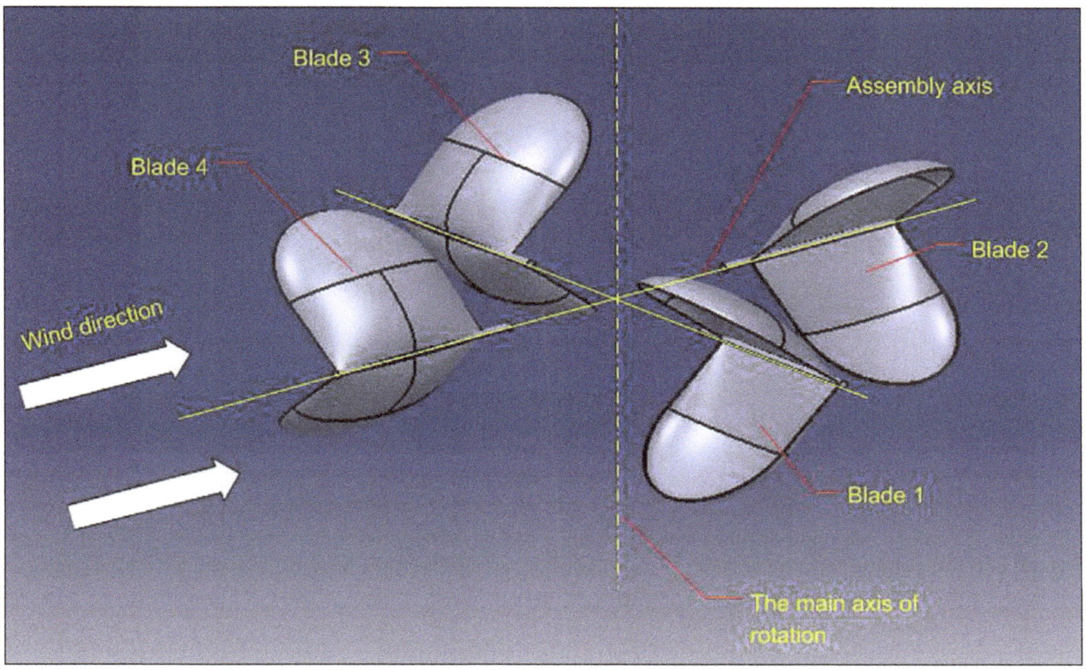

Fig. 2. Notations for the rotor under study

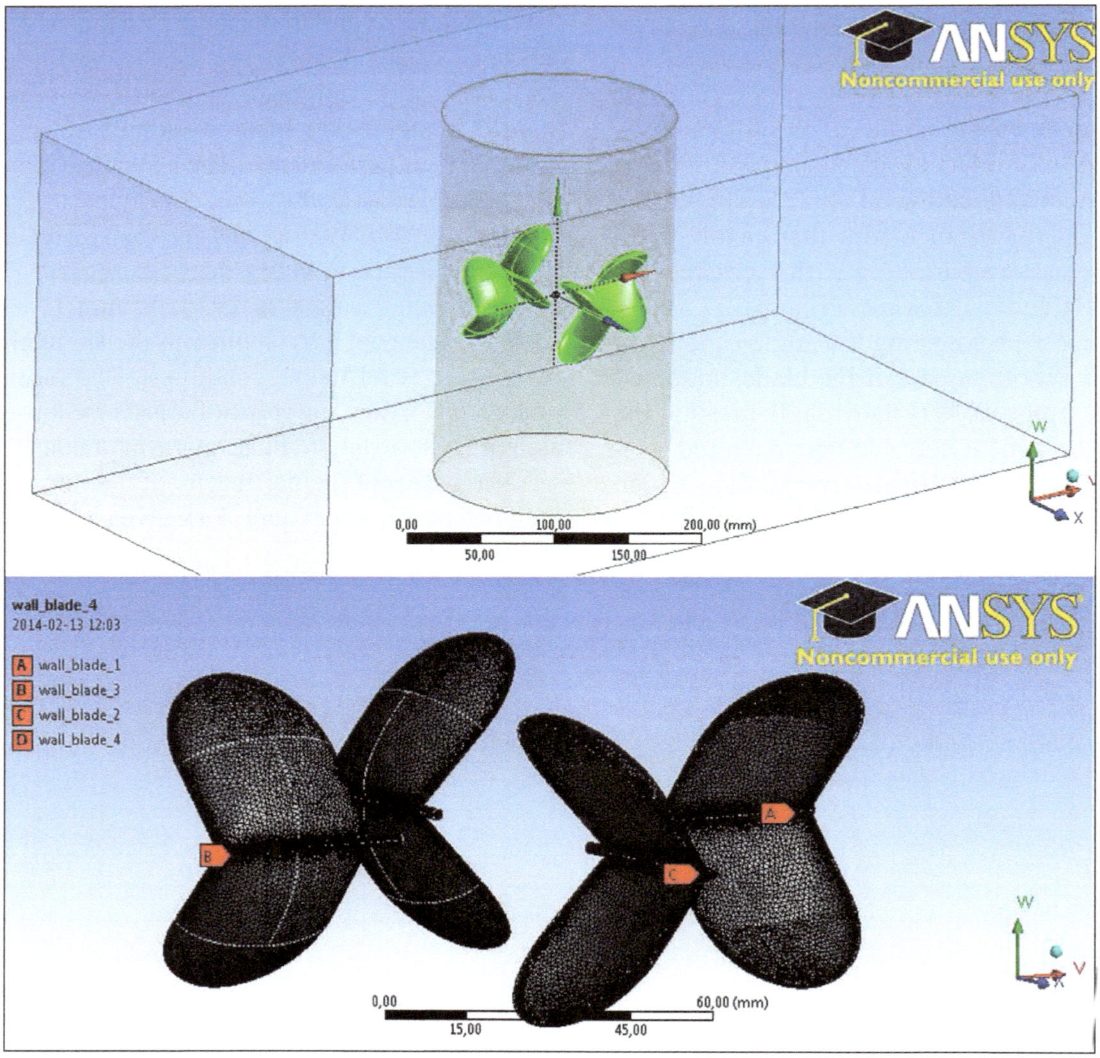

Fig. 3. A sample method of preparing a model of flow around a wind turbine (above) in the wind tunnel measurement chamber and the blade surface mesh density (below)

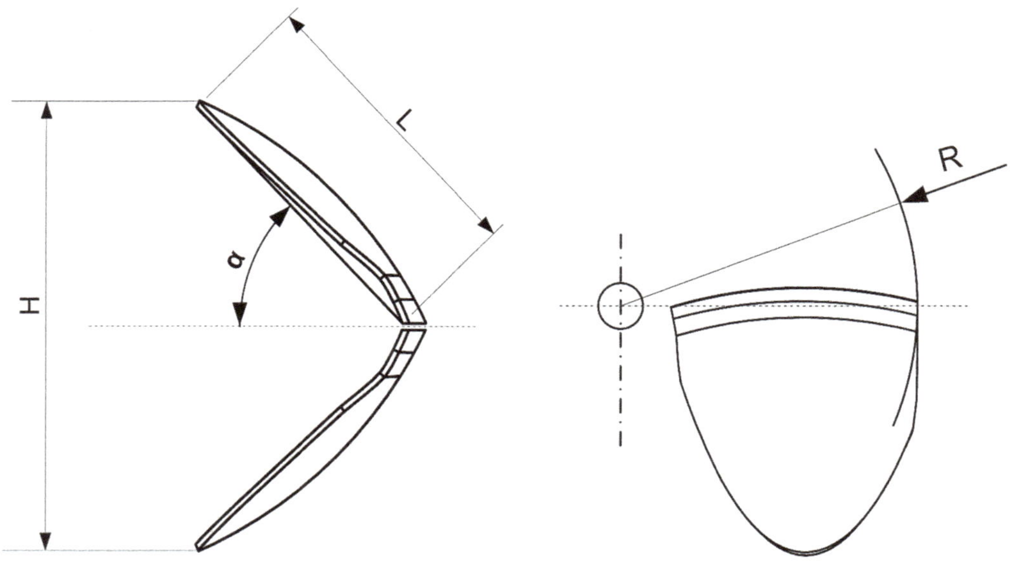

Fig. 4. Diagram of the geometric sub-model of the wind turbine rotor blade, **α** – angle of blade opening as the angle between the upper or lower blade part and the horizontal surface, R – blade radius [mm], L – blade length [mm]

Table 1. Volumes of rotor parameters

No.	Name	Indication	Value	Unit
1	Opening angle	α	45	[°]
2	Blade radius	R	65	[mm]
3	Blade pitch	H	55	[mm]
4	Blade length	L	39	[mm]

RESEARCH DESCRIPTION

The tests were done for several rotational velocities but identical wind velocity. Three complete rotations were simulated to record the changes in blade loading as a function of a rotation angle and rotation repeatability. Accordingly, Table 1 summarizes the auxiliary quantities which are the solver computing settings. If rotational velocity changes, time for a complete rotation changes, too. If the same number of time steps is to be maintained, the value of a single step for each velocity is calculated.

The research scope includes determining the coefficients of the torque for the blades (for each blade) and the entire rotor based on the sum of components. The torque coefficient is defined by the relation in (3), [3].

$$C_m = \frac{T}{0{,}5 \cdot \rho \cdot v^2 \cdot A \cdot Lc} \tag{3}$$

where: T – torque [Nm],

ρ – air density [kg/m^3],

v – wind velocity [m/s],

A – reference surface [m^2],

Lc – reference length [m].

Range of the quantities under study that describe the turbine rotor:

- torque coefficient – blade 1,
- torque coefficient – blade 2,
- torque coefficient – blade 3,
- torque coefficient – blade 4,
- torque coefficient – rotor,
- torque [Nmm] – rotor
- rotor power [mW],
- wind power [W]
- efficiency [%].

The calculations refer to the reference quantities that were used to calculate torque coefficients C_m. The values of those coefficients were obtained from the calculation solver. Having converted them according to (3), the change in torque at time T(t) for the entire rotor system was specified. The rotor power P was determined from equation (4).

$$P = (T \cdot n_m)/9550 \tag{4}$$

The performance of the system under study was calculated from the comparison of the value of rotor power P and wind power P_w for a given rate flow [4].

$$P_w = 0.5 \cdot \rho \cdot A\, v^3 \tag{5}$$

The numerical calculations were done with computational solver Ansys/Fluent. Turbulence model k-ω SST was employed in the calculations because it is best for this type of research as claimed in [6]. It combines two other models. The basic and most common k-ε model fails in near-wall flow modeling and reflecting the phenomena that occur in the near-wall layer. The

Table 2. Auxiliary data for the numerical computations

Rotational velocity n_m [rpm]	Rotational frequency ω [rad/s]	Rotational velocity n_s [rps]	Time for three rotations [s]	Number of time steps [-]	Value of a time step [s]
0	0	0	0	540	0
50	5.233	0.833	3.600		0.00667
100	10.467	1.667	1.800		0.00333
150	15.700	2.500	1.200		0.00222
200	20.933	3.333	0.900		0.00167
250	26.167	4.167	0.720		0.00133
300	31.400	5.000	0.600		0.00111
350	36.633	5.833	0.515		0.00095
400	41.867	6.667	0.450		0.00083
450	47.100	7.500	0.400		0.00074
500	52.333	8.333	0.360		0.00067

Table 3. Reference quantities in the simulations

Name	Symbol	Value	Unit
Reference surface	A	32.6	[cm²]
Reference radius	L	46.5	[mm]
Air density	ρ	1.2257	[kg/m³]
Air flow rate	v	6	[m/s]

results obtained from that model are reliable when the flow is slow with a turbulent area in the whole area that is far away from the near-wall layer. Wilcox k-ω model can replace the k-ε model because it can give satisfactory results for the near-wall flows. Combining those two models, a new one, i.e. k-ω SST is created which uses Wilcox k-ω model for the near-wall flows and k-ε model for flows in areas far away from the walls of the object under study so the numerical results are more reliable and comparable with the real results.

RESEARCH RESULTS

First, the values of torque coefficients for each of the blades as a function of time are compared. Notice that time [s] is the reference quantity. Despite the fact that the calculations were done for the same number of time steps, their values vary depending on velocity. Hence, the time courses of quantities such as power, torque and performance given in Figures 10, 11 and 12 are shorter for higher velocities.

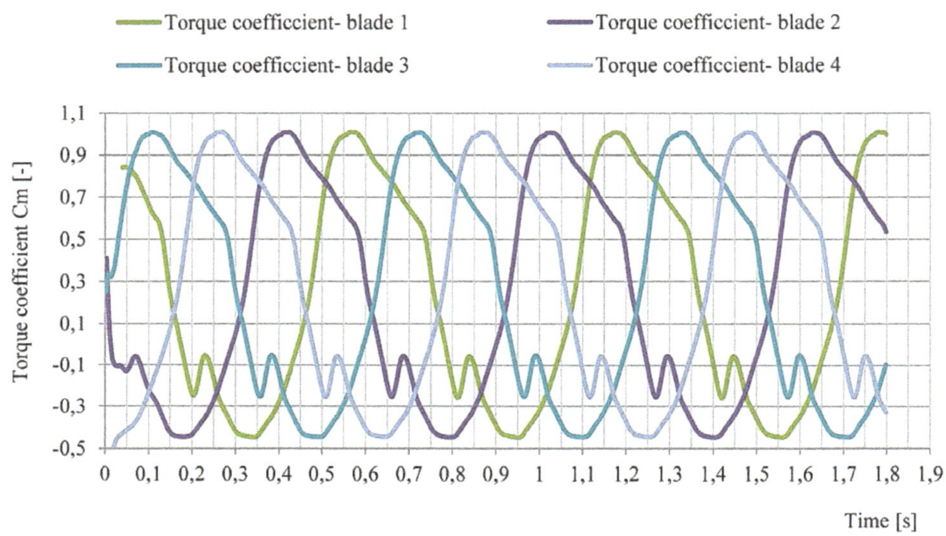

Fig. 5. Time courses of torque coefficient C_m [-] as a function of time for each wind turbine rotor blade for three complete rotations at an angle of 90° and a rotational velocity of 100 rpm

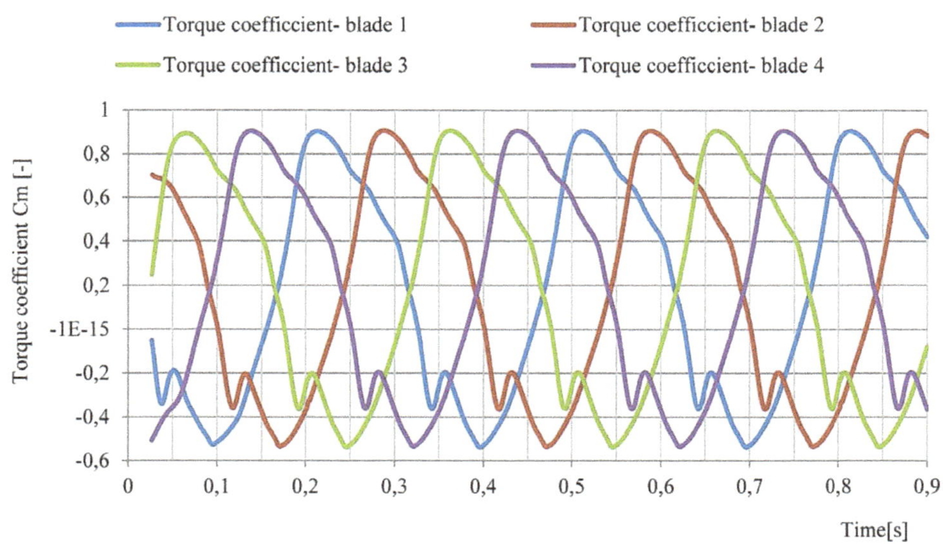

Fig. 6. Time courses of torque coefficient C_m [-] as a function of time for each wind turbine rotor blade for three complete rotations at an angle of 90° and a rotational velocity of 200 rpm

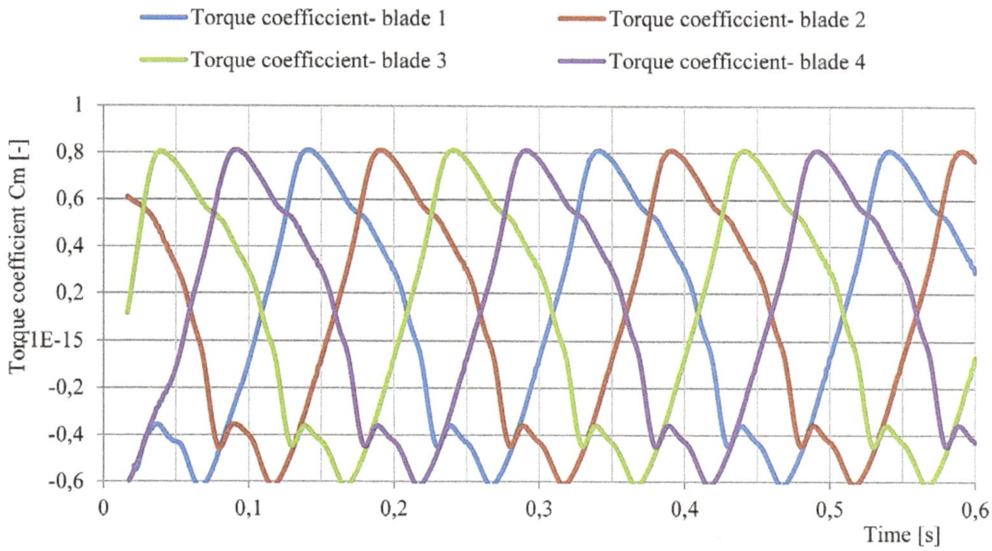

Fig. 7. Time courses of torque coefficient C_m [-] as a function of time for each wind turbine rotor blade for three complete rotations at an angle of 90° and a rotational velocity of 300 rpm

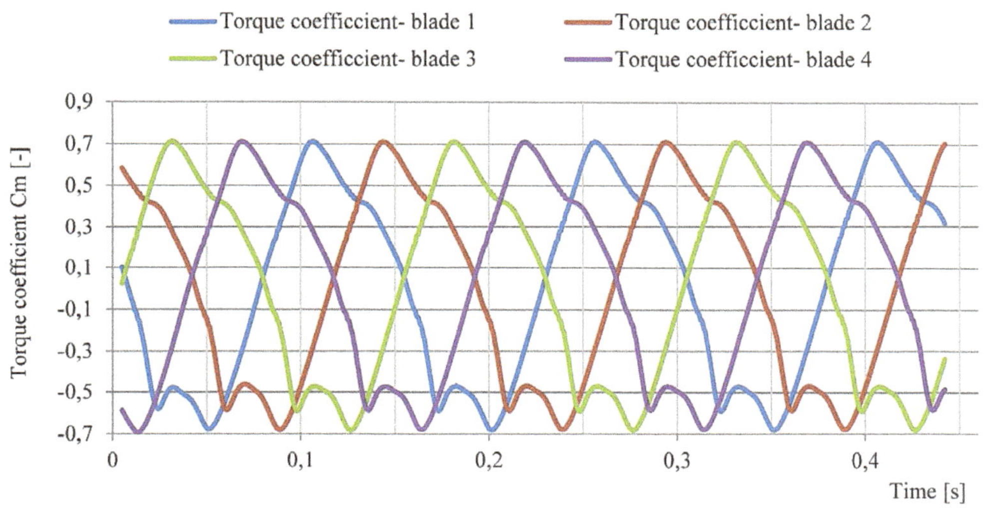

Fig. 8. Time courses of torque coefficient C_m [-] as a function of time for each wind turbine rotor blade for three complete rotations at an angle of 90° and a rotational velocity of 400 rpm

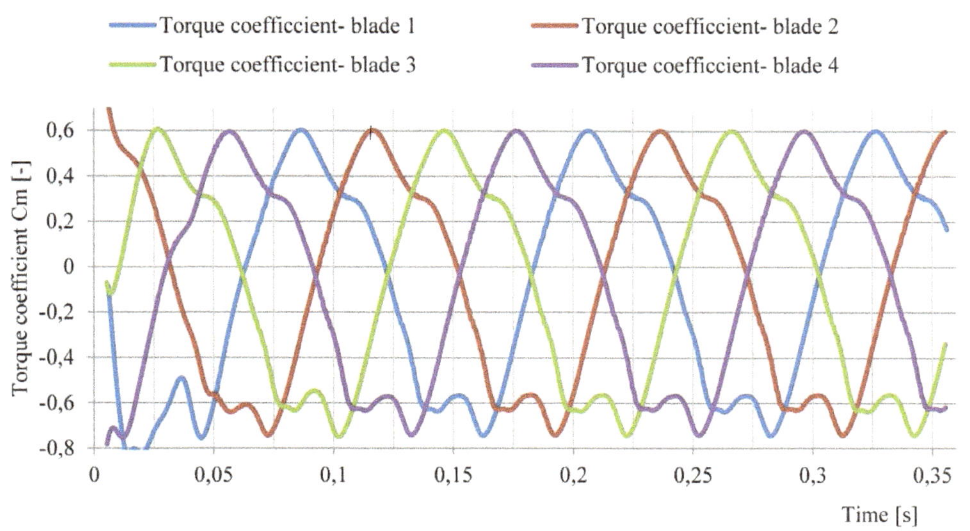

Fig. 9. Time courses of torque coefficient C_m [-] as a function of time for each wind turbine rotor blade for three complete rotations at an angle of 90° and a rotational velocity of 500 rpm

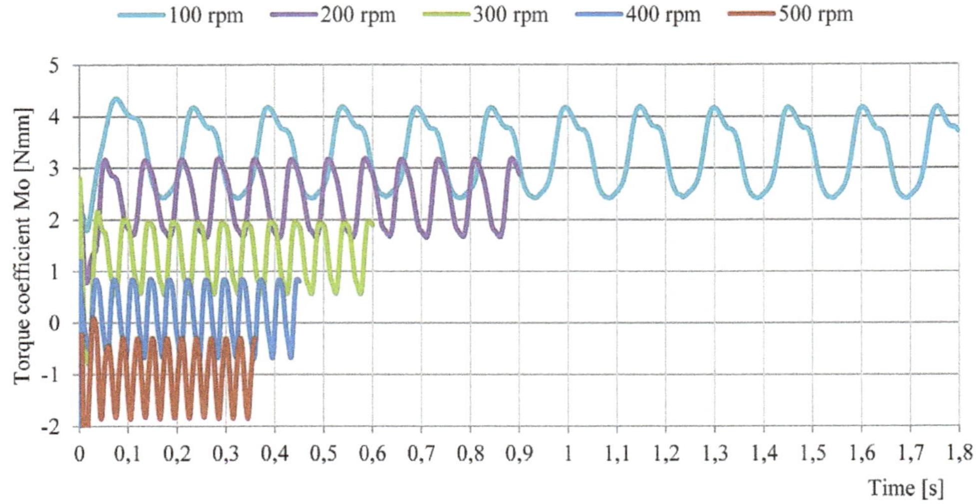

Fig. 10. Torque coefficient for the entire wind turbine rotor [Nmm] as a function of time for three complete rotations at an angle of opening of 90° and varied velocities

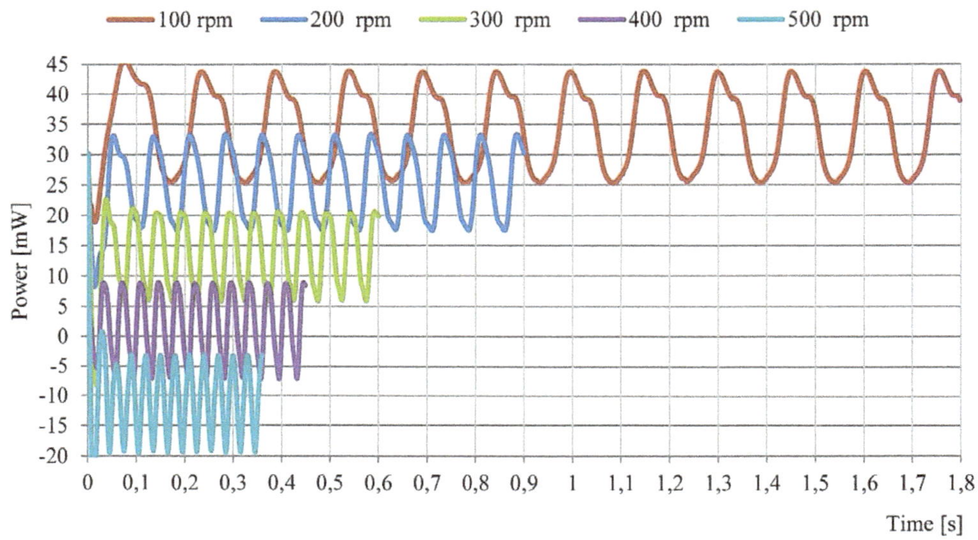

Fig. 11. Power generated by the entire wind turbine rotor [mW] as a function of time for three complete rotations at an angle of opening of 90° and varied velocities

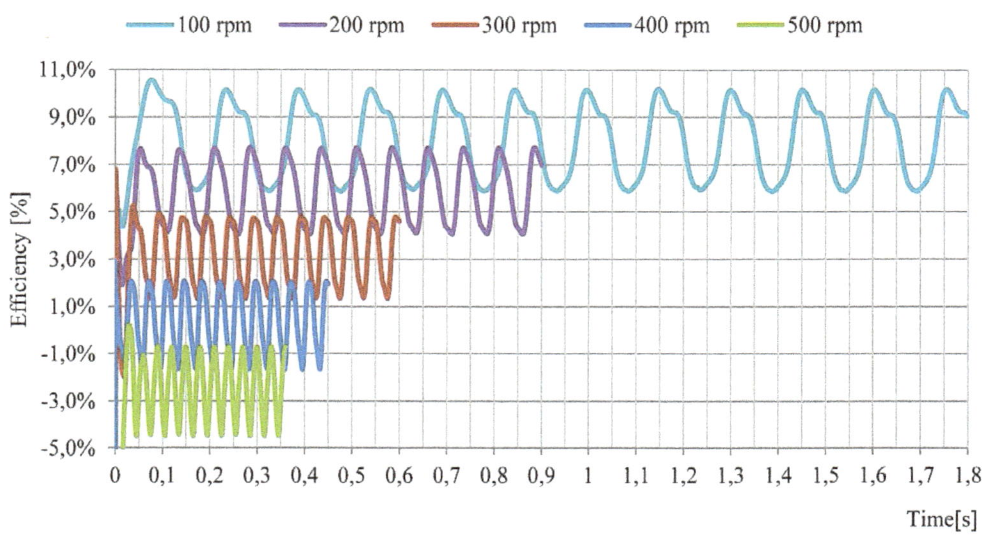

Fig. 12. Efficiency e of the entire wind turbine rotor [mW] as a function of time for three complete rotations at an angle of opening of 90° and varied velocities

SUMMARY

Torque values over time oscillate with a certain amplitude whose boundary values depend on rotor rotation frequency. Also, the coefficient of torque decreases with increasing velocity, which is triggered by a rise in oncoming blade drag. As soon as the velocity boundary value is exceeded, the coefficient C_m reaches a negative value, which means that the oncoming blade at such high speed reaches the instantaneous drag force larger than that of the leaving blade. Actually, this comes from the sum of the linear velocity of a point on the blade and wind velocity. The leaving blade loses its drag as the result of the reduced wind velocity by the linear velocity of a point on the blade. C_m of each blade over time reaches different values depending on the angle of rotation. The largest absolute values are achieved by the extreme blades, i.e. 2 (leaving) blades and 4 (oncoming) blades as in Figure 2. For the oncoming blades, those values are negative (adverse effect). The total coefficient C_m of the entire rotor needs to be positive. That coefficient is almost 0 at a velocity of about 400 rpm, whereas at higher velocities it begins to be negative. This means that it should be given power to run at such high velocities. That need is also confirmed by the negative rotor performance. The maximum instantaneous performance of the system studied is more than 10% at a velocity of 100 rpm while the average one is about 8%. The suggested changes in the structure to continue the investigation on improving the performance refer to avoiding losses due to pressure leak near the assembly axis. Also, a more concave shape in that area is recommended in order to achieve high ratio drag coefficient of the leaving blade to oncoming blade.

REFERENCES

1. Czyż Z., Kamiński Z.: Badania symulacyjne wirnika turbiny wiatrowej o regulowanej powierzchni roboczej, Nowe trendy w naukach inżynieryjnych 3, pod redakcją dr. inż. M. Kuczera, Tom II, Creative Science, Monografia 2012, Creativetime, Kraków 2012, 243-251.
2. Digraskar D.A.: Simulations of Flow over Wind Turbines, Master Thesis. University of Massachusetts – Amherst, 2010.
3. FLUENT 13, User's Guide, Fluent Inc.
4. Jagodziński W.: Silniki wiatrowe. PWT, Warszawa 1959, 84-89.
5. Wendeker M., Kamiński Z., Czyż Z.: Urząd Patentowy RP nr zgłoszenia: P.402214, 2012.
6. Wilcox D.C.: Turbulence Modeling for CFD- Third Edition. DCW Industries, United States 2006.

EVACUATION SYSTEMS OF SCREW-TYPE WATER TURBINES IN SMALL HYDROPOWER PLANT

Jacek Caban[1], Leszek Gardyński[1]

[1] Faculty of Mechanical Engineering, Lublin University of Technology, 36 Nadbystrzycka Str., 20-618 Lublin, Poland, e-mail: j.caban@pollub.pl; l.gardynski@pollub.pl

ABSTRACT

This paper presents evacuation solutions of hydroelectric power set applied in a screw-type turbine in Small Hydropower Plant (SHP). The article compares the solutions applied in Poland and other European countries in the yield of green energy from the water sector. Innovative solutions of the turbine safety problem in hydro building development in case of flood waters of rivers in Poland.

Keywords: construction machinery, water turbine, Small Hydropower Plant.

INTRODUCTION

A major problem for hydraulic structures is a periodic occurrence of floods leading to material and social losses. Quite a common and very dangerous phenomenon in dry areas is the so called flash flood. Flooding is a transient hydrological phenomenon involving a stage of increased water level in rivers or seas in watercourses which cause flooding of large areas of land, coastal areas or depressions when the water is beyond the shoreline. The fight against flooding is an ever present problem known worldwide. A major impact on the occurrence of floods is the existing system of rivers and occurring in different periods of the year hydrological and meteorological situation. The main causes of flooding include: heavy rain, spring thaw, river damming by ice jams or landslides, damage to hydraulic structures (e.g. interruption of the dam), and others.

This paper presents an evacuation solution of the hydroelectric set used for the functioning of a screw-type turbine in a Small Hydropower Plant (SHP). New-approach solutions to the problem of turbine safety, in case of flood in the hydro building development at national rivers, are presented. Attention is paid to the solutions used in other European countries where green energy yield of the water sector is much higher than in Poland.

PROBLEMS OF SMALL HYDROPOWER PLANT

One of the most important ways to protect the environment is the gradual introduction of energy efficient technologies in industry and in power production replacing conventional energy sources – renewable energy [8]. The renewable energy sources (RES) include: water, wind, solar, geothermal energy and bioenergy. According to Directives 2001/77/EC and 2003/30/EC (OJ EU L 140, 05.06.2009, p.16), the Member States of the European Union are required to increase the level of energy production from renewable sources up to 20% of the total energy balance by 2020. In Poland, this share is expected to be 15%. In the interwar period in Poland there were about 6500 small hydro plants. Unfortunately, by 1982, only about 100 were left, and now there are approximately 700. In 2011 the share of electricity produced from renewable energy sources in the national gross electricity consumption in Poland was 7.9% [2] and increased from 2.58% in 2005.

Hydropower, the energy from moving water, is one of the oldest renewable energy sources and the total global electric power capacity of hydropower, including large hydropower, small hydropower, and ocean power, was approximately 820 GW in 2005, which accounted for almost 20% of the renewable energies [10]. In the field of hydropower, the SHP are important, characterized by an installed capacity of up to 5 MW. The theoretical hydropower resources in Poland – depending on the evaluation method – are estimated to be: 23,000 – 25,000 GWh/a. [6]. In Poland, in 2009, hydropower plants produced 2683 GWh, which represents 1.8% of the total electricity generation of 151,697 GWh [1]. Each hydropower plant is individually designed with close fit to local conditions [4]. The decisive criteria for choosing a solution include minimal construction costs, short delivery time and broad possible use of river flow [9]. In the domestic setting, details of the objects which have been implemented and the new ones are prototypical [6].

In Figure 1, a comparison of the efficiency of turbines are presented: Archimedes screw – 1, Kaplan – 2 and Francis – 3. At low heads the most ideal solutions are: Kaplan turbine, Archimedes screw, Banki-Michell's turbines and the so-called siphon turbine [3]. Banki-Michell's turbine does not apply to low heads because of the loss induced by the lack of a suction tube.

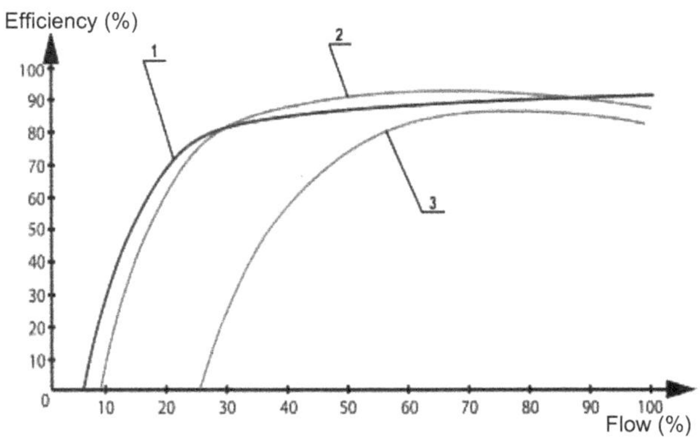

Fig. 1. Performance chart of individual turbines, (description in the text) [11]

To simplify and reduce the cost of SHP solid turbine blades are used. It is important when working with constant load or constant transfer to the turbine. In a SHP more often, also in Poland, power plants based on Archimedes screw turbine [3] are built. Archimedes screw uses the kinetic energy of flowing water, usually provided by the river, and even a small stream [11]. Its simple design manifests high reliability of the device; the main elements are arranged in a rotor mounted gutter. The turbine wheel is connected mostly to the asynchronous generator by means of a gear, with a low speed of 500–700 rpm. According to some manufacturers [13, 14, 17–19], the efficiency of these turbines is up to 90% and the obtainable capacity is up to 500 kW. Small hydraulic structures have a great potential for drawing energy from natural flowing water [17].

DEVELOPMENT OF EXISTING HYDRO BUILDINGS

Exploitation of hydraulic structures is considered in two ways. The first one is the operative exploitation, which aims to make the best use of flow and head to obtain the highest possible energy production and distribution of the most favourable time [7]. The second one is a technical operation designed to provide a normal and trouble-free use of power as long as possible, maintaining safety regulations.

Threats of SHP may occur as a result of flow operation exceeding reliable flows. The operation of hydrological or geological conditions with exceptional features, such as a very thick ice, improper passing of excessive water or ice flows, moving parts' freezing, etc. The passage of excess water by water structures of small power plants, especially passing the great flood water should be carried out in a manner to ensure safety and to minimize the damaging effects of water [7]. Huge congestion and water thrust affect the hydro building as well.

WAYS TO EVACUATE TURBINE

Very often a mobile crane is used with the first turbine unit installation . So far turbine installation and removal were made with the use of lifting equipment, gantry cranes or, for smaller turbines hoists embedded in the frames were used. These solutions of disassembly or partial disassembly by crane has its drawbacks, of which the principal one is the time in which the operation is carried out and which may last a few hours. In case of flood, time is a critical parameter and in specific weather conditions driving up a specialized equipment may not be possible. Evacuation system is created in order to pass the flood waters including Q = 1 [16] (so called „century water" – that is one that theoretically occurs once every 100 years).

The main way to prevent damage caused by flood is locating an object on the side channel of the river bank and not on the main channel (mill builders have used such a strategy for a long time). By-pass channels are also used. A diagram of such a solution is shown in Figure 2.

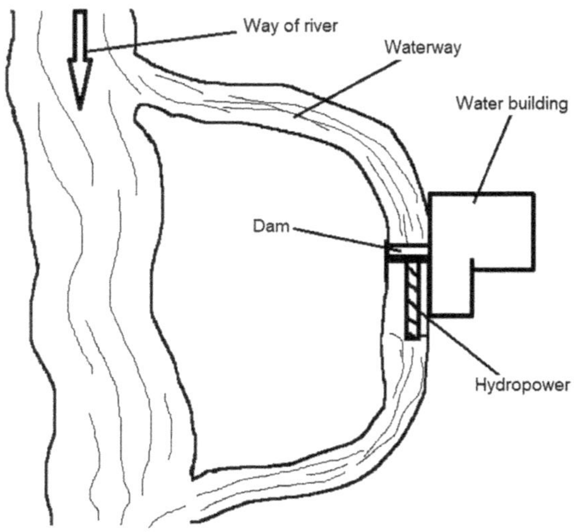

Fig. 2. Diagram of hydraulic engineering structures built on the bypass channel

Evacuation systems are good in situations where a dam or other hydro-technical structure, equipped with only a single path for water are electrified. Then, in the event of water excess, it can be let through easily, quickly and without resistance, protecting the device from damage. One of the evacuation systems of the entire turbo system with an Archimedes screw above the surface of the water is the solution proposed by a company called EMNET. It involves installing a rail construction aside, with a crane, , at the edge of the dam or other hydro structure. This device is also equipped with a flap that lets the flood water through after being opened in the place of turbinemachinery, which is then put aside (Fig. 3). The principle of operation of the turbine set for working with maximum and minimum water flows is shown in Figure 4. Unfortunately, this type of solution has not been implemented so far on the river, it is still only in a draft form.

These risks can be eliminated by appropriate preventive measures, such as responding in an appropriately short time. Another solution allows for very rapid changes in the Archimedes screw turbine operation and evacuation of the whole turbo machinery above the water surface using a specially designed evacuation equipment construction. Czech unit GESS [14] uses a hydraulic cylinder system that can extend over the surface of the water, the turbinemachinery embedded in the gutter of the turbine (Figures 5–7). This design is aimed at rapid evacuation and protection of the turbine from damage in danger of flooding. The entire assembly mechanisms are normally submerged in water and can eas-

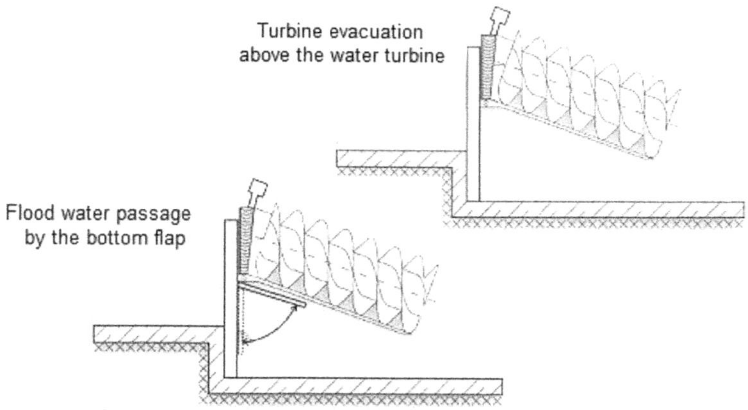

Fig. 3. The way of turbine evacuation over the surface of water and passing through of the flood water [12]

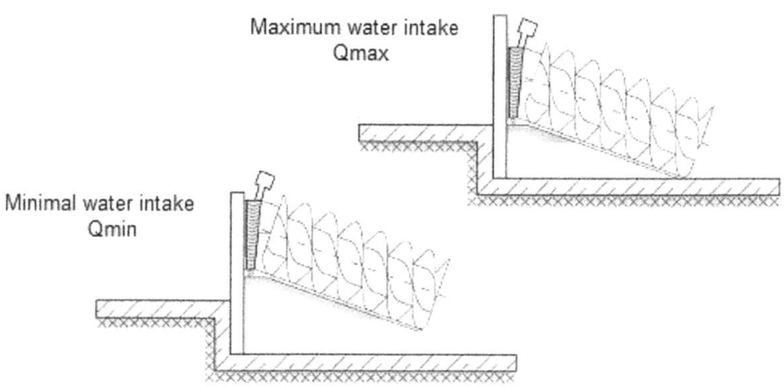

Fig. 4. The method of turbine operation at different levels of water [12]

ily be raised above the upper level of water as it is rotatably mounted on a special hinge system. Body design also allows greater deflection of the bearing during installation and operation of the turbine [14]. Such solutions already exist in our country. Figure 7 shows the solution of GESS turbine during normal operation, i.e, the actuator is in the stowed position.

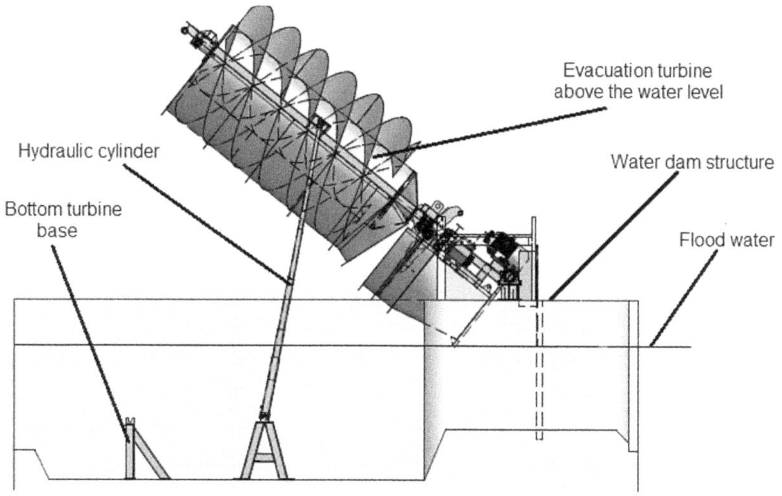

Fig. 5. GESS Turbinemachinery in the evacuation position after stopping over the top water [14]

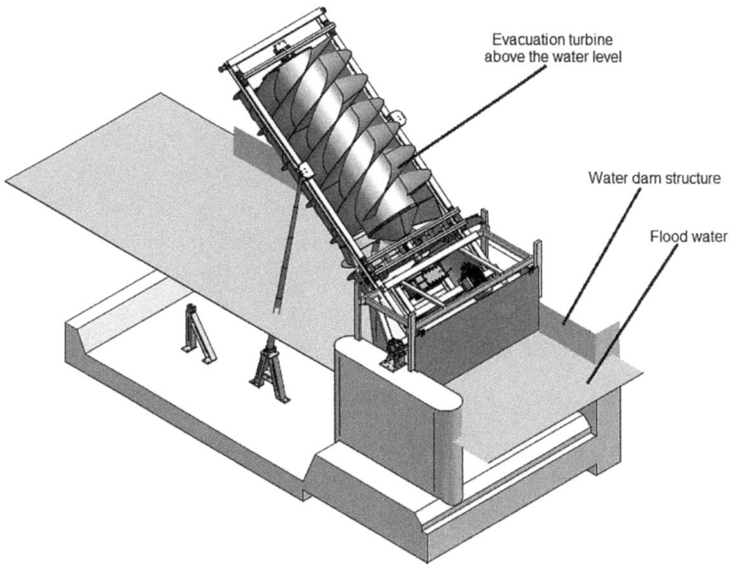

Fig. 6. GESS turbine machinery in the evacuation position in the view from the top water [14]

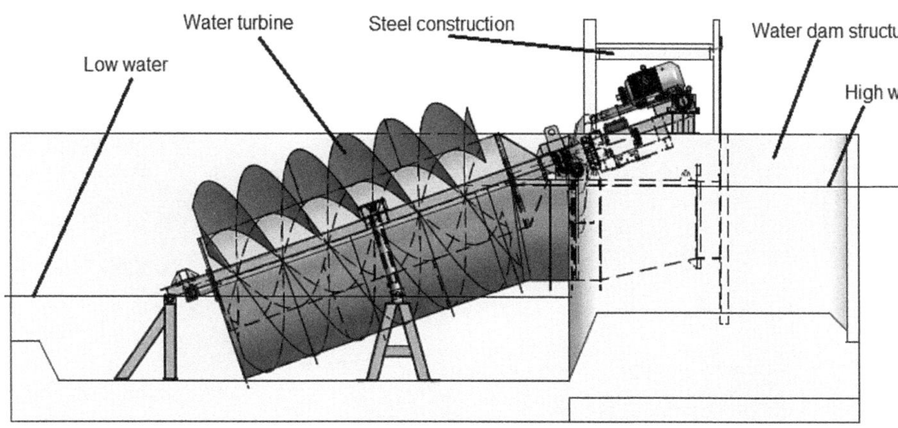

Fig. 7. GESS turbine machinery in the normal work position [14]

In the second half of September 2011 construction and assembly SHP works in Bieleckie Młyny in Świętokrzyskie Voivodship were completed [20]. This is the third hydro power plant in Poland which uses this type of solution. And it is also one of the first hydromachinery of this kind in the world. Two other small hydro turbines were installed on the Radomka river (Fig. 8) [11, 13, 15, 20] by the end of June 2011 and on the river in Luboszyce on the Mała Panew river [11, 20].

Mass production allows the use of special solutions, using high-quality materials, in advanced - and also cost competitive technology in comparison to individual solutions [6]. The presented machine concept design will eliminate the problem of flood and water resource management in the area of the lowest damming. This approach significantly improves the technical parameters associated with the evacuation of the Archimedes screw turbine over the surface of the water. The advantage is the increase of device functionality in hydroclimate of Poland. The solution offered by Czech company GESS is new to the market and significantly different from other products for SHP. An example of the evacuation system is a very creative combination of the known technologies used in transport systems (e.g, tipper trucks) adopted to the evacuation of Archimedes screw turbine self-supporting structure. It should be mentioned that it is not identical solution, it works in entirely different specific environmental conditions, but operates on the basis of a similar technical concept.

Among European (Austria, the Netherlands, Germany) manufactures of Archimedes screw turbines [17–19] there are no reports on the production of evacuation systems for turbine unit. As it is clear from the conducted analysis, the evacuation system was developed exclusively by GESS company also with a view to the market and the Polish customers.

As it has been pointed out, the hydropower plants, driven by an Archimedes screw turbine have now permanently attached turbine unit. In practice, this means that each amount of water greater than the nominal swallowing capacity of turbine is not energetically used. The solution allows moving the turbine inlet up and down generating the energetical use of large water flows, thus significantly improving the SHP technical parameters along with the Archimedes screw turbine. It is estimated that the energy yield from the water additionally introduced onto the turbine can increase the efficiency of the turbine by about 20%, compared to the energy from the water in the nominal swallow capacity of turbine. The problem could be only the rotor flooding. To avoid it, effective decrease is reduced by a half of the diameter of the rotor plus the distance for the low water at large flows.

Separation the turbine from the turbine unit allows the use of larger amounts of water indicated on the rotor. The technical solution also allows rapid evacuation of Archimedes screw turbine on the water surface. As mentioned ear-

Fig. 8. First in Poland Archimedes screw turbine on the Radomka river [11]

lier, none of European manufacturers which are global providers does not have such a range of products (for the Archimedes screw turbines). The example of evacuation system based on the technical concept known from transport systems (e.g. lifts, winches) was used to evacuate self-supporting structure of Archimedes screw turbines. Figure 9 shows the turbine during normal operation. While working, the turbine unit is based on the lower bearer and manoeuvring rolls placed on the abutment and the pillar of the dam and this is so called operational phase of the operation of the plant.

Evacuation is carried out by using two mechanical- and hand-driven chain hoists mounted on a crane beam of a steel structure. A scheme of the evacuation way of the turbine above the surface of the water by means of a chain hoist is shown in Figure 10.

CONCLUSIONS

It has been pointed out that depending on the type of management section of the river and the present SHP technical solution, there are various ways of quick turbine evacuation in danger, and hence, prevention of damage to the turbine. The solutions presented in this paper concerning prevention from damage to the turbine in the event of flood waters are an innovative approach for SHP investors.

The concept of separation turbine unit and control system of water flow onto the turbine rotor, while maintaining a constant level of the top water, allows for greater energetic use of the flowing water.

The presented solutions can be an inspiration to renovate the existing hydro buildings or building or the reconstruction of existing hydro-engi-

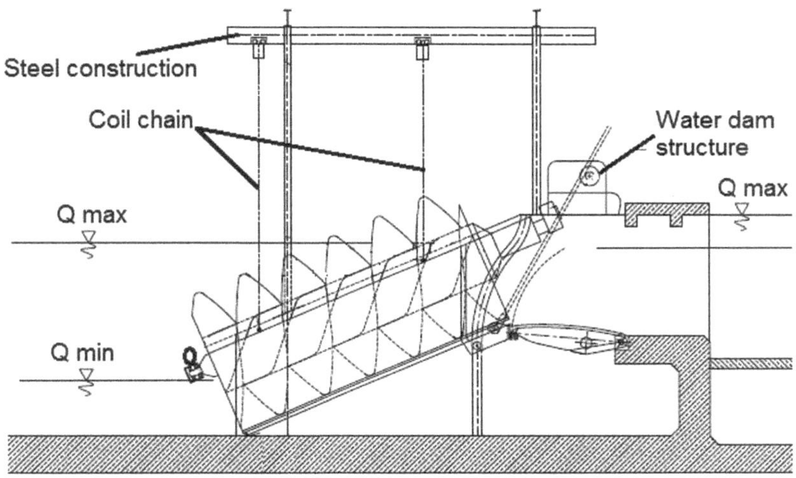

Fig. 9. Archimedes screw turbine scheme of evacuation system at work

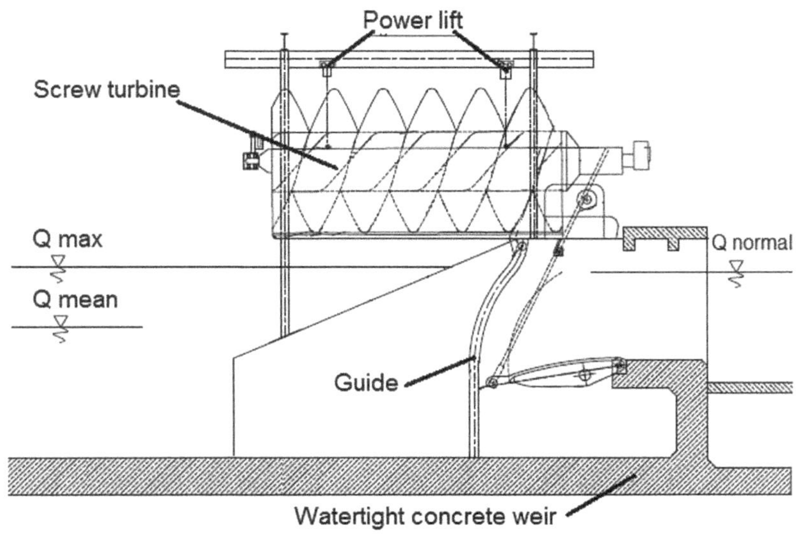

Fig. 10. A scheme of the evacuation way of the turbine above the surface of the water by means of a chain hoist

neering facilities. In a view of the possibility to obtain funding from various structural and ministerial sources and funds, the investment in SHP becomes more favourable. Hydroelectric power plants operate at a predictable time and power and are the most reliable source of electrical energy from renewable sources. The advantage of hydroelectric power is the ability to stabilize river flow and prevent flooding (e.g. Czorsztyn-Niedzica dam on the Dunajec river). Small hydropower systems have been increasingly used as an alternative energy source so that a small system is installed in small rivers or streams with little environmental effect [5]. In addition, investment in turbines with fish-friendly technology "fish friendliness" allows the use of energy in protected areas, such as "Nature 2000".

REFERENCES

1. Biuletyn Urzędu Regulacji Energetyki, 3(71), 4 maja 2010.
2. Biuletyn Urzędu Regulacji Energetyki, 2(80), 29 czerwca 2012.
3. Caban J., Gardyński L.: Problemy eksploatacyjne węzłów łożyskowych turbin wodnych. Autobusy, Technika Eksploatacja, Systemy Transportowe, 10, 2011, 99–105.
4. Caban J., Gardyński L.: Ocena działania łożyska ślizgowego w maszynie przepływowej. Proceedings of the Institute of Vehicles, Warsaw University of Technology, 5(86), 2011, 19–28.
5. Hwang I.S., Lee Y.H., Kim S.J.: Optimization of cycloidal water turbine and the performance improvement by individual blade control. Applied Energy, 86, 2009, 1532–1540.
6. Iwan J.: Studium badawczo-rozwojowe problemów turbin wodnych małej energetyki. Wydawnictwo Politechniki Gdańskiej. Gdansk 2006.
7. Juniewicz S., Michałowski S.: Elektrownie wodne. PWT, Warszawa 1957.
8. Lewandowski W.M.: Proekologiczne odnawialne źródła energii. WNT, Warszawa 2007.
9. Przekwas M.: Efekty gospodarcze elektrowni wodnych i przykłady modernizacji obiektów istniejących. Małe Elektrownie Wodne, nr 2. Towarzystwo Rozwoju Małych Elektrowni Wodnych, Gdańsk-Oliwa 1995.
10. Renewables Global Status Report, 2006.
11. http://dobraenergia.info/page8.php January, 2012,
12. http://www.ekomew.pl/ January, 2012,
13. http://www.ewzeneris.pl/index.php?site = wspolpraca February, 2012.
14. www.gess.cz/cz February, 2012.
15. http://www.mew.pl/Forums/viewtopic/t=1540.html January, 2012.
16. http://mewsa.pl January, 2012.
17. http://www.andritz.com/oi-andritz-atro-gmbh February, 2012.
18. http://www.spaansbabcock.com/hydro_power/advantages.aspx February, 2012.
19. http://www.rehart.de/ February, 2012.
20. http://ioze.pl/energetyka-wodna/aktualnosci/pierwsza-turbina-archimedesa-wyprodukowana-w-polsce-mew-bieleckie-mlyny January, 2012.

POLISH EMOTIONAL SPEECH RECOGNITION USING ARTIFICAL NEURAL NETWORK

Paweł Powroźnik[1]

[1] Institute of the Foundations of Electrotechniqu and Electrotechnology, Faculty of Electric and Information Technologies, Lublin University of Technology, Nadbystrzycka 38A, 20-618 Lublin, Poland, e-mail: ppowroznik@gmail.com

ABSTRACT
The article presents the issue of emotion recognition based on polish emotional speech analysis. The Polish database of emotional speech, prepared and shared by the Medical Electronics Division of the Lodz University of Technology, has been used for research. The following parameters extracted from sampled and normalised speech signal has been used for the analysis: energy of signal, speaker's sex, average value of speech signal and both the minimum and maximum sample value for a given signal. As an emotional state a classifier fof our layers of artificial neural network has been used. The achieved results reach 50% of accuracy. Conducted researches focused on six emotional states: a neutral state, sadness, joy, anger, fear and boredom.

Keywords: emotional speech, artificial neural network, communication.

INTRODUCTION

The recognition of the emotional state of a speaker, based on the analysis of speech signals, is a relatively new issue, however, its significance is increasing rapidly. One of the reasons of such a direction of changes is the dynamic development of systems based on the human – computer type of communication. Other applications involve processing of the speech signal, wherein speaker's emotions may play a certain significance. Among the potential applications of algorithmically modified emotional speech those associated with marketing and telephone contact with the client should be replaced [1]. Another group of applications involves the use of driver's emotional state by onboard computers. This kind of systems may be installed in vehicles, which can initiate appropriate safety procedures based on collected data [5].

The research carried out so far has been mostly based on databases in which every speech sample is matched with a specific emotional tone of voice [3]. However, the achieved results are mostly acceptable. This is due to the fact that for an average person it is possible to recognize another person's emotional state only in 60% of all cases [4].

There are several research centres in Poland which investigate the matters of emotional speech recognition (in the Polish language) [3, 5, 6]. The basic classifier applied to this type of research is the support vector machine (SVM), and the k-Nearest Neighbours algorithm (or k-NN for short).

The subject of this research is to determinate the optimal parameters for an artificial neural network allowing an effective recognition of the emotional states of a speaker. The second objective is the determination of the characteristics of the Polish emotional speech.

The article is divided into three parts. The first one characterizes the discussed matter. The second part treats the database used in the research. The third part includes the analysis of the available algorithms, research methods, and parameters for the Polish emotional speech and presents the obtained results and suggestions for improving the adapted research methods.

ANALYSIS OF ISSUES

The analysis of speech signals is connected with possessing a proper database of sound files. One of the available collections of emotional speech sound samples is the Berlin Database of Emotional Speech [7]. This database contains recordings of speech in seven emotional states, that is, a neutral state, joy, sadness, fear, anger, boredom, and disgust, prepared by 10 actors of both sexes [8]. Another group of sound files collections are databases prepared on the basis of recordings from TV and radio programs. The Lodz University of Technology has prepared and shared their own database of the Polish emotional speech. The basic problem of automatic emotional speech recognition is the choice of a proper feature vector. Descriptors, which are commonly used in this particular case do not vary from the ones used the analysis and processing of speech signals. A set of that kind of descriptors contains parameters like the signal's energy and a basic frequency [9]. Nowadays, the standards in voice recognition are *Linear Predictive Coding (*LCP), *Perceptual Linear Predictive (*PLC) [10], and *Mel-frequency Cepstrum (*MFCC) [11, 12], which are also used in the analysis of emotional speech recognition.

Based on the determined parameters of the speech signal the sets of attributes are created. These sets are then used as the starting vector for classification algorithms. As a classifier Polish emotional speech k – nearest neighbours (k – NN) algorithm [6] and support vector machine (SVM) [13] are used. However, global trends help draw the assumption of equally good performance of artificial neural networks in the analysis of the above issues [14].

The Polish database of emotional speech, prepared and shared by the Medical Electronics Division of the Lodz University of Technology, has been used for research. The collection consists of 240 recordings prepared by 8 actors: 4 women and 4 men. Each of the speakers pronounce five different sentences in 6 emotional tones, that is: boredom, fear, anger, sadness, joy, or without any emotional tone [3]. The database contains sound files in the 'wav' format sampled with 44.1 kHz frequency and the bit rate of 16 bps. The database includes the following statements: 'They have bought a new car today', 'His girlfriend is coming here by plane', 'Johnny was today at the hairdresser's', 'This lamp is on the desk today' and 'I stop to shave from today on'.

SPEECH SIGNALS PARAMETERS AND PROPOSED CLASSIFIER

The process of emotion identification based on a speech signal requires a distinction of characteristic parameters in the voice. Research carried out until now has not resulted in the determination of a uniform and universal set of features so far. As a consequence, a heuristic approach has been adopted [13]. It involves identifying, according to the signal, as many parameters describing the signal as it is possible, and selecting, by means of experimenting or using algorithms, those parameters which describe the researched matter best. Among the parameters extracted from the signal the most useful ones are: the laryngeal tone [13], energy of the signal [13], formant values, and MFCC, LPC, PLP factors [3].

ENERGY AND AVARAGE VALUE OF SPEECH SIGNAL

The research which has been carried out focuses on the energy, and the average value of the speech signal. The energy of the signal is defined as the integral of the square of the signal, that is, energy emitted with unitary resistance. For digital signals it is described by the following formula [15]:

$$Ex = \sum_{n=0}^{N} x^2(n), \tag{1}$$

where: n – sample number,
$x^2(n)$ – square value of sample signal.

The average value of the whole signal is defined in the following way [15]:

$$\overline{x_N} = \lim_{n \to \infty} \frac{1}{2N+1} \sum_{n=-N}^{N} x(n),$$

where: $x(n)$ – value of n-sample,
N – total mumber of samples.

For the sake of the research, before the abovementioned parameters have been set, the values of particular samples have undergone the process of normalization. A standard algorithm of signal processing is based on three basic stages: preparation of the data set, the designation of the feature vector and classification. Thanks to the high quality of recordings the first stage has been limited to normalization process.

The second stage is the determination of the feature vector describing the analyzed matter as precisely as possible. In this study, this stage was limited to five elements: the energy of the signal, the average value of the whole signal, the sex of the speaker (1 – woman, 0 – man), and both the minimum and maximum sample value for a given signal. This set is the entry vector in a neural network. The last stage is classification. A five entry neuron network has been proposed as a classifier, twelve neurons in the first, hidden layer, six in the second one, and six entry neurons. The model of the network is depicted in Figure 1.

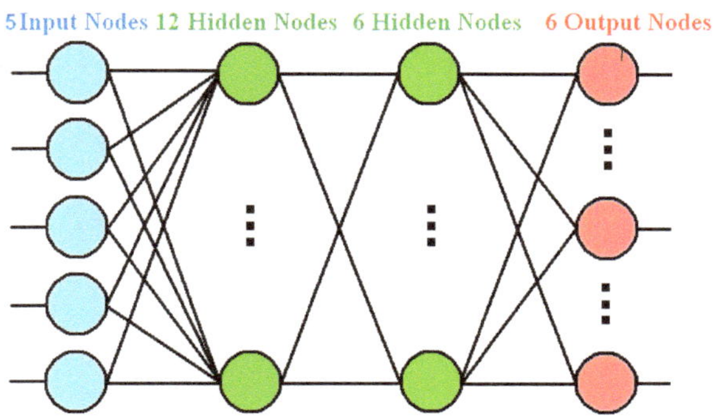

Fig. 1. The model of used Artificial Neural Network

The sigmoidal function has been used as an activation function. The research has been carried out in MatLab, where the network has been trained with the backpropagation with momentum and factor adaptation (traingdx). The learning process ended with either the achievement of the given number of epochs, that is, 1500 in the analyzed cases, or the achievement of a normalized result different from the expected one by no more than 0.1. The achieved results have been depicted in Figure 2. The confusion matrix has been shown in Table 1.

CONCLUSIONS

The recognition of emotions in speech signal is a difficult task, and the achieved results are far from ideal. Research carried out by Swiss scientists show, that the evaluation of an emotional state is difficult even for a human being. Nevertheless, it is possible to improve the achieved results [4]. The first step is to expand the fea-

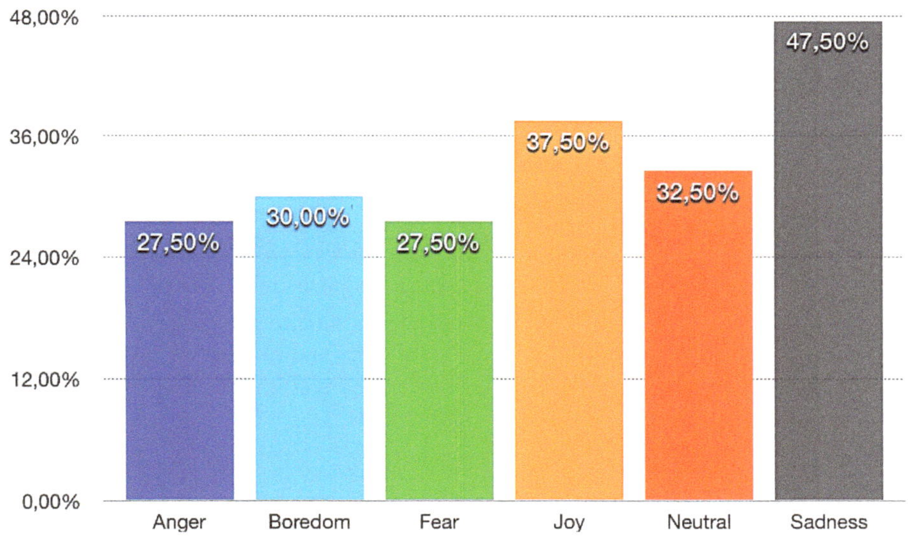

Fig. 2. The effectiveness of the recognition of individual emotional states

Table 1. Confusion matrix

Predicted class \ *Actual class*	Anger	Boredom	Fear	Joy	Neutral	Sadness
Anger	**11**	5	10	12	1	1
Boredom	4	**12**	6	1	12	5
Fear	10	8	**11**	6	3	2
Joy	11	1	8	**15**	4	1
Neutral	4	10	4	1	**13**	8
Sadness	1	8	6	1	5	**19**

ture vector. The signal energy and its average value are not enough to achieve good results. It is absolutely necessary to expand above set of parameters with MFCC and LCP parameters. Constructed artificial neural network (ANN) has also failed to accomplish the expectations. It is necessary to expand the ANN or to transform it into, for instance, the Kohonen network. The effectiveness of emotion recognition can be also increased by combining a voice analysis system with semantic analysis [16]. A natural direction of development is, first and foremost, the application and testing of the suggested solutions in both the process of abstracting signal and classifier parameters.

REFERENCES

1. Ramakrishnan S.: Recognition of emotion from speech: A review. Speech Enhancement, Modeling and Recognition – Algorithms and Applications, March 2012.
2. Kamaruddin N., Wahab A.: Driver behavior analysis through speech emotion understanding. Intelligent Vehicles Symposium (IV), IEEE, 2010, 238–243.
3. Kamińska D., Pelikant A.: Zastosowanie multimedialnej klasyfikacji w rozpoznawaniu stanów emocjonalnych na podstawie mowy spontanicznej. IAPGOŚ 03, 2012.
4. Scherer K.: Vocal communication of emotions: A review of research paradigms in speech communication, 40, 2003, 227–256.
5. Database of Polish Emotional Speech, available: http://www.eletel.p.lodz.pl/bronakowski/med_catalog/ (Accessed 10.08.2014).
6. Ślot K., Rozpoznawanie biometryczne, WKiŁ, Warszawa, 2010.
7. Berlin Database of Emotional Speech, available: http://www.expressive-speech.net/ (Accessed 10.08.2014).
8. Polzehl T., Schmitt A., Metze F.: Approaching multi-lingual emotion recognition – from speech – on language dependency of acoustic/prosodic features for anger recognition. Proc. of Speech Prosody, Chicago 2010.
9. Yeqing Y., Tao T.: An new speech recognition method based on prosodic analysis and SVM in Zhuang language. Proc. 2011 International Conference on Mechatronic Science, Electric Engineering and Computer, 2011, 1209–1212.
10. Shauka A., Chen K.: Emotional state recognition from speech via soft-competition on different acoustic representations. Proc. Neural Networks (IJCNN), 2011, 1910–1917.
11. Plutchik R.: The nature of emotion. American Scientist, Vol. 89, July-August 2001, 344–350.
12. Niewiadomy D., Pelikant A.: Implementation of isolated words boundaries recognition. Proc. XII International Conference System Modeling and Control SMC, Zakopane 2006.
13. Janicki A., Turkot M.: Rozpoznawanie stanu emocjonalnego mówcy z wykorzystaniem maszyny wektorów wspierających (SVM). KSTiT, Bydgoszcz 2008.
14. Soltani K., Ainon R.: Speech emotion detection based on neural networks. Proc. Signal Processing and Its Applications, 2007, 1–3.
15. Zieliński T.: Cyfrowe przetwarzanie sygnałów. Od teorii do zastosowań. WKŁ 2009.
16. Wang Y., Guan L.: Recognizing human emotional state from audiovisual signals. Proc. IEEE Transactions on multimedia, Vol. 10, 2008, 659–668.

26

NON-DESTRUCTIVE TESTS OF LOCK TONGUES USED IN ATR-72 AIRCRAFT LANDING GEAR BASED ON MAGNETIC METHOD

Mirosław Malec[1], Tomasz Cieplak[2], Sławomir Walczuk[3]

[1] Fundamentals Technology Faculty, Lublin University of Technology, 38 Nadbystrzycka Str., 20-618 Lublin, Poland , e-mail: m.malec@pollub.pl

[2] Management Faculty, Lublin University of Technology, 38 Nadbystrzycka Str., 20-618 Lublin, Poland , e-mail: t.cieplak@pollub.pl

[3] Graduate student of Fundamentals Technology Faculty, Lublin University of Technology, Poland

ABSTRACT

The purpose of this work is to highlight the opportunities of using and analyzing process progression of Non-destructive Testing in aeronautical industries and technologies. This paper concentrates on magnetic-fluorescent method, which is used to showcase the practical test of lock tongue installed in ATR-72 aircraft landing gear.

Keywords: NDT research, magnetic-fluorescent method, wear and tear.

INTRODUCTION

Non-destructive Testing (NDT) techniques enjoy widespread use in the field of aeronautical technologies due to meeting safety requirements for the duration of aeronautical activities as well as fulfilling the need to provide high degree reliability of operations and the highest possible utility readiness. In addition to these factors, manufacturers and users of the aircraft have started to put significant emphasis on economic considerations recently. Non-destructive tests conducted over the course of operating the aircraft allow for technical inspection of multiple components without the need to disassemble the craft or with a limited disassembly. This proves to be of great significance as NDT methods enable engineers to take advantage of these tests at any point during the utilization of aircraft and they do not interfere with their exploitation and prevent conducting aeronautical operations only to a marginal degree.

This paper highlights the process and results of the conducted NDT tests using magnetic-fluorescent method. The tests have been carried out in compliance with current legal provisions and norms.

TEST OBJECT CHARACTERISTIC

The object of this test was an aircraft lock tongue responsible for blocking the landing gear hatch of ATR-72 airplane. It cooperates with a capstan and an electromagnetic lifter. Each opening of the lock is signaled by the pressure on the capstan and relay contacts, the joining of which is further relayed in the cockpit and interpreted as the opening of the hatch.

The tongue performs the function of a lock block, preventing the hatch from opening during the flight. It is made of the tool steel NC11X-210Cr12 and its dimensions are 150×60×20 mm. Due to the function of this device any damage to this element preclude it from further use. Figure 1 presents this component. All tests of this element have been conducted in NDT research laboratory, furnished with tools and specialized equipment necessary to conduct the tests based on magnetic-fluorescent method.

A device critical to perform these tests is a magnetic defectoscope Contrmag CM-152 manufactured by Srem company. For the purpose of this research a number of other devices were used: a demagnetizing device, Berthold's model, Block A model, sedimentation containers ASTM D 96,

Fig. 1. The tested component – a lock tongue of a landing gear

magnetic field meter, a source of white light, an ultraviolet radiator UV/A, a lighting intensity meter, UV/A radiation measuring tool, ultrasonic cleaner, magnetic suspension and cleaning liquids.

PREPARATIONS AND THEIR PROGRESS

The preparations for the main test have been opened by confirming the expiration date of magnetic suspension Fluxo 6C used in the test procedures and the quality of the tested liquid. To this end 100 ml of Fluxo 6C liquid was poured into a sedimentation container ASTM D 96 which is used to determine to level of concentration of the magnetic fraction in the liquid carrier – in this case, kerosene. The test took 30 min, and the measured value was set between 0.1–0.7% magnetite corpuscles. The final result amounted to 0.1% which signifies that the tested liquid fulfilled all usage requirements. In the course of the test the intensity of day lighting and ultraviolet radiation was measured, the latter of which was generated by ultraviolet radiator installed on a test stand. The test was conducted using a universal measuring device Pollux 707. Considering currently accepted norms the intensity of ultraviolet light should be greater than 1500 $\mu W/m^2$. The final result of UV radiation is 2500 $\mu W/m^2$. In addition, the intensity, which does not to exceed 20 lx, was estimated to value 5 lx.

VERIFICATION OF CM – 152 DEFECTOSCOPE PERFORMANCE

The next stage of the preparation for the testing process is the verification of the defectoscope performance used in NDT. In this case, the subject of inspection is CM – 152 magnetic defectoscope. The aforementioned inspection is twofold: it involves assessing the direction of magnetization and its thoroughness. The direction of magnetization was verified with Berthold Magnetic Field Indicator (Figure 2). This indicator is a simple measuring device, which enables researchers to determine approximation of measurements in the test conducted with the use of magnetic powder and helps to verify the direction of magnetic field. Berthold's Magnetic Field Indicator comprises: a ring, which is to be free of any magnetic residue, and a cylinder of soft steel set into it, which consists of four segments. These segments are separated from each other by non-magnetic facets which create two perpendicular fissures. A membrane with a thin brazen lid is screwed onto the cylinder. As the membrane turns the lid's position relative to the ground becomes shifted. In order to conduct the measuring the indicator must be placed on a magnetized object; as a result a fraction of magnetic field lines will flow through the cylinder made of soft steel. After the powder of magnetic suspense is applied, depending on the direction of the field the membrane reflects one or two lines of cross fissure. As the membrane is moved away upwards without the loss of measurements, the measurements taken using this method are proved to be more accurate as a result of higher current intensity or the quality of the magnetic suspense. The model component rotates slowly around the cylinder's axis until the fissure becomes clearly reflected. In this position the

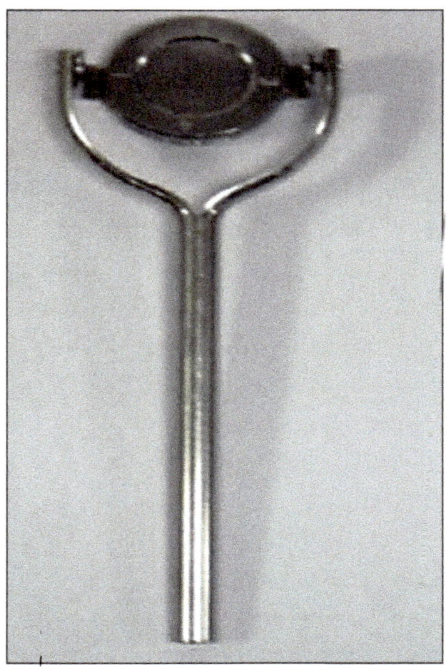

Fig. 2. Berthold's Magnetic Field Indicator

magnetic field goes perpendicularly in relation to reflected fissure. The estimation of the direction of magnetization using the abovementioned indicator has been supplemented with the evaluation of the thoroughness of magnetization with Block A model. Block A model is a steel cuboid measuring 30×110×20 mm, with multiple holes 1.5 mm in diameter each.

If the location of damage is known it is easy to determine the thoroughness of magnetic test. The abovementioned model – Block A – has been placed into the clip of the defectoscope CM – 152 after which it has been subjected to magnetization three times, with a 2.7 kA/m current within the duration of 5 seconds each time (Figure 3).

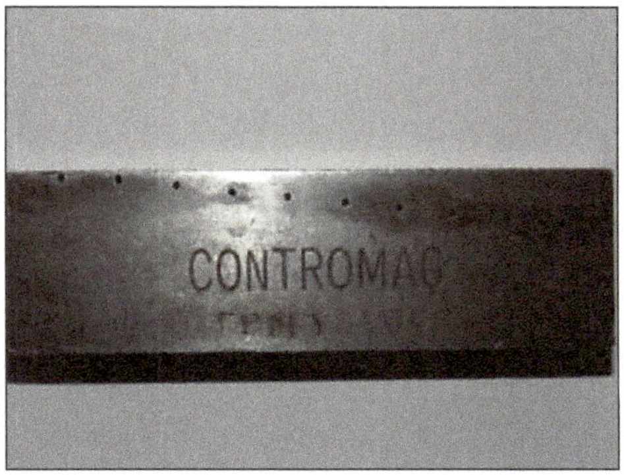

Fig. 3. Model – Block A

Furthermore, in order to perform additional check of the direction of magnetization a Berthold's Magnetic Field Indicator was applied onto Block A. Using MDC3 and MF300 H gauge the value of magnetic field was measured. The value of intensity of the magnetic field in an appropriately conducted test should be between 6400 –12800 A/m. This value in a given test amounted to 9000 A/m. It is of paramount importance that during this stage of the test the gauge probe was placed in accordance with the direction of the forces of magnetic field, as this ensures correct readings of its values. Otherwise there is a chance of distortion and the tested item may become undermagnetized or overmagnetized, which will result in defectograph inaccurate. In the course of the magnetization procedure a magnetic suspension solution was applied on Block A. After magnetization was completed the results were acquired; The results are presented in Figure 4.

Circular magnetization is the next step in test preparations. The process involves a shift of the magnetization angle by 90° in relation to the former position and evaluation of the efficiency of the process. In order to assess the thorough-

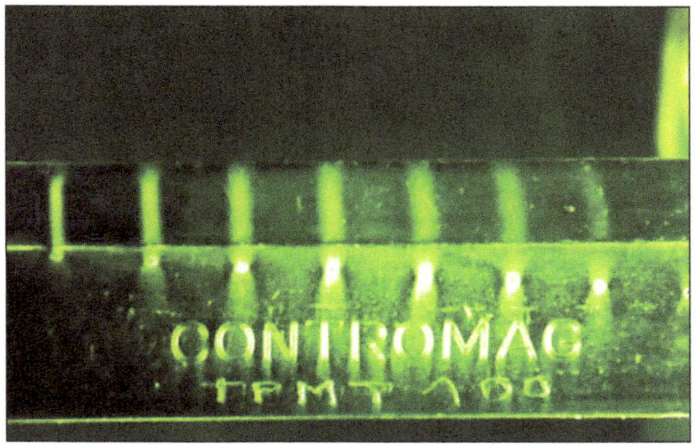

Fig. 4. The model Block A in UV lighting after completed magnetization and application of magnetic suspension

ness of magnetization a Ketos Ring was placed on a copper electric conductor, after which the current of 8 KA was switched on for the duration of 2 seconds – this procedure was repeated three times. Ketos Ring is a metal ring, 127 mm in diameter with drilled holes located 1.78–21.34 mm from the outer rim. Just as during model – Block A check, in this case the quality of defectograph was determined. After the test had come to an end it was possible to pinpoint deliberately damaged segments. A number of lines appeared on the rim of the ring, which signified that the test was conducted appropriately. The intensity of the field on the tested component was 1000 A/m. Figures 5 and 6 present, respectively, Ketos Ring and the way it looked after magnetization.

Cleaning the tested object from all kinds of waste, such as grease, dirt, oils and anticorrosive countermeasures is a crucial step in the testing process. In particular anticorrosive paints with ferromagnetic characteristics must not be left on the object. The object in question was first cleaned with a cloth wetted in special clearing salve, and then placed in an ultrasonic cleaner. Overall cleaning took 30 minutes. Next the researches proceeded to initial inspection of the tested object. The purpose of this test phase is to locate any defects which were revealed after the cleaning process was complete. The initial evaluation was conducted using magnifying glass in 5× magnification. The test did not reveal any defects so the researches proceeded to the next step. This stage involves demagnetization of the component in laboratory conditions using a demagnetizer so that the residual magnetism does not exceed 500 A/m. By default, residual magnetism of the tested object was 600 A/m. After demagnetization the levels of the object's magnetism decreased to 100 A/m.

Fig. 5. Ketos Ring

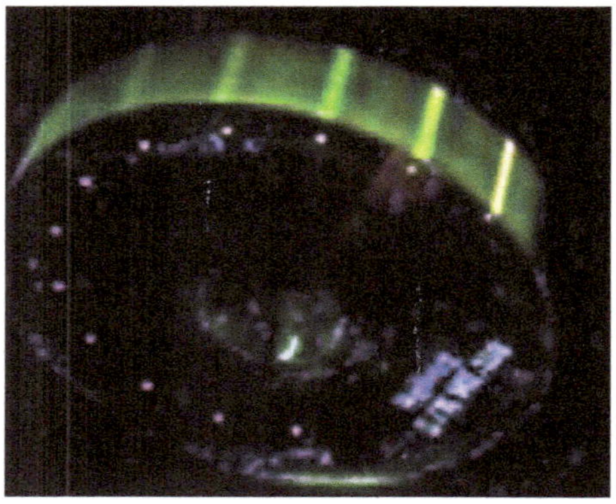

Fig. 6. Ketos Ring in UV lighting after magnetization and application of magnetic suspense

LONGITUDINAL MAGNETIZATION OF THE TESTED OBJECT

The next phase involves magnetization of the test tongue. First, longitudinal magnetization was performed. As a result, all defects located perpendicularly to the direction of the magnetic field should be visible once magnetization and application of the magnetic suspense has been completed before radiating the object with a UV radiator. To this end the object was inserted in the grip of CM-152 magnetic defectoscope (Figure 7). The object was subjected to three doses of magnetization with 400 A current, which took 5 seconds each time. The intensity of the magnetic field in the object was 11000 A/m. During the two first cycles of magnetization the tested object was poured in magnetic solution, while the third magnetization did not involve this process in order to stabilize the defectograph. Once magnetization was finalized, the object was removed from the grips and subsequently underwent inspection in UV lighting.

No defects were found. Once the inspection was finished, in order to ensure accurate readings

Fig. 7. CM 152 magnetic defectoscope

the lock tongue was subjected to demagnetization. For this reason it was placed in a demagnetizer. As mentioned before it is important to ensure that the value of residual magnetic field does not exceed 500 A/m. The measuring was performed using Magnetis residual magnetic field gauge. The final value of the magnetic field during this test amounted to 95 A/m.

CIRCULAR MAGNETIZATION OF THE TEST OBJECT

Circular magnetization is the next stage of the test. The tested object is placed on a copper rod, which in turn is placed in the grips of a magnetic defectoscope. The object was once again subjected to magnetization with 400 A current for the duration of 5 seconds – the process was repeated three times. The intensity of magnetic field in the test object reached the value of 9000 A/m. The angle of direction of magnetic field had shifted in comparison with previous test by 90°. As in previous stage, this time during the first two cycles of magnetization the component was covered with magnetic solution, while the third cycle did not involve this process in order to stabilize the defectograph. Once magnetization was completed, the object was removed from the grips and it underwent inspection. The control component was once again subjected to UV radiation, which this time revealed accumulation of the ferromagnetic liquid at the joint between that tongue and control lifter. As referred above, during the inspection and after magnetization no defects were found, whereas after performing circular magnetization a major accumulation of ferromagnetic solution was revealed in one of the parts of the joining. The exact location of the suspected defect is visible in Figure 8. The spot underwent additional thorough scrutiny which confirmed the appearance of a crack across the lifter in the tested lock tongue.

Fig. 8. Location of the defect in the test lock tongue

In order to record the conducted test the documentation on performed activities is drafted. At this stage a report was filled in, which constitutes a record of accomplished tasks. Every aeronautical organization responsible for conducting Non-destructive Tests has a different set of documents, which are then validated by National Aviation Office. In the case Polish Civil Aviation Office is the competent body. Regardless of the country, the tests are conducted in, such a document ought to account for certain sets of rudimentary data, including:

- date of the test;
- test method;
- sketch of the tested item;
- name of the researcher;
- result of the test;
- evaluation skating whether the component was faulty or operational.

The last stage of the test based on magnetic method is the final demagnetization of the test component (Figure 9), and clearing it from the suspension. The final demagnetization is very important, because the value of magnetic field intensity must not exceed 240 A/m. The test object – a damaged lock cap – was isolated from other, flawless components used in the test and subjected to thorough clearing, which involved using cleaning salve and clearing in an ultrasonic cleaner. The object was supplemented with an information card, which contained rudimentary information concerning the damaged component, its origins (name, series no.), the date of the test and the name of the research responsible for the undertaking. Separation of the broken component is the last step in the test procedure.

Fig. 9. The final measurement of residual magnetism in the tested component

CONCLUSIONS

This paper concentrated on the procedure of Non-destructive test based on magnetic-fluorescent method, which is frequently used in aviation and aeronautic industry. In particular the work highlighted stages of conducting thorough test of a lock tongue installed in the a landing gear of ATR-72 airplane. The conclusions accumulated after successful completion of the procedure are twofold. First, there are the tangible benefits stemming from the application of this method in aeronautic technologies. The authors of this paper conclude that the most important of them are: the reliability of the accumulated results of the test, high level of detection of defects, low expense threshold of conducting the test, capacity to perform the tests directly on board of the aircraft using portable defectoscopes and fluorescent powder. Second, there are practical conclusions which came about over the course of the test: approachability of the method and simplicity of the test procedure. It is important, however, to pay particular attention to certain aspects of the tests which, if done inappropriately, may result in inaccurate readings. One must account for: checking expiration date and the quality of magnetic liquid, controlling the intensity of white light and UV/A radiation, controlling the intensity of magnetic field in the test object during magnetization, demagnetizing the test object between longitudinal and circular magnetization, and correct choice of direction of magnetization, which depends on the shape of the object.

If all of the aforementioned requirements are met the resulting research may be deemed as reliable - which means they give the true information on any defects, as well as wear and tear of a given object The subject of the analysis in this paper was the technical condition of the lock tongue installed in the landing gear of ATR-72 aircraft. In the given test case a 14 mm crack on the joining of the lifter was revealed, which precludes the component from any further use. In spite of apparent simplicity and approachable principles of the highlighted magnetic method it is important for further develop diagnostic equipment, which will ensure greater reliability and accuracy of the gathered test results. This, in turn, will improve safety of aeronautic undertakings and safeguard human lives, what is of paramount importance.

REFERENCES

1. Deputat J. Nieniszczące metody badania własności materiałów. Biuro Gamma, Warszawa 1997.
2. Lewińska-Romicka A. Badania magnetyczne. Podręcznik, t. 1 i t. II. Biuro Gamma Warszawa 1998.
3. Norma PN-EN ISO 9934-1 Badania nieniszczące. Badanie magnetyczno proszkowe – Część 1. Zasady ogólne.
4. Norma PN-EN ISO 9934-2 Badania nieniszczące. Badanie magnetyczno proszkowe – Część 2. Środki wykrywające.
5. Norma PN-EN ISO 9934-3 Badania nieniszczące. Badanie magnetyczno proszkowe – Część 3. Aparatura.
6. Norma PN-EN ISO 3059 Badania nieniszczące. Badania penetracyjne i magnetyczno-proszkowe. Warunki obserwacji.
7. Norma: PN-EN 571-1 Badania nieniszczące – badania penetracyjne. Zasady ogólne.
8. Pawłowski Z. Badania nieniszczące. Poradnik. Wydawnictwo SIMP, Warszawa 1988.

27

EFFECTIVENESS OF CELLULAR INJECTION MOLDING PROCESS

Tomasz Garbacz[1], Ludmila Dulebova[2], Volodymyr Krasinskyi[3]

[1] Lublin University of Technology, 36 Nadbystrzycka St., 20-618 Lublin, Poland, e-mail: t.garbacz@pollub.pl

[2] Technical University in Košice, 74 Mäsiarska St., 040 01 Košice, Slovakia, e-mail: lidmila.dulebova@tuke.sk

[3] Lviv Polytechnic National University, 12 St. Bandery St. , Lviv, Ukraine, e-mail: vkrasinsky82@gmail.com

ABSTRACT

In a study of cellular injection, molding process uses polyvinylchloride PVC. Polymers modified with introducing blowing agents into them in the Laboratory of the Department of Technologies and Materiase of Technical University of Kosice. For technological reasons, blowing agents have a form of granules. In the experiment, the content of the blowing agent (0–2,0% by mass) fed into the processed polymer was adopted as a variable factor.

In the studies presented in the article, the chemical blowing agents occurring in the granulated form with a diameter of 1.2 to 1.4 mm were used. The view of the technological line for cellular injection molding and injection mold cavity with injection moldings are shown in figure 1.

The results of the determination of selected properties of injection molded parts for various polymeric materials, obtained with different content of blowing agents, are shown in Figures 4–7. Microscopic examination of cross-sectional structure of the moldings were obtained using the author's position image analysis of porous structure. Based on analysis of photographs taken (Figures 7–9) it was found that the coating containing 1.0% of blowing agents is a clearly visible solid outer layer and uniform distribution of pores and their sizes are similar.

Keywords: thermoplastics polymers, cellular injection molding process, blowing agents, physical properties.

INTRODUCTION

Cellular structure is obtained by insertion of a blowing agent (porophor) in a form of an inert gas, low-boiling liquid or a solid body into the input plastic which, if in liquid or solid state, transforms into gas under determined conditions of the injection molding process. Inert gases and low-boiling liquids are introduced under pressure by means of special dispensing devices, directly into the feeding zone of the plasticizing system [1, 2, 3, 4]]. By contrast, some liquids and solid bodies, such as pigments, are mixed with polymer plastics in a typical way, i.e. prior to their delivery to the injection molding machine dispenser, or they are introduced into the plastic already during the extrusion process. If the process of gas emission is started in the proper temperature, numerous microspheres, which had been generated, dissolve in the ambient plastic material due to the operation of pressure and surface expansion.

Cellular plastic in a liquid form is still not a stable system because as a result of interfacial tension at the plastic – gas contact point, and as a result of diffusion, the number of cells in the plastic is reduced, whereas their dimensions enlarge, what comprises an undesirable effect. The obtained cells keep enlarging until gas pressure and interfacial tension are balanced. Favourable small-cell structure can be maintained in the plastic product by its immediate cooling and solidification. During the injection molding process, blowing agents are exposed to exactly the same conditions as the processed plastic, compression, homogenizing treatment and carrying already before gas emission [2, 3, 6].

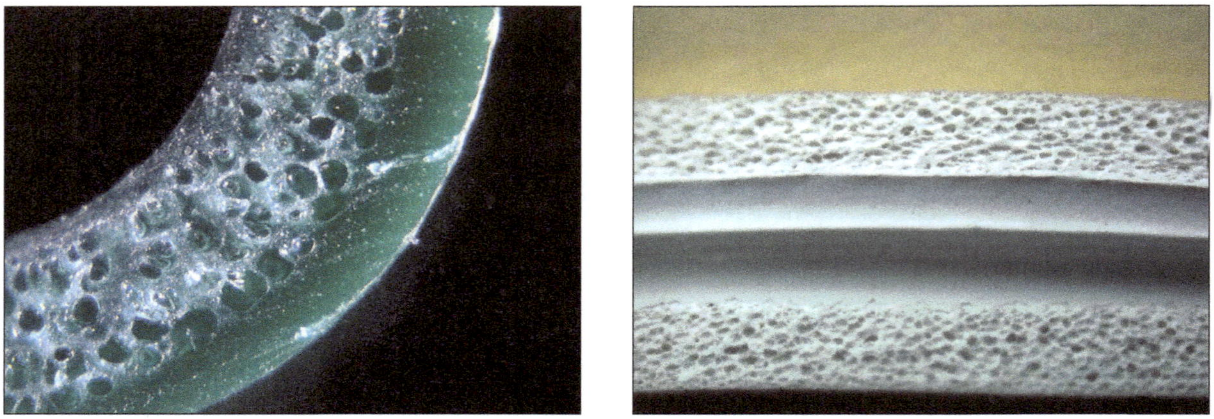

Fig. 1. Cross section of porous structure of polymer products made from the examined polymers

Cellular of thermoplastic materials may be conducted with the use of chemical blowing agents (to maintain low cost of technical equipment and machines used in processing) and physical blowing liquids (gases), to obtain high degree of porosity.

By contracts, chemical blowing agents are added to the granulated polymer as solids (powder, granulated product) fed to the plasticizing system, most often by means of gravimetric or volumetric dispensers. Plastic mixed with the blowing agent is then processed in the plasticizing system and shaped with a traditional tool (injection mould, extruder head etc.). As it is the case with physical blowing agents, gas dissolved in the polymer is obtained [7, 8, 9, 10].

The type and volume of the added blowing agents affects final form of the cellular plastic and determines cellular method. Proper feeding system is the basic condition of effectiveness. Blowing agents should be dissolved in a polymer.

The obtained product may have a fully solid or cellular structure, or it may be cellular in its entirety, or it can have the cellular topcoat and solid surface [5, 7, 12, 13]. Proper distribution of temperature in the plasticizing system of the injection molding machine is required to obtain favorable cellular structure in the process. Temperature in the first, second and alternatively third functional zone of the plasticizing system should be lower than the initial decomposition temperature of the blowing agent. Only in the fourth zone of the plasticizing system, temperature should be higher than decomposition temperature of the blowing agent. Cooling of the cellular product (scope and intensity) determines creation of the cellular structure of the product. Insufficient cooling may lead to the emergence of a non-uniform cellular structure [2, 6, 7, 13, 14].

EXPERIMENTAL PROCEDURE

The polymer used in the experiments was poly(vinyl) chloride PVC, marketed under the trade name ALFAVINYL - GFM/4-31-TR, with a density of 1230 kg/m^3 and tensile stress at yield of 21 MPa, according to the data provided by the manufacturer. In the experiments, two types of chemical blowing agents were used: Hostatron P1941, Hydrocerol 530, both manufactured by Clariant Masterbatches GmbH, ADCOL-blow UP-0X-B-X11020 manufactured by Clariant and Expancel 950MB120 manufactured by AkzoNobel.

Based on the adopted research program, the polymeric material for cellular injection molding was modifed in such a way that the blowing agents were fed in during the mechanical mixing process. The blowing agents used in the cellular injection molding process were fed into the polymers being processed in the following quantities: 0,2%, 0,4% and 2,0% by mass (w/w). In order to obtain high process efficiency for the above blowing agents, the processing temperature range was 170–190 °C.

The test stand consist of a screw injection molding machine, Demag 3-25t, equipped with an injection mold, as shown in Figure 1a. The machine has a single-screw plasticizing system, with a screw diameter of 30 mm. To the moveable table of the machine a moveable subassembly of the injection mold (Figure 1b) is mounted, in which two mold cavities are located. The mold cavity has the shape and dimensions of a tensile specimen for static tension tests, the so–called specimen type II [15].

The mold cavity dimensions are as follows: length 150 mm, width from 10 to 20 mm, thick-

Fig. 2. Test stand used in the experiments: a) injection molding machine Demag 3-25t, b) injection mold unit and molded specimens

ness 4 mm. Directly in the mold cavities there are point ejector pins mounted, with a diameter of 6mm. In the fixed subassembly of the mold, there are flow system channels supplying the polymeric material to the mold cavities; the channels have a direct contact with an injection nozzle of the plasticizing system.

The following parameters of the cellular injection holding process were adopted: injection and clamping time – 2 seconds; molded piece cooling time – 30 s; injection pressure in the hydraulic system – 10 MPa. In the experiments, the temperatures of the investigated PVC polymers were set in specific heating zones of the plasticizing system in the following way: in zone I – 160 °C, in zone II – 170 °C, III – 180 °C, IV – 190 °C.

The applied shape and dimensions of the specimens complied with the relevant norm. The specimen thickness corresponded to the injection molded part thickness and it was measured and registered each time together with the width of the measuring length right before the tests.

The porous products were examined in terms of their selected physical, mechanical and functional properties in accordance with appropriate international norms. Also, tests of tensile strength, hardness, and porous structure were made. The applied shape and dimensions of the specimens complied with the relevant norm. The specimen thickness corresponded to the injection molded part thickness and it was measured and registered each time together with the width of the measuring length, right before the tests.

Strength properties of the injection molded parts under static tension were determined using a testing machine manufactured by TIRA test 2300. The machine was equipped with 10 kN screw wedge chucks together with the accessories. The measurements were done at a tensile speed of 0.83 mm/s (50 mm/min) and under the measuring load range 0–500 N. The hardness tests by Shore's method were conducted using an Affri-manufactured hardness tester, type ART13 – Shore's method D. Such choice was due to the

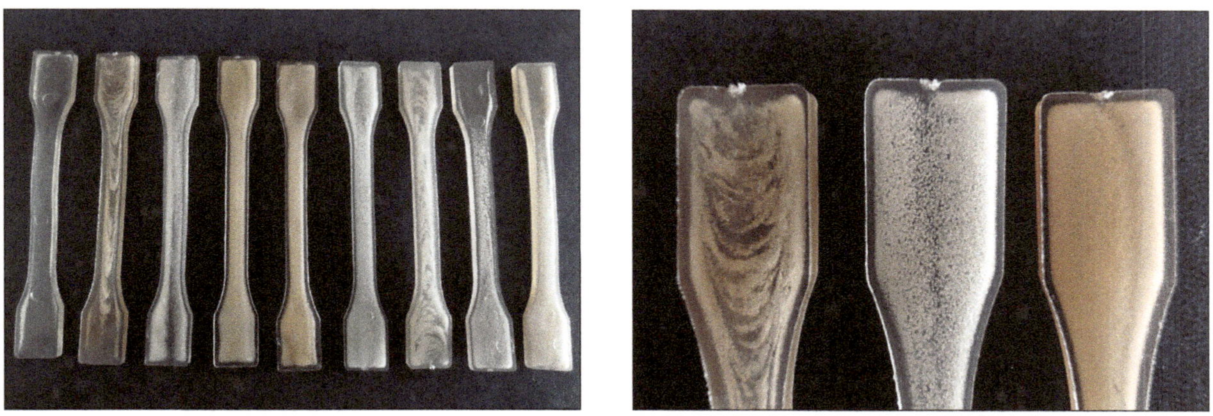

Fig. 3. Porous PVC injection molded parts with an ingot and a fragment of the molded part with a visible porous structure

employed method for determining hardness of solid polymers [16].

The investigation and analysis of the porous structure of the produced injection molded parts were conducted using authors' stand for porous structure image analysis. Porous structure image analysis is used to determine geometric characteristics of pores, based on the images of porous structure taken. The images are taken using a metallographic microscope, optical camera with an electronic amplifying system and then analyzed in a computer stand equipped with special graphics software.

RESULTS AND DISCUSSION

As a result of the process, the degree of porosity of the obtained porous products ranged from 0% to 45%. The value of the degree of porosity is defined by the percentage of cellular surface in relation to the total tested surface . In density testing of the cellular PVC coat samples, the pycnometrical method was applied.

The results of determining selected strength properties of the injection molded parts, obtained at different contents of the blowing agent in the processed polymers, are shown in Figures 4–5. It has been observed that increasing the blowing agent content in the product decreases the value of tension at break, in a non-linear manner in the whole content range of the blowing agent in the polymer. The distribution of the curves demonstrates that at high content of the blowing agent (over 0.4%) in the polymer, the intensity of the decrease in tensile strength is lower, irrespective of the type of the processed polymer and the type of the blowing agent used. The results of the tensile strength tests, of the specimens made from the polymers modified by the blowing agents correspond to the results of the conducted hardness tests. The injection molded parts made from porous polymers have decreased mechanical properties, strength properties included.

Significant differences in mechanical strength of the porous injection molded parts can be observed. For example, blowing agent P 1941, with the endothermic decomposition characteristics, dosed in the range from 0 to 2.0% causes a decrease in tensile strength by 25–30% on average in the case of PVC (Figures 4, 5).

The results of hardness of the injection molded parts are graphically represented in figure 6. The type of a blowing agent used, be it with the endothermic or exothermic decomposition characteristics, has no effect on the surface hardness of the injection molded parts. The changes in hardness illustrated in the figures predominantly depend on the type and properties of the polymeric materials used in the cellular injection molding process. It has been observed that increasing the blowing agent content in the product decreases the value of hardness, in a non-linear manner in the whole content range of the blowing agent in the polymer. The pore structure becomes finer and finer, the number of pores increases, while their size is decreased. Moreover, such a large content of blowing agent is economically ineffective.

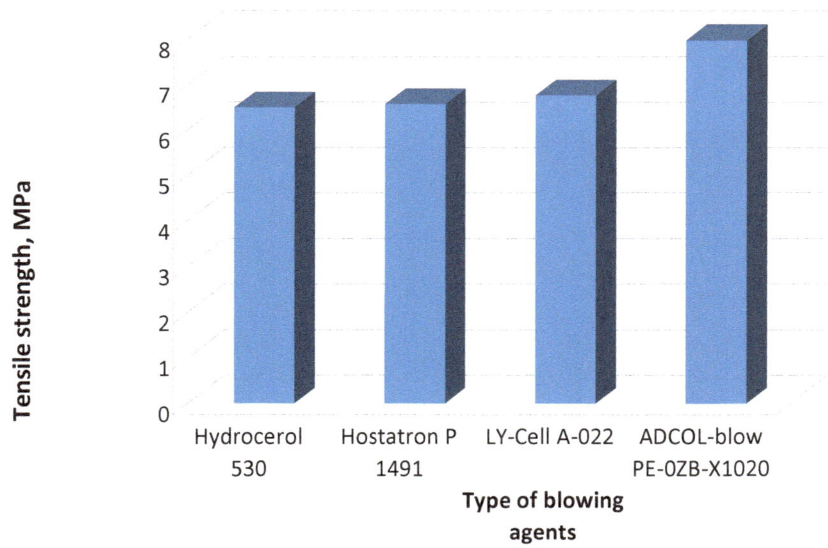

Fig. 4. Dependence of tensile strength of the injection molded parts made from the injected polymers on the content of blowing agents 0,4 % by mass

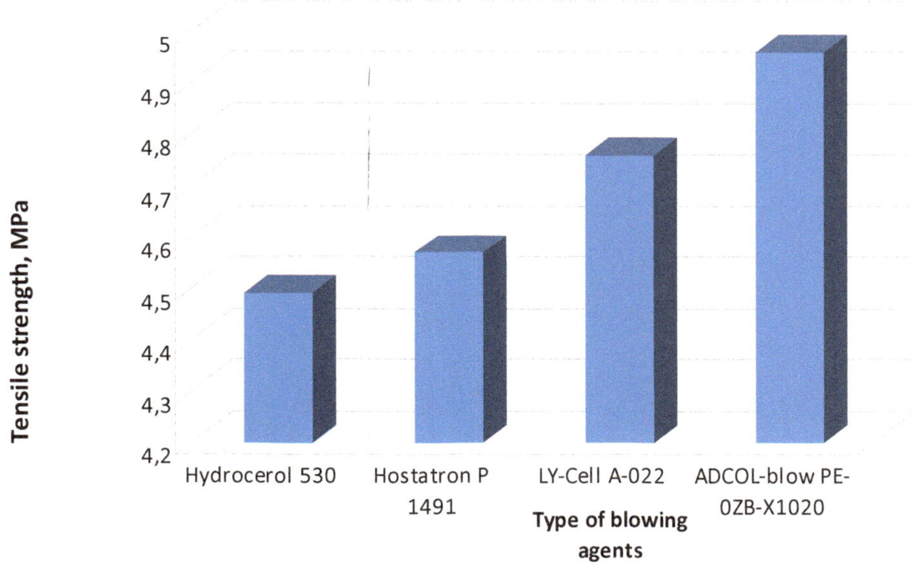

Fig. 5. Dependence of tensile strength of the injection molded parts made from the injected polymers on the content of blowing agents (2,0 % by mass)

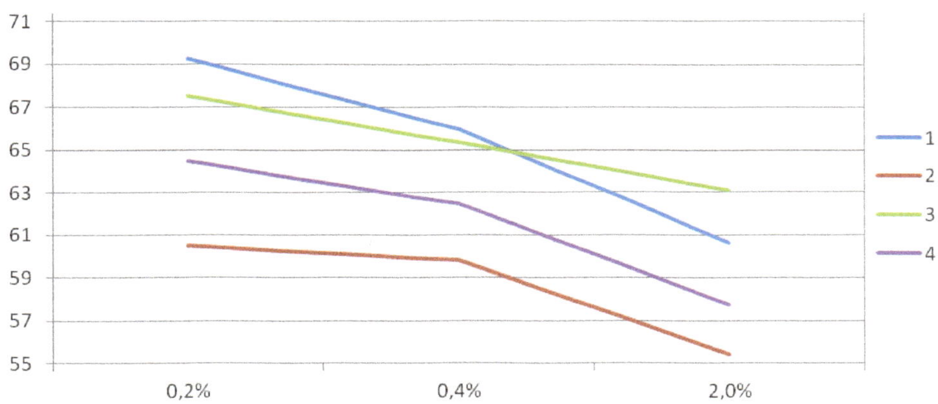

Fig. 6. Dependence of hardness of injection molded parts examined with Shore's method on the content of blowing agents, 1 – Hydrocerol 530, 2 – Hostatron P 1491, 3 – LY-Cell A-022, 4 – ADCOL-blow PE – 0ZBX1020

A discernible effect of the blowing agent used, its type and effect characteristics on the obtained morphology of porous molded pieces have been observed. Based on the analysis of the photographs taken, it has been found that the injection molded parts with 0.2% content of the blowing agent (Figure 7) have a clearly visible solid outer layer and their pore distribution is non-uniform to the greatest extent.

The change in the number of pores and their surface quantity in the cross section of the molded parts can also depend on cooling intensity. Fast cooling hampers the occurrence and growth of pores, especially of those located closer to the surface layer. The outer layer of the molded part undergoes direct and fastest cooling in the mold, which is why the number, size and surface quantity of the micro pores produced in this region are lower compared to the regions located closer to the core of the molded parts.

CONCLUSIONS

The blowing agents used in the cellular injection molding process give a possibility of creating products with new, modified properties, without causing important changes. The change in these properties offers new possibilities of applying this kind of products with simultaneous reduction in their production costs.

In the molded parts with 0.4% content of the blowing agent, the pore distribution is more uniform, and the visible pores are of similar sizes In the case of the molded parts with 2,0% content of the blowing agent (Figures 7–9), there is a sub-

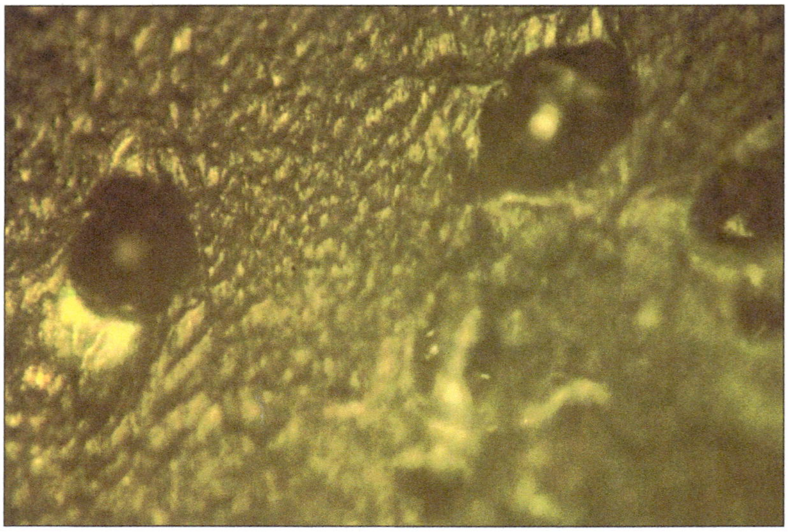

Fig. 7. Microscopic image of the cross section of the made of cellular tape polymer of the amount of the blowing agent (LY-Cell A-022) 0,2% mass, magnify 50×

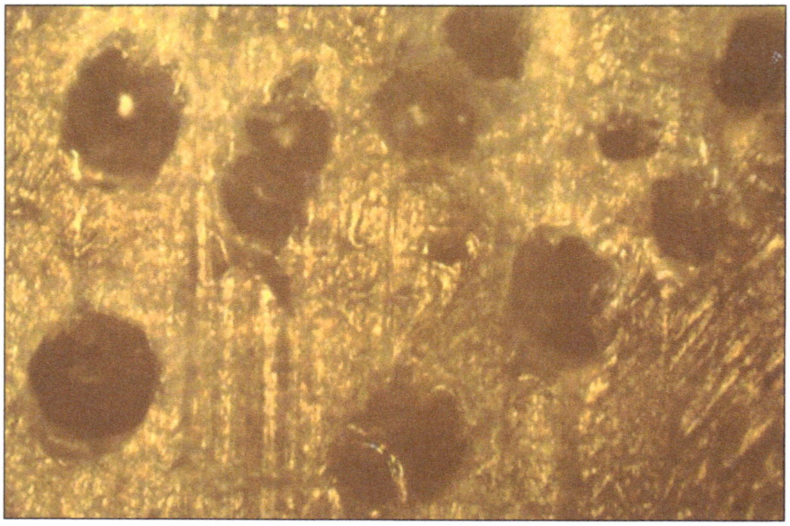

Fig. 8. Microscopic image of the cross section of the made of cellular tape polymer of the amount of the blowing agent (Hydrocerol 530) 0,4% mass, magnify 50×

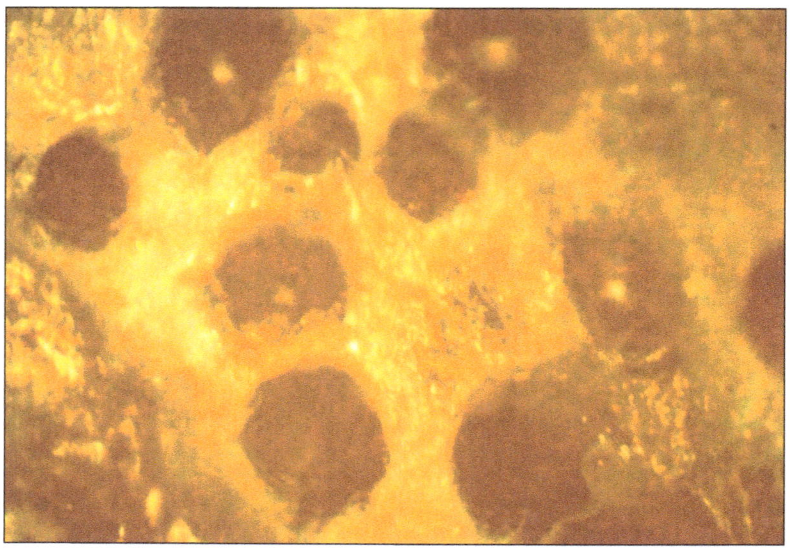

Fig. 9. Microscopic image of the cross section of the made of cellular tape polymer of the amount of the blowing agent (ADCOL-blow PE-0ZBX1020) 2,0% mass, magnify 50×

stantial concentration of pores of various sizes which, in extreme cases, can cause a discontinuity in the solid outer layer. For specific contents of the blowing agent in the polymeric material, the examined pore diameter increases, as the measuring capacity changes towards the product core. The smallest pore diameter was observed for the solid outer layer. The closer to the core the solid outer layer is, the more the pore size increases. The increase was most intensive at the lowest content of the blowing agent in the polymer being processed. As the dose of the blowing agent increases, the increase in the pore diameter for particular contents of this agent was smaller and smaller.

In the case of the porous parts produced using the blowing agents with the endothermic decomposition characteristics, the gas release in the course of processing comes to an end once the energy supply is stopped. The obtained porous structure is uniform; the pores have a spherical or quasi spherical shape. The pores have similar sizes, irrespective of their location in the product.

The porous structure is an advantage of injection molded parts produced in the cellular injection molding process, as it results in a decreased amount of the polymeric material needed in their production. Owing to the use of chemical blowing agents, porous parts have, among others, lower weight, enhanced damping properties, and lower processing shrinkage. Changing the injection molded part structure from solid to porous by using appropriate types and content of the blowing agents, brings numerous economical benefits without simultaneous decreasing the properties of such parts to values below expectations.

Acknowledgements

This paper is the result of the project implementation: Technological and design aspects of extrusion and injection molding of thermoplastic polymer composites and nanocomposites (PIRS-ES-GA-2010-269177) supported by The international project realized in range of Seventh Frame Programme of European Union (FP7), Marie Curie Actions, PEOPLE, International Research Staff Exchange Scheme (IRSES).

REFERENCES

1. Kelvin T. Okamoto.: Microcellular processing. Hanser Publishers, Munich, Germany 2003.
2. Bociąga E., Palutkiewic P.: Wpływ zawartości poroforu i warunków wtryskiwania na wybrane właściwości i stan powierzchni wyprasek z polipropylenu. Polimery, 1, 2012, 38-42.
3. Dulebová Ľ., Greškovič F.: Influence of regrind on properties of plastics processing by injection moulding. Materials Engineering - Materiálové inžinierstvo,18, 44, 2011.
4. Greškovič F., Dulebová Ľ., Varga J.: Technológie spracovania plastov. Vstrekovanie. Plastics Processing. Technologies. Injection Moulding, SjF TU v Košiciach, Košice 2010.
5. Matuana LM, Park CB, Balatinecz JJ.: Structures and mechanical properties of microcellular foamed polyvinyl chloride. Cellular polymers, 17, 1998, 1-16.
6. Bledzki K., Kirschling H., Steinbichler G., Egger P.: Polycarbonate microfoams with a smooth surface and higher notched impact strength. Journal of Cellular Plastics, 40, 2004, 489–496.
7. Garbacz T.: Properties of triple-layered PVC coatings synthesized in the micropore coextrusion method. Polimery, 56, 2011, 129-134.
8. Моравський В. С., Красінський В. В., Сікора Я., Кшижак А., Суберляк О. В.: Хімічна промисловість України, 6, 46, 2011.
9. Guo M.C., Heuzey M.C., Carreau P.J.: Cell structure and dynamic properties of injection molded polypropylene foams. Polymer Engineering and Science, 47, 2007, 1070–1081.
10. Klepka T., Dębski H., Rydarowski H.: Characteristic of high-density polyethylene and its properties simulation with use of finite element method. Polimery, 54, 2009, 668-673.
11. Zhou Q., Chuan-Bo C.: Exo-endothermic blowing agent and its blowing behavior. Journal of Cellular Plastics, 41, 2005, 225-234.
12. Sikora J.W.: Screw extrusion. In: Sabu T., Yang W. Advances in polymer processing. From macro to nanoscales. Woodhead Publishing Limited. Oxford-Cambridge-New Delhi, United Kingdom 2009.
13. Tor-Świątek A., Samujło B.: Use of thermovision research to analyze the thermal stability of microcellular extrusion process of poly(vinyl chloride). Maintenance and Reliability, 15, 2013, 58-61.
14. Garbacz T., Tor A: Effect of porophor content on the useful properties of external coatings of cables obtained by foaming extrusion. Polimery, 52, 2007, 286-293.
15. PN-EN ISO 527-1 (2010): Plastics. Determination of tensile properties.
16. PN-EN ISO 868 (2003): Plastics and ebonite. Determination of indentation hardness by means of a durometer (Shore hardness).

THE ISSUE OF ICING IN AVIATION. STUDY WITH THE USE OF NUMERICAL SIMULATION

Katarzyna Grzegorczyk[1], Ewa Wajszczak[2], Dominik Galas[3]

[1] Institute of Aviation, Al. Krakowska 110/114, 02-256 Warsaw, Poland, e-mail: katarzyna.grzegorczyk@ilot.edu.pl

[2] The State School of Higher Education in Chełm, Pocztowa 54, 22-100 Chełm, Poland, e-mail: e.wajszczak@gmail.com

[3] The State School of Higher Education in Chełm, Pocztowa 54, 22-100 Chełm, Poland, e-mail: dominikgalas91@gmail.com

ABSTRACT

Although it has been over 100 years since Wright brothers' first flight, icing has not been entirely investigated yet. The article is divided into 4 parts. The first one introduces the receiver of meteorological aspect of icing. Second part is devoted to physics aspect of ice accretion. The third and fourth parts describe icing with mathematical models and present a numerical simulation. We have completed our presentation with the results of numerical simulation.

Key words: icing, numerical simulation, ice accretion, meteorological aspect of icing, physics of icing.

INTRODUCTION

Icing is a process of creating an ice layer on a surface (e.g. of an aircraft). It can result in a change of aerodynamic characteristics of aerodyne or, worse, in a damage of the engine. The most exposed parts to icing are: leading edges of wings and stabilizers, engine inlets, leading edges of propeller blades and rotors. Other parts include: outer antenna, windshield or struts [8, 9].

It is generally accepted that aircrafts which move faster than 1000 km/h (540 kts) are practically resistant to icing due to frictional heat.

Considering causes of icing the following aspect are discussed: direct sedimentation of ice crystals or snow, freezing of supercooled water droplets when they hit the aircraft surface and water vapor sublimation on the aircraft surface [7].

Classification of icing, which is the most obvious and important for aviation study, is due to the shape of ice, therefore we can encounter [3]:

- Ice with a shape of a profile. It appears when the temperature is below -20 °C and the cloud is not very watery. When droplets hit the airplane, they freeze immediately and they do not change the shape of the airfoil (Figure 1). With small droplets ice accretion appears in the vicinity of the leading edge. With bigger ones the range of icing will grow. It's the most dangerous kind of ice accretion.
- Ice with a shape of a block. This type of icing forms in the temperature about -5 °C or lower when the cloud is very watery and an aircraft is moving at low speed. The droplets hitting the leading edge do not freeze, therefore, they are blown further where they do freeze (Fig. 2). It is less dangerous type of icing because vibration on flying airplane shatter the ice which in consequence falls off.
- Hoarfrost. It is possible to encounter it when the cloud has a small rate of aquosity or through sublimation of water vapor when the sky is cloudless. When an aircraft descends rapidly descending from high altitude (the surface of the airplane is colder than outside air temperature), the situation is perfect for creation of hoarfrost.

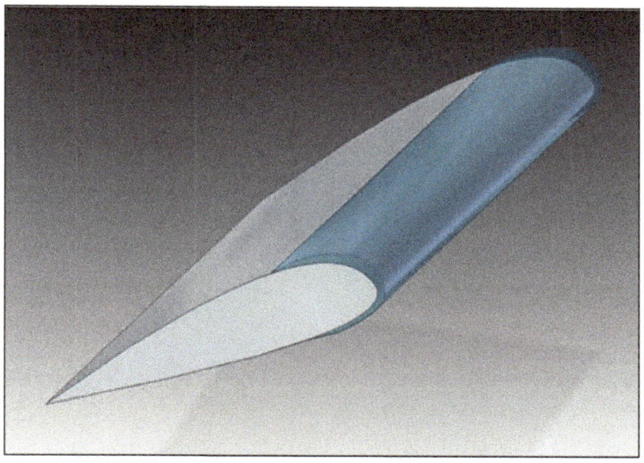

Fig. 1. Ice with a shape of the profile

Fig. 2. Ice with a shape of a block

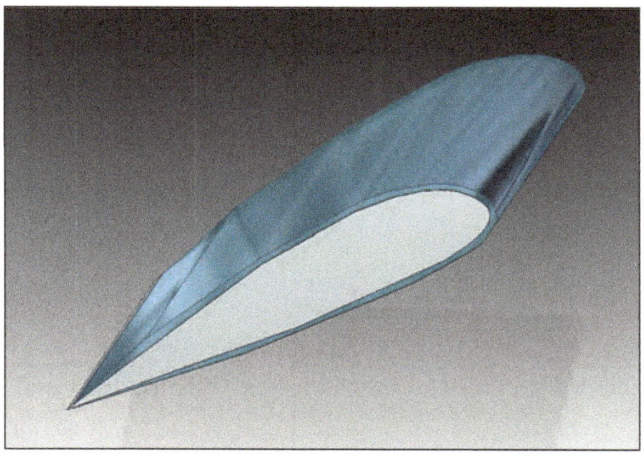

Fig. 3. Hoarfrost

The thickness of ice layer is normally slight and it does not change aerodynamic properties significantly, however, it remains dangerous due to the reduction of visibility from the cockpit (Fig. 3).

The speed of ice accretion provides other classification [4]:

- Trace – intensity of ice accretion is slightly bigger than rate of loss due to sublimation.
- Light – speed of ice accretion does not exceed 0.5 mm/min. It is assumed that this type of icing poses danger to flights in icing environment exceeding 1 hour.
- Moderate – the velocity is between 0.5 – 1 mm/min. With that rate of accretion it affects also flight of short duration.
- Severe – with intensity between 1 – 2 mm/min de-icing systems are not efficient enough.

It is relevant that preceding division is approximate, therefore, moderate conditions of icing for one aircraft could be severe for another.

According to ice structure we can name [2, 4]:

- Clear ice (glaze ice) – it occurs in the temperature of -20 °C – 0 °C, in watery clouds, from droplets which freeze at the moment when they hit the surface. It is quite dense and hard, therefore, it is very difficult to remove.
- Rime ice – the temperature for this type of icing is below -10 °C, in clouds formed by ice crystals or supercooled droplets with a diameter of 0.5 mm. Rime ice's most common characteristics are crystal structure, roughness of surface and lower density than clear ice due to content of air bubbles.
- Mixed ice – it is a combination of two above types of icing. It occurs in cumulus clouds (especially Cb clouds), where it is easy to find strong mixing of small and large water drops in the temperature of about -10,0 °C. It is also possible to encounter in Ns clouds.
- Hoarfrost – it is homogeneous layer on the aircraft's surface. Besides increase in drag, it causes reduction in visibility.

Icing is so hazardous because of consequences which it has in aviation. Among numerous effects the following should be emphasized [10, 13]:

- Ice accretion on leading edges of an aircraft disturbs the aerodynamic shape of wing. It results in reduction in lift, increase in drag, weight, stall speed or fuel consumption, which is the worst from the point of view of ergonomics.
- In conditions of severe icing the rate of ice accretion may amount to 1.25 cm/min. That would result in the loss of stability due to irregular distribution of weight. Furthermore, ice on the propeller may cause engine vibration. The moment it falls off it might damage the skin of the aircraft.
- As a result of icing it is possible to clog pressure sources (alike static and total inputs). Obviously, this would have an impact on instruments' readings (airspeed indicator, altimeter and vertical speed indicator).

- Deposition of ice causes (as mentioned) limitation in visibility from cockpit or precludes retraction of the undercarriage.

PHYSICS OF THE PHENOMENA

Considering icing from the physical perspective is an extremely complicated task due to multiplicity of processes which occur in the ice zone. Furthermore, there are phenomena with juxtapose the qualities of great power [5, 11, 12]. However, 3 basic conditions for ice creation can be named:

- water presence in the atmosphere;
- transport of water to the profile i.e. aircraft movement, fluid movement around the airfoil or diffusion;
- dissipation of heat leading to water freezing.

Basic processes seem to be obvious, but fragmentary issues are far more complex. Fragmentary issues include:

- Deflection of air stream induced by airfoil movement towards surrounding air;
- Water uptake by profile;
- Undercooling of water droplets in liquid state. Small water droplets remain in liquid state even in -40 °C.
- Kinetic heating of air. It occurs in a zone where the biggest air accumulation exists i.e. at the leading edge. It is remarkable that this process opposes icing but still leading edges are most susceptible to ice accretion.
- Kinetic heating produced by droplets. Water drops hitting the surface transfer their kinetic energy.
- Convective exchange of heat. It occurs between aircraft surface and the flowing air stream.
- Water evaporation or ice sublimation located on the surface of profile. This phenomenon contributes to heat dissipation.
- Heat emission related to crystallization (freezing of water which hits the surface).
- Heat emission related to friction. It is especially important considering high speed flow.

The above mentioned phenomena are not the only ones which have impact on icing but they are considered to be most essential ones.

Another aspect which complicates the comprehension of icing is the necessity consider many fields of classic physics such as atmospheric physics, meteorology, thermodynamics, aerodynamics and flight mechanics. With simplification of the process itself, it is possible to consider the phenomena based on fundamental laws and physical dependencies:

- The first law of thermodynamics. The change in internal energy of the system is equal to the amount of heat exchanged with the surroundings and the amount of work performed on or by the system. To simplify the problem – the temperature is a measure of the internal energy of a system. Considering icing using the first law of thermodynamics, it is possible to explain kinetic heating of air. In the stagnation point the kinetic energy of air particles is converted into the internal energy which increases the temperature.
- Temperature. As mentioned above, it may be defined as a measure of the internal energy of a system. The temperature lower than water freezing temperature is an indispensable condition for icing occurrence.
- Normal physics conditions. Normal temperature is equal to ice melting temperature under normal pressure. It provides a reference point to physical conditions in which icing occurs.
- Specific heat. According to the definition it is a measure of the amount of energy per unit of mass required to increase the temperature by one degree Celsius. Analyzing icing, it is important to remember that water has high specific heat.
- Isentropic gas transformation. In case of that transformation it is practicable to establish precisely the relation between variations in density, pressure and temperature of the considered system. For example, 100% increase of pressure causes 60% increase of density and 21% increase in absolute temperature. Considering 5% decrease in the temperature of melting that would mean 13.5 K drop. It is substantial for icing process .
- The relation between the velocity of gas stream and variation in the temperature and pressure. Taking the first law of thermodynamics into account and the decrease in gas speed, the increase in the temperature is observed and other way round. Pondering the isentropic gas transformation, one may designate changes of particular parameters.
- Continuity equation. If liquid or gas flows through a tunnel with a variable cross-section, mass flow remains constant. Considering the

flow around the airfoil, such as upper surface, one encounters the biggest growth of speed. Thus, the increase in speed is observed, there is also a decrease in internal energy and, hence, decrease in temperature.

Other factors should also be taken into consideration: water phase transitions (which has even higher specific heat than water - crystallization of one water drop supplies enough heat to warm a few other drops), humidity (cold air can hold smaller amount of water vapor than warmer), saturation pressure (under the ice layer it is bigger than the one you may find above the water layer). On the basis of the above facts, one may conclude that unsaturated air for water might be oversaturated for ice. This explains why ice occurs on airfoil without water presence in the atmosphere in a liquid from. Another factor which should not be omitted is water droplet uptake (which grows with an increasing drop diameter and flow speed while it decreases with the increase in distance from the leading edge). Heat exchange is a very important process . It is a basic way to interchange heat between the profile and surroundings. Moving fluid collects and carries energy from the body of higher temperature to the one with lower energy. For the aircraft flying in supercooled atmosphere the amount of lost heat by convection increases with the increase of speed and the decrease of dimensions. This explains why ice grows at first on antennas as well as small and protruding elements. In order to define icing, scientist had introduced icing coefficient β. It is a ratio of water amount that forms ice layer to all the water captured by the surface.

MATHEMATICAL ICING MODELS

There are the following mathematical icing models [2]:

- **Bilanin model.** a mathematical model of ice growing in time according to cloud aquosity, temperature and icing coefficient.
- **Dietenberger model.** It is a mathematical model which describes changes of lift coefficient C_z, drag coefficient C_x and critical angle of attack α_{cr} in correlation to the type and intensity (p) of contamination also maximum camber of its surface structure (k). Dietenberger model can be applied with under the following assumptions: ice layer appears at the whole leading edge (p = 1), wing chord equals MAC (mean aerodynamic chord), $k=\delta_l=\delta_{l\ max}(t)$, a and b coefficients are selected depending on the considered profile and contamination type.
- **Messinger model for icing.** In this consideration, the fundamental is the law of conservation of energy for equilibrium of temperature in isolated, unheated surface exposed to icing. The model has several limitations including: icy surface in Messinger model is smaller than the real one; the base is isolated; the model does not imply heat dissipation by the system; therefore, the only method of balancing the energy is to create hidden heat. In practice ice growth would be encountered [1].

NUMERICAL SIMULATION

For our simulation ICEM program was used for creating structural mesh and FLUENT as analysis system from ANSYS package [6]. The type of the simulation was 2D. The airfoil of ATR72 was used with average chord of 2.255 m [14]. The following parameters were set: air velocity v = 50 m/s, density ρ = 1.225 kg/m^3, angle of attack α = 0°. The computational domain dimensions were as follow: length – 22 m, height – 20 m, semi circle with radius of 10 m from the leading edge side. Mesh type for both cases was quad and the number of cells was around 30 000. Structural mesh with boundary layer for clean airfoil is displayed in figure 4.

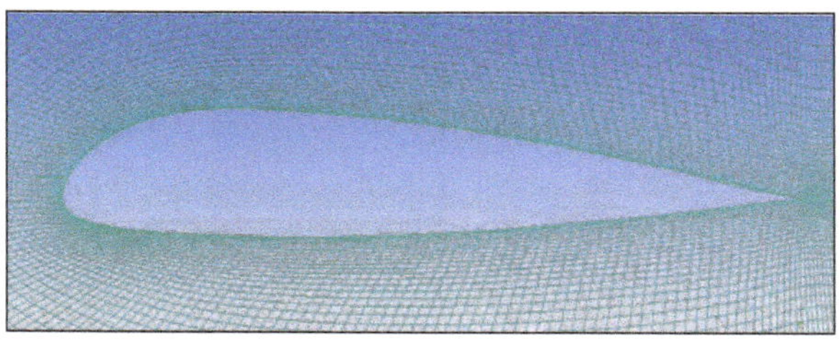

Fig. 4. Structural mesh for clean airfoil

Results which we obtained are presented below:

Airfoil	Drag	Lift
Clean	100%	100%
With ice accretion	212.729%	92%
Difference	+ 112.729%	-8%

Significant degradation of aerodynamics characteristic of the airfoil was observed. Drag increased significantly by 112.729% and lift decreased noticeably by 8%. The distribution of

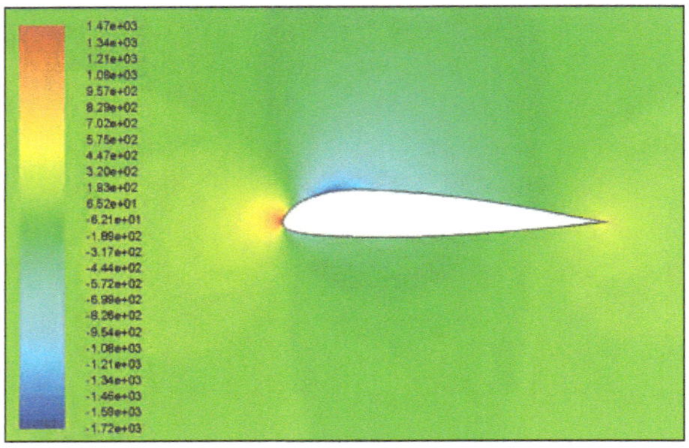

Fig. 5. Pressure distribution over clean airfoil

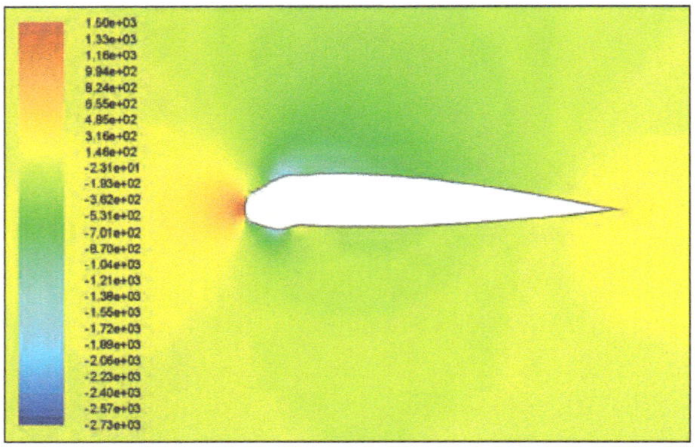

Fig. 6. Pressure distribution over airfoil with ice

pressure around both profiles is displayed in figures 5 and 6. Comparing both illustrations one can clearly see that the area of low pressure over clean airfoil is larger than the one over the airfoil with ice. What is more, the area of high pressure in stagnation point is significantly greater for airfoil with ice. Both of this observations explain the degradation of performances.

CONCLUSION

The study confirmed that icing is a very complicated phenomenon. Over the years since the beginning of aviation icing proved to be extremely dangerous. Even though the aeronautical industry is one of the most developmental branches; the 100% efficient method of counter-acting ice influence has not been discovered yet. For this reason, every pilot should respect the unpredictability and strength of nature and, therefore, when it is possible avoid the known icing conditions.

REFERENCES

1. Myers T.G. Extension to the Messinger Model for Aircraft Icing. AIAA Journal 39(2), 2001.
2. Al-Sharabi M.B.G. Modelowanie i symulacja numeryczna dynamiki obladzanego samolotu w fazie wznoszenia. Rozprawa doktorska, Politechnika Warszawska 2003.
3. Szefostwo służby hydrometeorologicznej sił zbrojnych RP. Meteorologia dla pilotów – poradnik. Warszawa 2011.
4. Meteorology, JAA ATPL Training, Jeppesen Publication.
5. Principles of Flight, JAA ATPL Training, Jeppesen Publication,
6. Jacobs S.J., Hospers J.M., Hoeijmakers H.W.M. Numerical simulation of ice accretion on multiple-element airfoil sections. 26th International Congress of the Aeronautical Sciences.
7. Jafernik H., Kozuba J. Oblodzenie samolotu: Część I. Przyczyny, wpływ na wykonywanie operacji powietrznych. Logistyka 3, 2011.
8. Meteorologia. Aeroklub Polskiej Rzeczpospolitej Ludowej, Wydawnictwa Komunikacji i Łączności, Warszawa 1975.
9. Schimidt M.T. Meteorologia dla potrzeb szybownictwa. Wydawnictwa Komunikacji i Łączności, Warszawa 1982.
10. Weinar J., Bonca Z. Zjawisko oblodzenia w układzie zasilania silnika lotniczego lekkiego samolotu. Technika Chłodnicza i Klimatyzacja, 6-7, 2007.
11. Abłamowicz A., Nowakowski W. Podstawy aerodynamiki i mechaniki lotu. Wydawnictwa Komunikacji i Łączności, Warszawa 1980.
12. Panas A., Kowaleczko G. Oblodzenie statków powietrznych. Rozdział II. Fizyka zjawisk oblodzeniowych statków powietrznych. ITWL, Warszawa 2005.
13. Smykla I. Wybrane zagadnienia z aerodynamiki i mechaniki lotu. Warszawa 2008.
14. Airfoil Coordinates Database, http://www.ae.illinois.edu/m-selig/ads/coord_database.html

29

THE ANALYSIS OF THE EFFECTS OF SCREENING HELICOPTER ENGINE COMPARTMENTS FOR INFRARED EMISSION

Stefan Fijałkowski[1]

[1] Lublin University of Technology, Nadbystrzycka 36, 20-608 Lublin, Poland, e-mail: s.fijałkowski@pollub.pl

ABSTRACT

The paper presents the research on using infrared screens protecting an engine in an engine gondola in PZL W-3 helicopter to limit infrared emission to the surrounding. The methods of numerical modelling and computing simulation (hover and horizontal flight forward) were applied to the analysis. The results enabled formulating the guidelines to construct screens.

Keywords: helicopter, infrared screen, infrared emission, egine gondola.

INTRODUCTION

The previous experimental tests of infrared radiation emitted by a helicopter during flight [4, 5, 6] prove that intense radiation of diffusers and exhaust gases released from them, the radiation from the hubcaps of air inlets and the screens of engine compartments are accompanied by quite intensive as presented in the images taken with thermographic camera Therma CAM S45 during the flight of the PZL W-3 Sokół helicopter (Fig. 1).

In order to limit infrared emission it is necessary to decrease radiation intensity and change radiation wave frequency. This can be done as follows [7]:

- by intensifying cooling the engine compartment, which can result in excessively decreased of engine temperature and worse engine performance.
- by choosing materials of appropriate emission indicators for engine surface and internal gondola surfaces, which is not always possible, to avoid the risk of engine and supplementary system overheating.

Lowering the emission can also be achieved by screening shields of engine compartments, which is the most effective method here.

An internal or external gondola surface can be screened. The latter method facilitates cooling

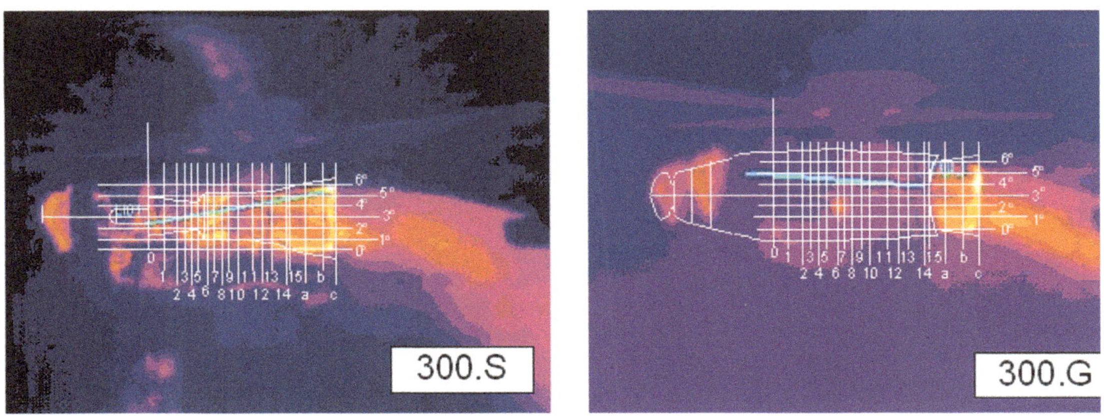

Fig. 1. Infrared emission by engine compartment screens to the surrounding during the flight of the PZL W-3 Sokół [4, 5, 6]

the screen and maintains the thermal conditions of the engine interior at the desired level (i.e. is avoided overheating the combustion chamber and engine turbine). Therefore, this paper discusses the external screening of engine compartment protection.

SCREENING INFRARED EMISSION FROM THE SURFACE OF A SOLID

An infrared screen in made of a substance that is impermeable for infrared radiation (zero transmission). It is not resistance for propagating radiation as such although it plays this role in radial systems [4]. Screen is to transform radiation energy that reaches it from the protected surface into energy radiated to the surrounding, as in the schematc in Fig. 2.

In the state of thermal and heat balance ($\dot{Q}_{1-E} = \dot{Q}_{E-O}$) the internal energy of a screen is defined at a certain level, which is manifested by the constant temperature of a screen.

Therefore, the role of screening as in [8, 9, 12] is to limit excessive infrared emission to the environment (decreasing the density of radial energy stream and decreasing the temperature). The functioning of a screen is demonstrated on the basis of a cylindrical-circular screen of unlimited length, without convective cooling. The schematic of screen geometry is in Fig. 3.

The screen is endlessly long without convectional cooling. In the system of transmitting solids 1-E-2, the non-concave surface of solid 1 is surrounded by the internal surface of screen E, and non-concave external surface of surrounded by solid 2. The surfaces of solids 1, E, 2 can differ in absorptiveness – a and – reflectiveness – r, at zero transmission – p, yet for each of them the following equations must be fulfilled:

- for total radiation:

$$a + r = 1 \tag{1}$$

- for monochromatic radiation (spectral):

$$a_\lambda + r_\lambda = 1 \tag{2}$$

The analyses of the technical screening of infrared emitting solids assume, in accordance with [8] and [9], that the solids are gray, and the reflectiveness of their surfaces is diffusion (dispersive). The reflectiveness of a concave screen surface may be of a mirror type (equality of angles of incidence and reflection) and then the equation is theoretically true:

$$r = r_\lambda = 1 \tag{3}$$

The above mentioned features of solids' surfaces are significant for the description of mutual emissiveness of two products. The description of

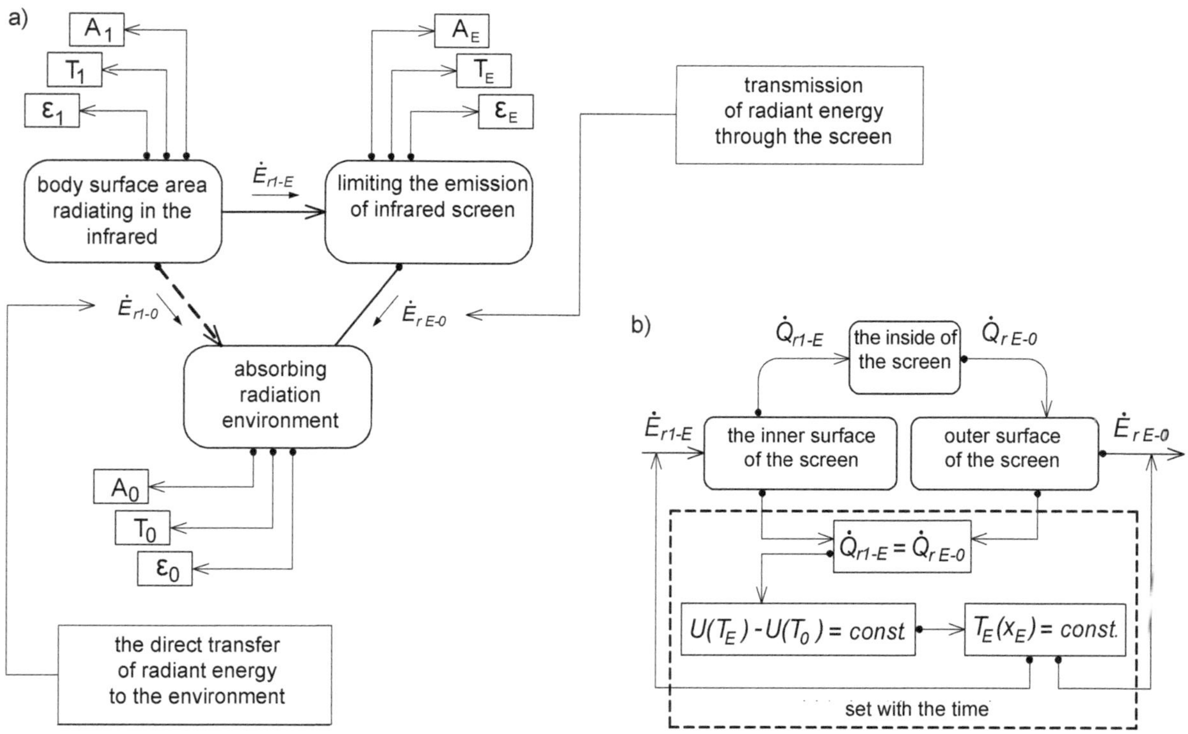

Fig. 2. Schematic of transforming the radiation energy in a screened system: a) a scheme of transforming energy, b) a scheme of balancing screen temperature

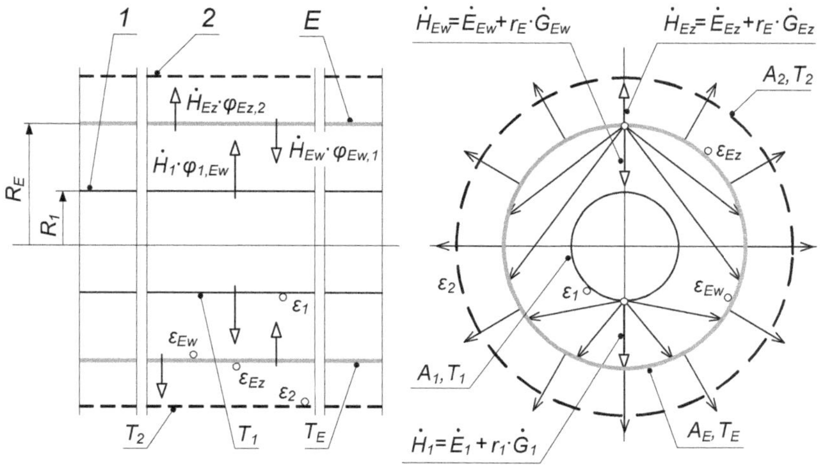

Fig. 3. Cylindrical-circular screen limiting infrared emission to the surrounding

a cylindrical-circular screen of unlimited length in the states of thermal and heat balance, without considering convection was presented with the method of lightness and radiating the surfaces of radiating solids [8]. These values refer to the energy exchange between two surfaces, one being non concave, closing a space filled with a diathermal factor (translucent for the infrared). The lightness and radiation of individual surfaces of the system (Fig. 3) is presented in Table 1.

The effective energy of radiation exchanged between the surfaces of system 1-E-2 is described by method following general equation, according the cited:

$$\dot{E}_{r_{k-l}} = \dot{Q}_{r_{k-l}} = \dot{H}_k \cdot \varphi_{k,l} - \dot{H}_i \cdot \varphi_{l,k} \qquad (4)$$

where: k, l – indexes attributed to particular surfaces 1, E, 2.

Measureless indicators φ in formula (1) are the relations to radiation, but to the isothermal surfaces, perfectly grey, radiating according to Lambert's law are, according to [8] equal to the relation of surface configuration.

The effective stream of radiant energy: emitted to screen $\dot{E}_{r1-E}$ aand to the environment given that $\dot{E}_{E-O}$, and that external and internal inequality of screen emission ($\varepsilon_{Ew} \neq \varepsilon_{Ez}$) is described in the following equation:

$$\dot{E}_{r1-E} = \dot{Q}_{r1-E} = \frac{A_1 \cdot C}{\frac{1}{\varepsilon_1} + \frac{A_1}{A_E}\left(\frac{1}{\varepsilon_{Ew}} - 1\right)}(\Theta_1 - \Theta_E) \qquad (5)$$

$$\dot{E}_{rE-O} = \dot{Q}_{rE-O} = \frac{A_E \cdot C}{\frac{1}{\varepsilon_{Ez}} + \frac{A_E}{A_2}\left(\frac{1}{\varepsilon_2} - 1\right)}(\Theta_E - \Theta_2) \qquad (6)$$

where:

$$\Theta_1 = \left(\frac{T_1}{100}\right)^4, \quad \Theta_E = \left(\frac{T_E}{100}\right)^4, \quad \Theta_2 = \left(\frac{T_2}{100}\right)^4$$

$C = 5{,}67 \left[\frac{W}{m^2 \cdot K^4}\right]$ – radiation constant of a perfectly black solid.

Using the condition of isothermicity and diffusiveness of all the surfaces, the state of thermal balance and the features of screen geometry (Fig. 3), are defined as:

Table 1. The description of lightness and radiation of specific elements of 1-E-2 system, according to Figure 3

Surface	Luminosity of surface	Radiation of surface
Solid 1 surface	$\dot{H}_1 = \dot{E}_1 + r_1 \cdot \dot{H}_{Ew} \cdot \varphi_{E,1}$	$\dot{G}_1 = \dot{H}_{Ew} \cdot \varphi_{E,1}$
Internal screen surface	$\dot{H}_{Ew} = \dot{E}_{Ew} + r_E \cdot \dot{H}_1 \cdot \varphi_{1,E}$	$\dot{G}_{Ew} = \dot{H}_1 \cdot \varphi_{1,E}$
External screen surface	$\dot{H}_{Ez} = \dot{E}_{Ez} + r_E \cdot \dot{H}_2 \cdot \varphi_{E,2}$	$\dot{G}_{Ez} = \dot{H}_2 \cdot \varphi_{E,2}$
Solid 2 surface	$\dot{H}_2 = \dot{E}_2 + r_2 \cdot \dot{H}_{Ez} \cdot \varphi_{2,Ez}$	$\dot{G}_2 = \dot{H}_{Ez} \cdot \varphi_{2,Ez}$

$$\Theta_E = \frac{\Theta_1\left[\frac{A_1}{A_E}\frac{1}{\varepsilon_{Ez}} + \frac{A_1}{A_2}\left(\frac{1}{\varepsilon_2}-1\right)\right] + \Theta_2\left[\frac{1}{\varepsilon_1} + \frac{A_1}{A_E}\left(\frac{1}{\varepsilon_{Ew}}-1\right)\right]}{\frac{1}{\varepsilon_1} + \frac{A_1}{A_E}\left(\frac{1}{\varepsilon_{Ew}} + \frac{1}{\varepsilon_{Ez}} - 1\right) + \frac{A_1}{A_2}\left(\frac{1}{\varepsilon_2}-1\right)}, \tag{7}$$

In this case, surface 2 represents the conditions that $\Theta_2 = \Theta_0$ and $\varepsilon_2 = \varepsilon_0 = 1$, the surface ratio is the configuration factors $\varphi_{E,1} = \frac{A_1}{A}$. The screen temperature in this case, is given by equation:

$$T_E = \left(\frac{T_1^4\left(\varphi_{E,1}\frac{1}{\varepsilon_{Ez}}\right) + T_0^4\left[\frac{1}{\varepsilon_1} + \varphi_{E,1}\left(\frac{1}{\varepsilon_{Ew}}-1\right)\right]}{\frac{1}{\varepsilon_1} + \varphi_{E,1}\left(\frac{1}{\varepsilon_{Ew}} + \frac{1}{\varepsilon_{Ez}} - 1\right)}\right)^{0,25}, \tag{8}$$

Equation (5) describes the temperature of screen knowing $\varepsilon_1, \varepsilon_E, \varphi_{E,1}$ and provided that the reflectiveness of the internal screen surface is diffusive.

If a surface is a mirror surface, [9], equation (5) changes and adopts a form:

$$T_E = \left(\frac{T_1^4 \cdot \left(\varphi_{E,1}\frac{1}{\varepsilon_{Ez}}\right) + T_0^4 \cdot \left(\frac{1}{\varepsilon_1} + \frac{1}{\varepsilon_{Ew}} - 1\right)}{\frac{1}{\varepsilon_1} + \frac{1}{\varepsilon_{Ew}} + \varphi_{E,1} \cdot \frac{1}{\varepsilon_{Ez}} - 1}\right)^{0,25}, \tag{9}$$

Limiting the stream of radial energy emitted to the surrounding through the screen depending on the internal surface quality, is described in the following way:

- for a diffusive surface:

$$K_{EK,D} = \frac{\dot{E}_{r1-E-O}}{\dot{E}_{r1-O}} = \frac{\dot{Q}_{1-E-O}}{\dot{Q}_{1-O}} = \frac{1}{1+\varphi_{E,1}\left(\frac{1}{\varepsilon_{Ew}} + \frac{1}{\varepsilon_{Ez}} - 1\right)\cdot\varepsilon_1} < 1 \tag{10}$$

- in case of mirror surface:

$$K_{EK,Z} = \frac{\dot{E}_{r1-E-O}}{\dot{E}_{r1-O}} = \frac{\dot{Q}_{1-E-O}}{\dot{Q}_{1-O}} = \frac{1}{1+\varepsilon_1\cdot\left(\frac{1}{\varepsilon_{Ew}} + \varphi_{E,1}\cdot\frac{1}{\varepsilon_{Ez}} - 1\right)} < 1 \tag{11}$$

Equation (6) implies that the decrease of energy emitted to the surrounding can be obtained by increasing configuration coefficient $\varphi_{E,1}$ and decreasing screen emission ε_E.

DESCRIPTION OF IN AIR-COOLED SCREEN

An air-cooled screen is described referring to a screen protecting a gondola of a helicopter drive motor (Fig. 4). Figure 5 presents a functional schematic of a module limiting infrared emission from a gondola surface of a helicopter drive to the environment with the use of a convection air-cooled cooling screen. The experimental research [4], [5], [6] enabled a description of a screen with two-sided convection cooling where:

- the field of temperature on the gondola surface is unevenly distributed along the gondola's length $T_G = T_G(x)$, whereas along the perimeter the changes are mathematically insignificant;
- the temperature in the immediate vicinity of the helicopter depends on flight altitude and the temperature at the ground $T_H = T_H(T_0, H)$;

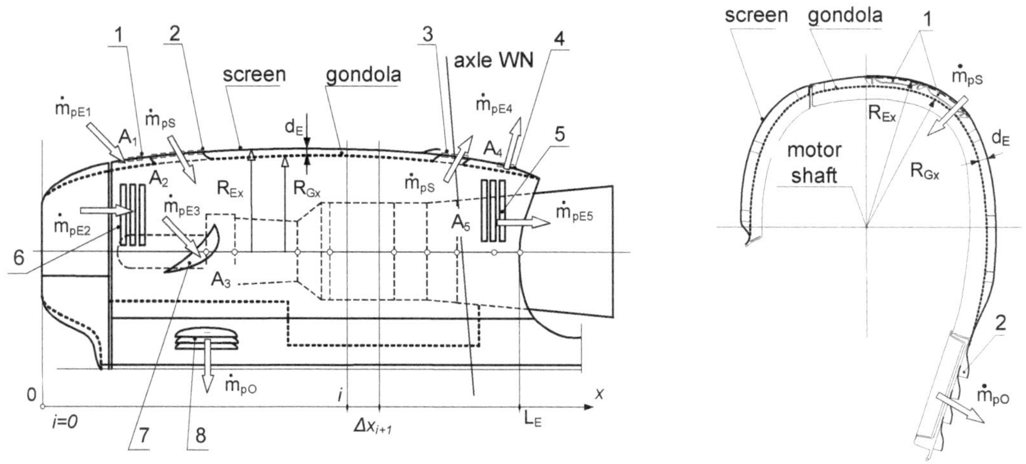

Fig. 4. Schematic of a screen limiting infrared emission through the gondola cover in the drive motor of the PZL W-3 Sokół. The scheme presents the location of the screen and air intakes and outlets cooling the screen interior. Own elaboration.

- the radial coefficients of gondola and screen surfaces and air do not depend on temperature in the discussed range, thus they are described by average values;
- air as a mixture of two-atom gases is treated as transparent for infrared radiation (diathermic medium) [4].

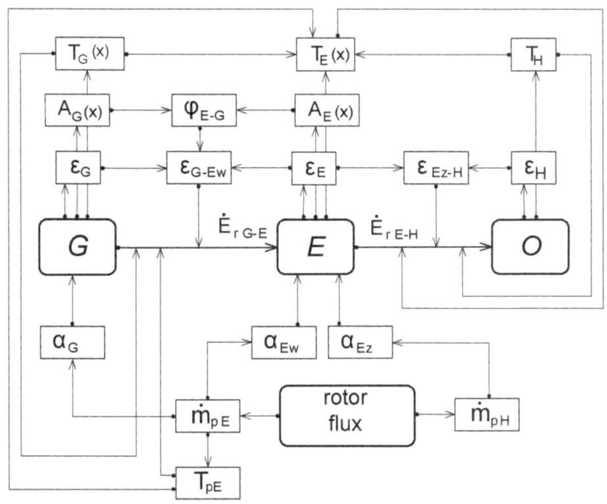

Fig. 5. Schematic of transforming radial energy in a system with convectional cooling from the side of gondola and from the outside, provided that $\varepsilon_H = 1$

Additionally, it was assumed that the screen is from the inside with air passing between the surfaces: gondola covers and screen interior; and from the outside with the rotor stream flowing around the gondola covered with the screen.

Air which flows inside the screen is generated by pressure distribution in the post-rotor stream in the areas of air intakes and outlets of the screen.

The description of the cooling screen was formulated, alike in point 2, by the method of surface luminosity and radiation, given that the internal and external screen emissions are not equal ($\varepsilon_{Ew} \neq \varepsilon_{Ez}$), and the surrounding emission ($\varepsilon_H = 1$). The coefficients of heat absorption are neither equal for the internal and external screen surfaces ($\alpha_{Ew} \neq \alpha_{Ez}$) (Fig. 6.)

The radial energy emitted as radiation to the screen assumed to be transmitted as radiation luminosity to the environment and as heat to the cooling air. The schematic of the model of the screen with two-sided convection cooling was presented.

Simultaneous equations describing the model of the screen: the equations of enthalpy of air inside the screen and the energy balance equation for the screen are as follows:

$$dT_P = \left(T_{Gx} - T_{Px}\right) \cdot \frac{\alpha_{Gx} \cdot O_{Gx}}{\dot{m}_{pE} \cdot c_p} dx + \left(T_{Ex} - T_{Px}\right) \cdot \frac{\alpha_{Ewx} \cdot O_{Ex}}{\dot{m}_{pE} \cdot c_p} dx \tag{12}$$

$$\left(\frac{T_{Ex}}{100}\right)^4 \cdot \left(1 + \frac{O_{Gx} \cdot \varepsilon_{GEw}}{O_{Ex} \cdot \varepsilon_{Ez}}\right) + T_{Ex} \cdot \left(\frac{\alpha_{Ewx} + \alpha_{Ezx}}{C_c \cdot \varepsilon_{Ez}}\right) - T_{px} \cdot \left(\frac{\alpha_{Ewx}}{C_c \cdot \varepsilon_{Ez}}\right) - \left(\frac{T_{Gx}}{100}\right)^4 \cdot \left(\frac{O_{Gx} \cdot \varepsilon_{GEw}}{O_{Ex} \cdot \varepsilon_{Ez}}\right) +$$
$$- \left(\frac{T_H}{100}\right)^4 - T_H \cdot \left(\frac{\alpha_{Ezx}}{C_c \cdot \varepsilon_{Ez}}\right) = 0 \tag{13}$$

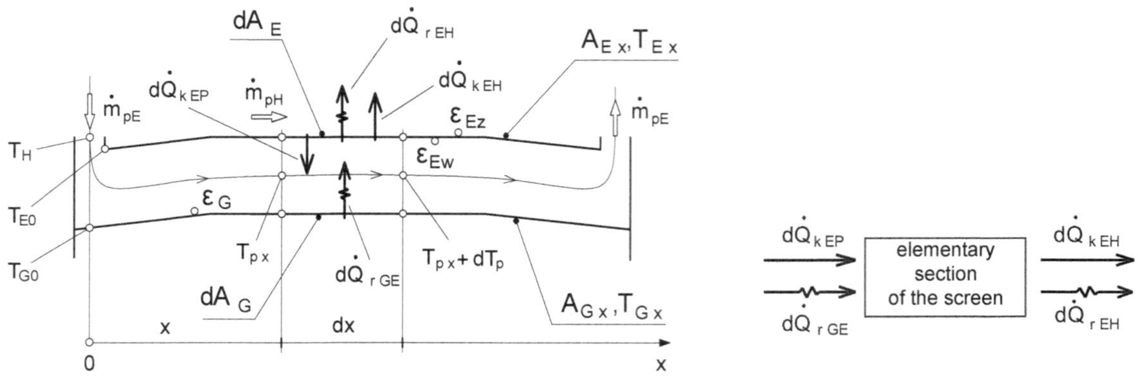

Fig. 6. Schematic of model of convection cooling module covering the motor drive gondola

with the boundary conditions:

$$\text{for } x = 0 \text{ is } T_{Gx} = T_{G0},\ T_{Ex} = T_{E0},\ T_{Px} = T_H \tag{14}$$

where: $\dot{m}_E$ – stream of the mass of air colling the screen interior,
C_c – constant of black body radiation,
ε_{GEw}, ε_{Ez} – emission coefficients: mutual emission of gondola and screen inside surface and external screen surface, respectively,
O_{Gx}, O_{Ex} – current perimeters of gondola and screen,
α_{Gx}, α_{Ewx}, α_{Ezx} – heat absorption coefficients: from the gondola surface, internal and external screen surface

The mutual emission is described by the following equations:

- inside the screen

$$\varepsilon_{GEw} = \frac{\varepsilon_G \cdot \varepsilon_{Ew}}{\varepsilon_{Ew} + \left(\varepsilon_G - \varepsilon_{Ew} \cdot \varepsilon_G\right) \cdot \varphi_{GEw}} \tag{15}$$

- outside the screen

$$\varepsilon_{EzH} = \varepsilon_{Ez} \tag{16}$$

Equation (9) is a first order ordinary differential equation with a variable coefficient strongly non-linear, whereas equation (10) is an algebraic non-linear equation. Both of the equations are coupled with air temperature T_{Px} changing along the air path inside the screen. The solution of the equation is the function of temperature distribution in the screen T_{Ex}. The solution should be computed numerically because of those difficulties.

Thus, a discreet system of space variable x, yet, for a changeable grade Δx, the gives a method for experimental data processing from measuring the surface temperatures on gondola and motor drive surfaces:

$$x_{i+1} = x_i + \Delta x_{i+1},\quad i = 0, 1, 2, 3, \ldots \tag{17}$$

The discretisation of numerical solution is given in Figure 7.

The considered area of the gondola cover, screen, space between the gondola and the screen and external immediate surrounding were divided with the i-indexed planes into elementary parts of unequal thicknesses (uneven grades of surface discretisation). Therefore, i^{th} layer from the direction of flow ends with i^{th} cross section.

It was also assumed that the air flowing through i^{th} layer absorbs heat of average temperature T_{GGi} from the gondola wall and the heat of average temperature T_{EEi} from internal screen surface as specified in the following equations:

$$T_{GGi+1} = \frac{T_{Gi+1} + T_{Gi}}{2} \tag{18}$$

$$T_{EEi+1} = \frac{T_{Ei+1} + T_{Ei}}{2} \tag{19}$$

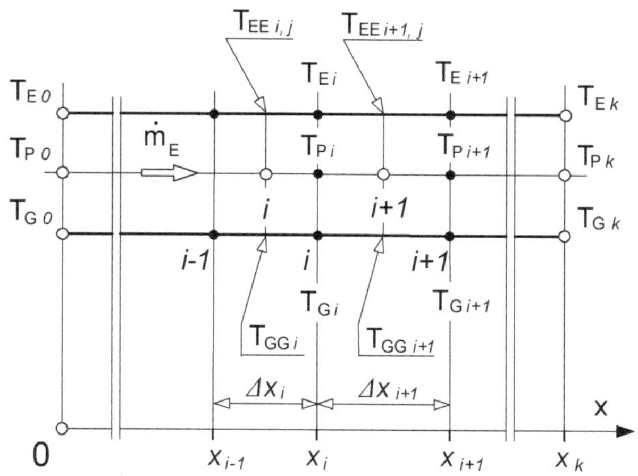

Fig. 7. Schematic of discretisation of the numerical solution

Equation (6) with the temperature derivative $\frac{dT_p}{dx}$ substituted with a difference quotient with "a grade forward" has the following form:

$$T_{Pi+1} = T_{Pi}\left[1-\left(S_G + S_E\right)_{i+1}\right] + T_{GGi+1}\cdot S_{Gi+1} + T_{EEi+1}\cdot S_{Ei+1}\ , \tag{20}$$

and equation (7) has the following numerical form:

$$\left(\frac{T_{EEi+1}}{100}\right)^4 \cdot E_{Ai+1} + T_{EEi+1}\cdot E_{Bi+1} - T_{Pi+1}\cdot E_{Di+1} - E_{Ci+1} = 0 \tag{21}$$

Individual equation coefficients (15) and (16) are described in the following way:

$$S_{Gi+1} = \frac{O_{Gi+1}\cdot \alpha_{Gi+1}}{\dot{m}_E \cdot c_p}\cdot \Delta x_{i+1}, \tag{22}$$

$$S_{Ei+1} = \frac{O_{Ei+1}\cdot \alpha_{Ewi+1}}{\dot{m}_E \cdot c_p}\cdot \Delta x_{i+1}, \tag{23}$$

$$E_{Ai+1} = 1 + \frac{O_{Gi+1}\cdot \varepsilon_{GE}}{O_{Ei+1}\cdot \varepsilon_{Ez}}, \tag{24}$$

$$E_{Bi+1} = \frac{\alpha_{EPi+1} + \alpha_{Ezi+1}}{\varepsilon_{Ez}\cdot C_C}, \tag{25}$$

$$E_{Di+1} = \frac{\alpha_{EPi+1}}{\varepsilon_{Ez}\cdot C_C}, \tag{26}$$

$$E_{Ci+1} = \frac{O_{Gi+1}\cdot \varepsilon_{GE}}{O_{Ei+1}\cdot \varepsilon_{Ez}}\left(\frac{T_{GGi+1}}{100}\right)^4 + \left(\frac{T_H}{100}\right)^4 + \frac{\alpha_{Ezi+1}}{\varepsilon_{Ez}\cdot C_C}\cdot T_H \tag{27}$$

The boundary conditions corresponding to the numerical solution i+0 are as follows:
for $i = 0$

$$\begin{gathered} T_{Pi} = T_{P0} = T_H, \\ T_{Gi} = T_{G0},\ T_{GGI} = T_{G0} \\ T_{Ei} = T_{E0},\ T_{EEI} = T_{E0} \end{gathered} \tag{28}$$

Due to the coupling of the two equations (16) and (17), their solution for temperature T_E was completed with the subsequent approximations by Newton-Raphason method:

$$T_{EEi+1,j+1} = T_{EEi+1,j} - \frac{F\left(T_{EEi+1,j}\right)}{F'\left(T_{EEi+1,j}\right)}, \tag{29}$$

for:

$$F'\left(T_{EEi+1,j}\right)=\frac{d\,F}{d\,T_{EE}} \tag{30}$$

Function F and its derivative F' are described by the following equations:

$$F\left(T_{EEi+1,j}\right)=\left(\frac{T_{EEi+1}}{100}\right)^{4}\cdot E_{Ai+1}+T_{EEi+1}\cdot E_{Bi+1}-T_{Pi+1}\cdot E_{Di+1}-E_{Ci+1} \tag{31}$$

$$F'\left(T_{EEi+1,j}\right)=4\cdot\left(\frac{T_{EEi+1}}{100}\right)^{3}\cdot E_{Ai+1}+E_{Bi+1}-E_{Di+1}\cdot S_{Ei+1} \tag{32}$$

The initial value of $T_{EEi+1,0}$, due to the effect of cooling, the first approximation was assumed in a form of solution without convectional cooling, yet for a limited length of the screen. Then, for $j = 0$ the equation is:

$$T_{EEi+1,0}=\left(\frac{T_{Gi+1}^{4}\left(\varphi_{EG}\dfrac{1}{\varepsilon_{Ew}}\right)+T_{H}^{4}\left[\dfrac{1}{\varepsilon_{G}}+\dfrac{1}{\varepsilon_{Ez}}-1\right]}{\dfrac{1}{\varepsilon_{G}}+\varphi_{EwGi+1}\left(\dfrac{2}{\varepsilon_{Ew}}-1\right)}\right)^{0,25}, \tag{33}$$

Coefficients φ_{EwGi+1} corresponding to the radiation relations were expressed by the relations of configuration (according to previous reservations) as for cylindrical and circular surfaces of limited length.

Such an approach is justified for comparing the distances between the screen and gondola with the thickness of individual layers and for constant temperatures in i[th] layers of gondola surfaces and corresponding screen surfaces. In the first approximation coefficient φ_{EwGi+1} was described by the equation as for two concentric cylinder of limited length (labels as for Fig. 7). The equation has the following form:

$$\begin{aligned}\varphi_{Ew-Gi+1}=&\frac{1}{ER_{i+1}}-\frac{1}{\pi\cdot ER_{i+1}}\left\{\arccos\frac{EB_{i+1}}{EA_{i+1}}-\frac{1}{2E\Lambda_{i+1}}\times\right.\\&\left.\times\left[\sqrt{\left(EA_{i+1}+2\right)^{2}-\left(2ER_{i+1}\right)^{2}}\arccos\frac{EB_{i+1}}{ER_{i+1}\cdot EA_{i+1}}+EB_{i+1}\arcsin\left(\frac{1}{ER_{i+1}}\right)-\frac{\pi\cdot EA_{i+1}}{2}\right]\right\}\end{aligned} \tag{34}$$

where:

$$ER_{i+1}=\frac{R_{Ei+1}}{R_{Gi+1}},\quad E\Lambda_{i+1}=\frac{\Delta x_{i+1}}{R_{Gi+1}},$$

$$EA_{i+1}=E\Lambda^{2}_{i+1}+ER^{2}_{i+1}-1,\quad EB_{i+1}=E\Lambda^{2}_{i+1}-ER^{2}_{i+1}+1.$$

CONVECTIONAL SCREEN COOLING

A helicopter motor drive gondola screen is convection cooled from the inside and from the outside [7]. From the inside, the screen is cooled by the stream of air flowing into the space between the screen and gondola through intakes and flowing out through the outlets onto the side of screen due to the reaction of post-rotor stream (distribution of velocities and pressures in this stream). The external surface of the screen is also cooled by a post-rotor stream.

Table 2 presents the criteria equations that are used to described the coefficients of heat absorption from the internal screen surface to the stream of air cooling the space between the gondola and screen and from the external screen surface to the stream of air flowing down the screen in the post-rotor stream.

Table 2. Criteria equation to define the coefficient of heat absorption from the external screen surface to the stream of air flowing down the screen in the post-rotor stream

Form of equation	Condition	No
$Nu_{Ez} = 0{,}224 \cdot Re_Z^{0{,}612} \cdot (1 - 0{,}54 cos^2 \beta)$, as in [9] $Nu_{Ez} = \frac{\alpha_{Ez}\, d_{Hz}}{\lambda_{pz}}$ Re_Z – Reynolds number for the post-rotor flow of gondola, β – angle of flow cutting.	$3 \cdot 10^3 \leq Re \leq 1{,}5 \cdot 10^4$	(34)

SIMULATION CALCULATIONS AND THEIR RESULTS

Simulation calculations concerning the infrared screen covering the gondola of motor drive PZL 10W in the PZL W-3 Sokół were made with author's own developed software INFRED-EKRAN_1_0.exe, Part 2 written in Fortan 95 language. The block scheme of the programme is depicted in Fig. 9.

In the numerical calculations, the results of calculations made in Part 1 of the programme were used (these results transfer automatically from Part 1 to Part 2 of the programme) and the experimental results concerning the distribution of temperatures in drive T_S and gondola surface T_G with thermographic camera Therma CAM S45.

The sample results and simulation calculations are depicted in the graphs. The heat absorption coefficients for the external gondola surface and for the internal screen surface were determined from equation (35) table 3, where as the heat absorption coefficients for the external gondola surface from equation (34). Emission coefficients for the external gondola surface, internal and external screen surface were assumed as in [11] and [13].

The results of experimental tests and temperature distribution are given in Figure 10.

Table 3. Criteria equation to define the coefficient of heat absorption from the internal surface of the screen to the stream of air cooling the space between the gondola and the screen

Item.	Form of equation	Condition	No
1.	$Nu = 0{,}26 \cdot Re_p^{0{,}8} \cdot Gr_p^{-0{,}15}$, (for the air flow), as in [3, 14]	$10^3 \leq Re \leq 10^5$ $3 \cdot 10^3 \leq Gr \leq 10^6$	(35)
2.	a) internal surface of the screen $Nu_{Ew} = 0{,}023 \cdot Re_p^{0{,}8} \cdot Pr_p^{\frac{1}{3}} \cdot \left(\frac{\mu_p}{\mu_s}\right)^{0{,}14}$ b) external surface of the screen $Nu_{Gz} = 0{,}017 \cdot Re^{0{,}8} \cdot Pr^{\frac{1}{3}} \cdot \left(\frac{d_z}{d_w}\right)^{0{,}18}$	$Re > 10^4$; $0{,}7 < Pr < 16700$ $\frac{d_z}{d_w} = 1{,}2 \div 14$	(36)

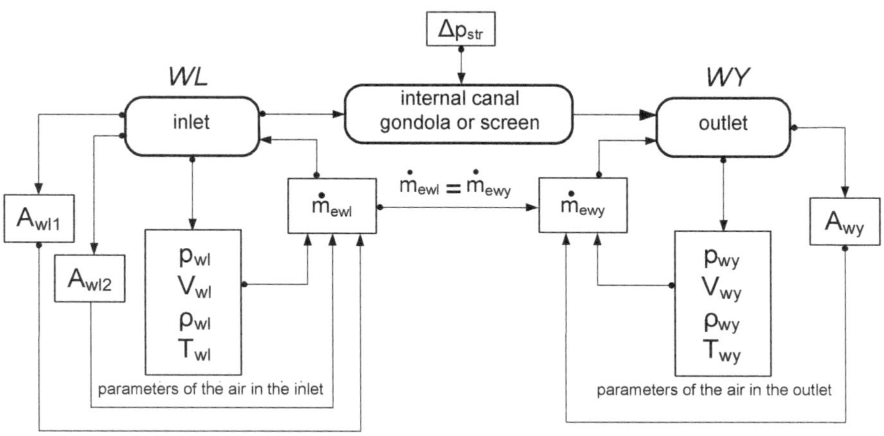

Fig. 8. Schematic of discretisation of the solved numerical task

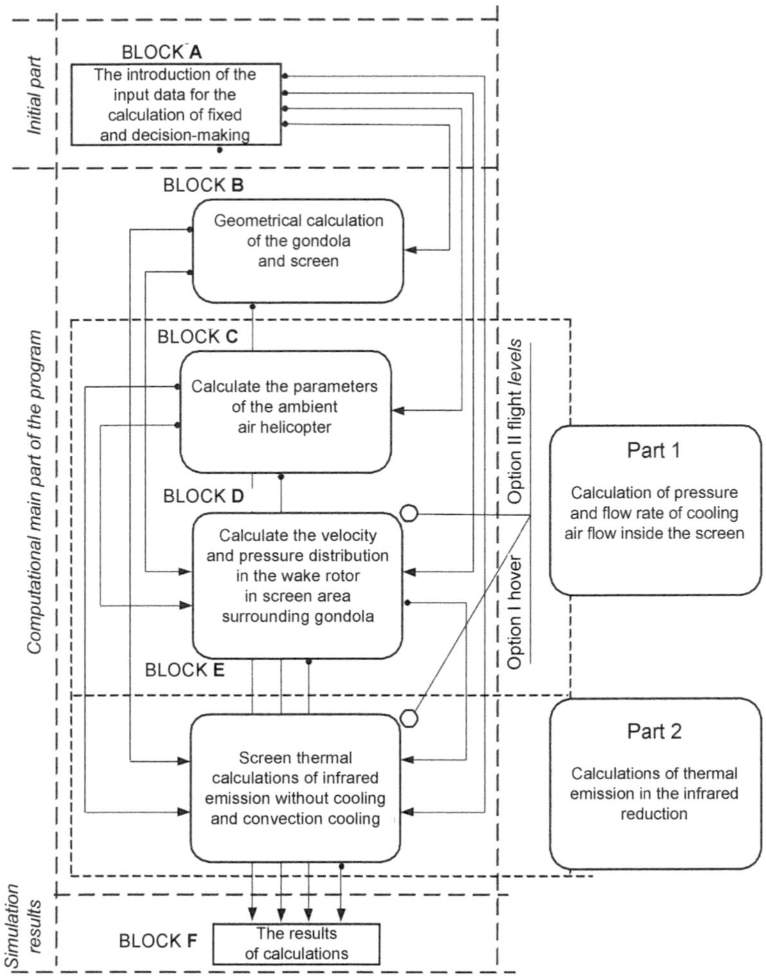

Fig. 9. Block schematic of calculation programme INFRED-EKRAN_1_0.exe, Part 2 written in Fortan 95 language

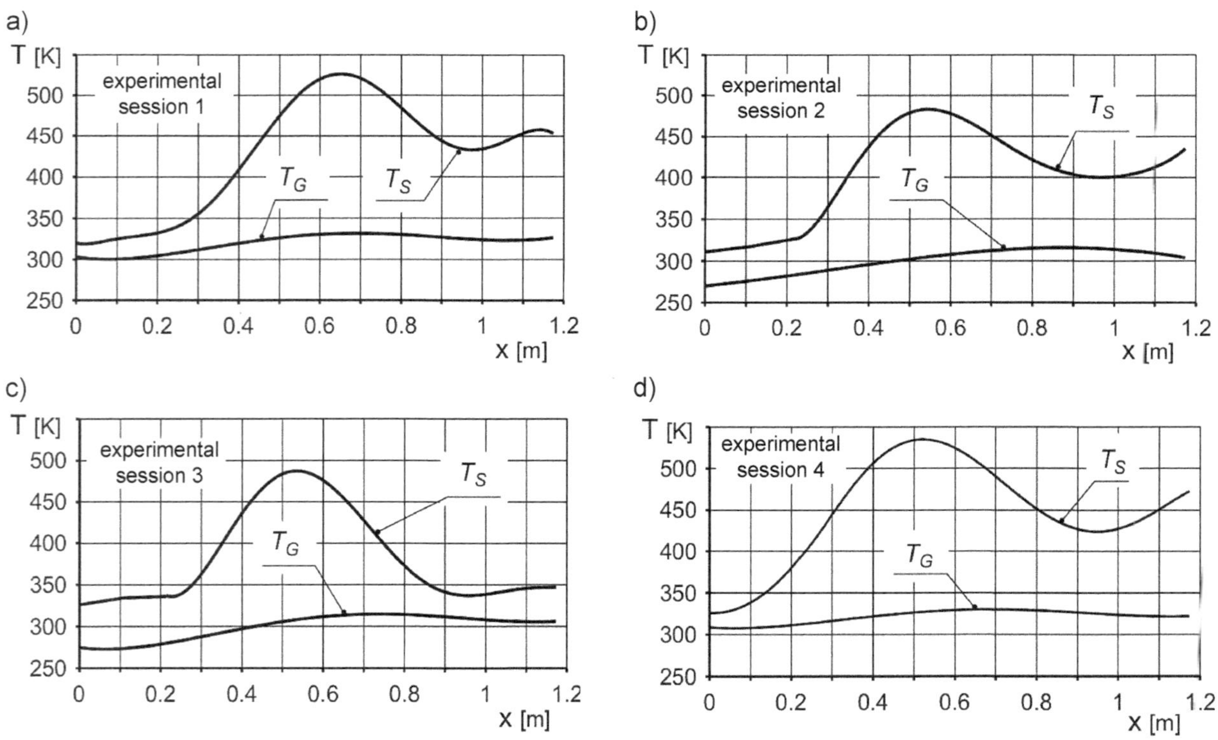

Fig. 10. Distribution of temperatures in drive T_S and gondola surface T_G along the path determined in four experimental sessions

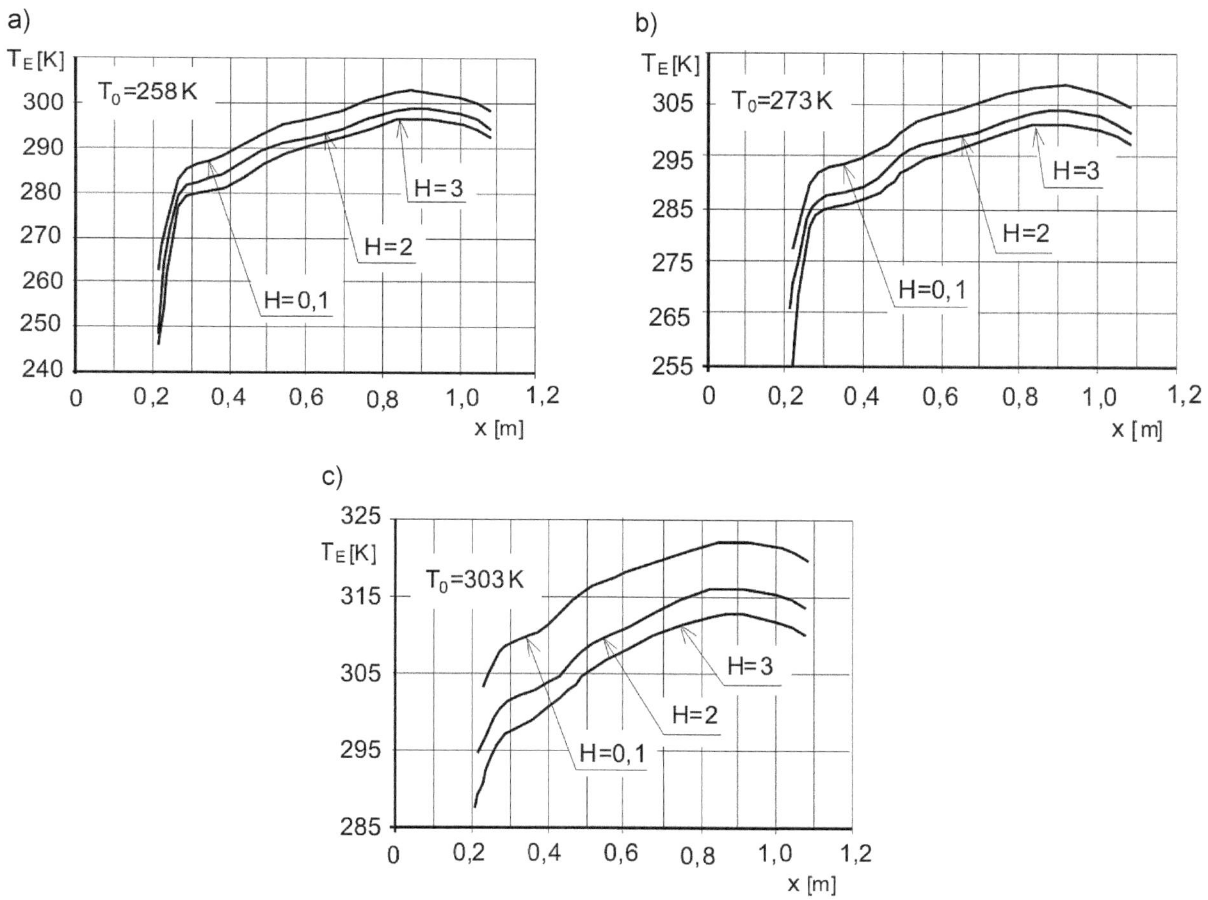

Fig. 11. Distribution of temperatures on the screen surface along the path of the gondola protecting the drive in helicopter PZL W-3 Sokół at different flight altitudes, for different initial temperatures at the ground level: a) T_0 = 258K, b) T_0 = 273K, c) T_0 = 303K

Figure 11 presents the results of numerical solutions of screen temperatures at different altitudes and for initial temperatures at the ground level T0.

CONCLUSIONS

The results of modelling and simulations allow for the following conclusions:

1. The simulation calculations concerning the functioning of infrared screens protecting a helicopter drive gondola confirm that the screens in helicopter PZL W-3 Sokół are necessary. The same applies to other types of helicopters with similar engine compartments.
2. The internal screen surface should manifest reflectiveness similar to the reflective body, which increases the efficiency of screening.
3. The internal cooling of the screen by the post-rotor air stream in the proximal surrounding of a gondola increases the efficiency of infrared screening. At the same time, cooling the space between gondola and the screen decreases the danger of overheating the gondola and helicopter turbine drive.

REFERENCES

1. Bielecki J., Suchenek M.: FORTRAN dla zaawansowanych. PWN, Warszawa 1983.
2. Bronsztejn I.N, Siemiendajew K.A.: Matematyka. Poradnik encyklopedyczny. PWN, Warszawa 1995.
3. Chmielniak T, Kosman G.: Obciążenia cieplne turbin parowych. WNT, Warszawa 1990.
4. Fijałkowski S.: Analiza emisji podczerwieni przez śmigłowiec w locie na podstawie badań eksperymentalnych. Prace Instytutu Lotnictwa, Zeszyt 211, Warszawa 2011.
5. Fijałkowski S.: Badanie i analiza emisji promieniowania podczerwonego nieosłoniętego silnika napędowego i strumienia spalin śmigłowca PZL Sokół w warunkach przedstartowych i przy starcie przy użyciu kamery termowizyjnej. Nr S38/M/2008 zad.2.3. Lublin 2008.

6. Fijałkowski S.: Badanie i analiza emisji promieniowania podczerwonego śmigłowca PZL Sokół w warunkach lotu przy użyciu kamery termowizyjnej. Nr S38/M/2009 zad.2.3. Lublin 2009.
7. Fijałkowski S.: O uwarunkowaniach chłodzenie przedziałów silnikowych śmigłowców z napędem turbinowym w różnych warunkach lotu. VI Konferencja nt. „GIS, GPS i Technika Lotnicza w Praktyce", Chełm, 25-26.10.2012.
8. Kostkowski E: Promieniowanie cieplne. Warszawa 1993.
9. Madejski J.: Teoria wymiany ciepła. PWN. Warszawa 1963.
10. Madura H.: Pomiary termowizyjne w praktyce. Agenda Wydawnicza PAKu, Warszawa 2004.
11. Raznjevic: Tablice cieplne z wykresami. WNT, Warszawa 1966.
12. Rudnicki Z.: Modelowanie matematyczne radiacyjnego przepływu energii. Gliwice 2003.
13. Sala A.: Radiacyjna wymiana ciepła. WNT, Warszawa 1982.
14. Zisina-Mołożien i in.: Tiepłoabmien w turbomaszinach. Izd. Maszinostrojenije, Leningrad, 1974.

30

TESTING ENERGY CONSUMPTION IN THE TROLLEYBUS AND THE BUS ON A CHOSEN PUBLIC TRANSPORT LINE IN LUBLIN

Ewa Siemionek[1], **Mieczysław Dziubiński**[1]

[1] Mechanical Engineering Faculty, Lublin University of Technology, Nadbystrzycka 36, 20-618 Lublin, Poland, e-mail: e.siemionek@pollub.pl; m.dziubinski@pollub.pl

ABSTRACT

One of the most important elements determining the assessment of the quality of trolleybuses and buses in public transport is the operating cost which comprises mainly the cost of power or fuel consumption. The knowledge of economic properties of new vehicles is justified when the decision about their purchase is being made. The information about vehicles which have not been used in the company so far comes mainly from the protocols of homologation tests as well as the opinions of other users. In both cases such knowledge is insufficient against the expected information providing for specific driving conditions in urban traffic. Chosen driving tests applied for the evaluation of energy consumption of trolleybuses and buses operation are presented in the article. Working conditions of the trolleybus were analysed on the basis of the comparative tests of the bus and the trolleybus for a chosen communication line in Lublin. On the basis of the determined characteristics describing working conditions of the vehicles constituting public transport in Lublin, the MPKPL test was drawn up and the analysis of findings was carried out for a chosen communication line.

Keywords: energy consumption, public transport, trolleybuses.

INTRODUCTION

The cost of power consumed by public transport vehicles constitutes a considerable part of balance sheets of communication companies. Economic properties of trolleybuses, including power consumption, are one of the basic factors determining the assessment of the trolleybus quality. Great importance is also attached to the cost of purchase, service, reliability and durability.

There are many representative driving cycles drawn up mainly by automotive organizations and institutions or European circles for environmental protection. The cycles are mostly designed for light vehicles, motor cars, and those designed for goods vehicles include only information about engine worktime at a given load. All the cycles are intended to test emissions of car exhaust fumes and do not reflect the real driving conditions of the car and especially of the trolleybus or the city bus driving cycles on a road without a special bus lane.

Fuel consumption in buses is examined in driving tests with assumed speed courses. A relevant problem is synthesizing the driving test taking into account the criteria of the resemblance of operational use conditions and tests carried out for buses and trolleybuses for a given network of communication lines.

Tests carried out for the Municipal Public Transport Company in Lublin showed that the conditions of trolleybuses use, including the kind of communication line determining the character of their functioning, had a significant effect on power consumption in trolleybuses.

The source of information concerning economic properties is mainly the protocols of homologation tests as power consumption was analysed for trolleybuses not used in public transport companies. It is also possible to gain knowledge about power consumption in a given type of a trolleybus from other users [1, 2, 5].

METHODOLOGY AND COURSE OF EXAMINATION

In the course of the study, road measurements were taken, including geographic coordinates, times of stops and rides, speeds, acceleration of buses and trolleybuses driving on chosen transport routes. A selected number of 10 stops and landform features were aimed at reflecting in an accurate way the movement of the bus and the trolleybus in the conditions of normal work of the carrier for both types of vehicles. In Figures 1 and 2, chosen time characteristics of speed for the tested vehicles have been presented [8].

Measurements were taken for a random number of passengers, changing at each bus stop. They were taken repeatedly on separate driving routes and the findings were averaged. On the basis of road measurements of real city vehicle driving cycles were carried out in the area of Lublin city, a representative driving cycle called MPKPL was developed. The developed representative driving cycle reflects real driving conditions of buses, including such parameters as achieved speeds and speeding up as well as the duration of stops and driving at a constant speed. The speed profile of the representative MPKPL cycle is presented in Figure 3.

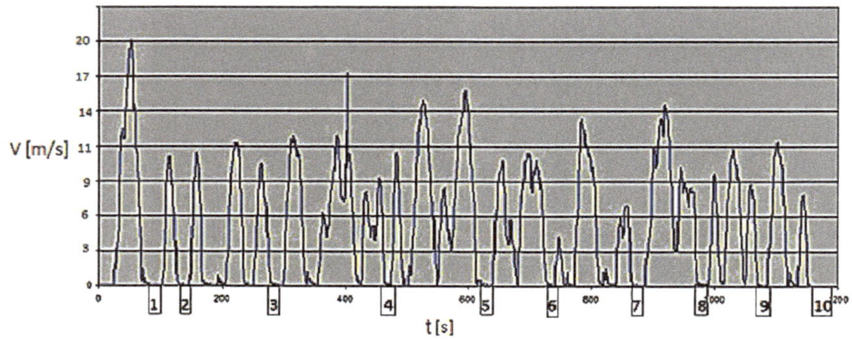

Fig. 1. Graph presenting speed in the function of time for the trolleybus during the ride Mickiewicza stop – Honorowych Krwiodawców roundabout

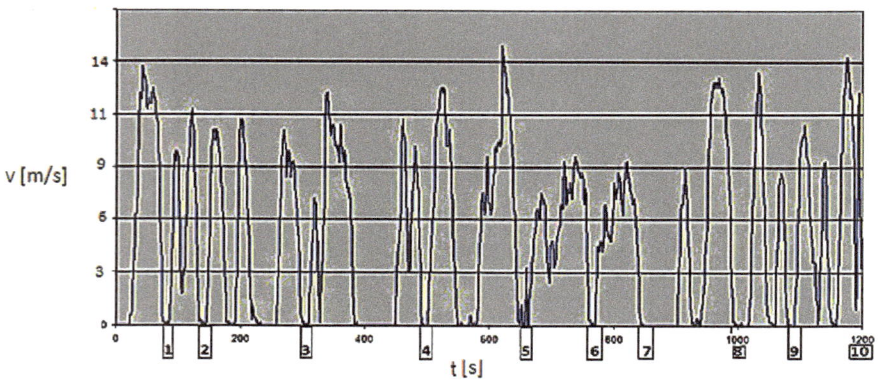

Fig. 2. Graph presenting speed in the function of time for the bus during the ride Honorowych Krwiodawców roundabout stop – Mickiewicza

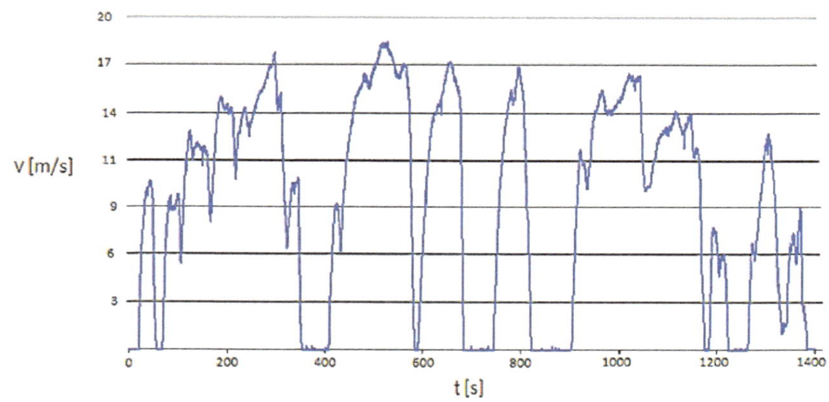

Fig. 3. Velocity profile of the representative MPKPL cycle

The length of the route is 8000 m, the average speed 15m/s, the maximum speed 18 m/s. The average driving speed includes not only the motion but also breaks related to passengers' getting on and off – from the moment of starting at the first stop to the moment of finishing at the last stop.

ANALYSIS OF ENERGY CONSUMPTION

Energy balance of the representative driving cycle was prepared on the basis of data for the trolleybus Solaris Trollino 12 and the bus Solaris Urbino 12.

Regarding weekly energy or fuel consumption, driving conditions on weekdays are more significant than those on weekend days. It results from a much bigger number of weekdays and realized transport routes.

Figure 4 shows the dependence of unitary energy consumption on the load of city transport vehicle after the percentage of seats taken on buses and trolleybuses have been analyzed. It can be concluded from this relationship that at the load equal to the number of seats, the consumption is about 50–100% higher than for a nominal load, however, at the little load, unitary consumption rises disproportionally.

Regarding distances between bus stops, unitary energy consumption is inversely proportional to the distance between stops *L* which is shown in Figure 5. The influence of little road inclination is practically of no importance for energy consumption, and with proper placing of stops and braking with energy recovery, the consumption could be even reduced.

The efficiency of converting energy in electric and combustion drives depends largely on the

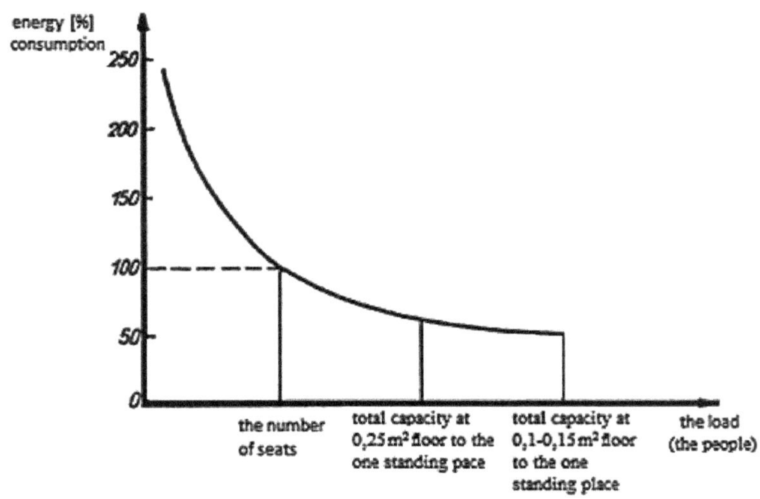

Fig. 4. Dependence of unitary energy consumption on the load of city transport vehicle [4]

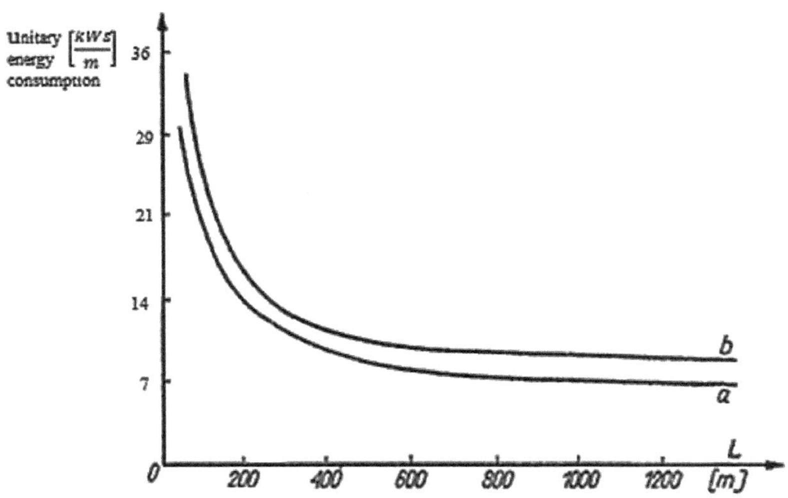

Fig. 5. Unitary energy consumption depending on the distance between stops *L* [4]

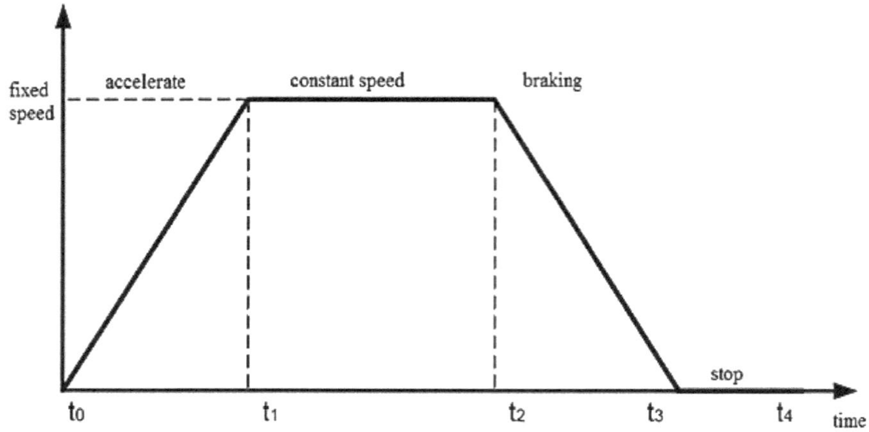

Fig. 6. Elementary car driving cycle [7]

profile of a given driving cycle. Figure 6 shows a basic car driving cycle.

An elementary driving cycle consists of acceleration phases until the set speed is reached, driving at an established speed, braking and a stop. When the vehicle accelerates, energy is used to realize effective work (overcoming resistance to motion and increasing kinetic energy of the vehicle) and making up for the loses in the power transmission system. The power needed to accelerate the vehicle to a set speed is much bigger than the one needed to drive it at a constant speed. However, when braking, the kinetic energy accumulated in the vehicle mass and rotating inertia masses is lost in a form of thermal energy in the vehicle friction brakes. The abrasion of friction surfaces causes the wear of brake lining as well as brake disks and drums, whereas the rubbed dust pollutes the environment. The energetic efficiency of drive systems was analyzed assuming that the trolleybus and the bus are of the same mass and have the same traction parameters, move on the same road and in the same driving cycle (a stop, accelerating, driving at the established speed, braking) [3, 6].

CONCLUSIONS

On the basis of professional literature, it can be stated that one of the basic criteria in evaluating means of city transport is their operating cost, dependent on energy consumption, in the conditions of real-life operational use.

Empirical research of buses and trolleybuses on selected communication lines has been conducted as part of the project 'Multidimensional analysis of means of city transport with regard to their operation and energy consumption'. On the grounds of the results of the conducted analysis and experiments, a representative driving cycle MPKPL was drawn up, making it possible to carry out energy balance for the tested group of vehicles.

The results of measurements and calculations enabled determining: total path, total time, mean speed of driving time, driving time, established motion time, slowing down time as well as elements of energy consumption balance between particular stops on the route.

For the comparative ride of the bus/trolleybus on the route Mickiewicza street – Honorowych Krwiodawców roundabout – Mickiewicza street:

- during the ride on the flat surface the bus is characterized by smaller cycle speedup averages than the trolleybus;
- mileage energy consumption for the bus Ψ = 6 kJ during the ride on the flat surface is bigger than for the trolleybus Ψ = 2.6 kJ;
- mileage energy consumption for the trolleybus during an uphill ride Ψ = 4 kJ is bigger than for the bus Ψ = 2.9 kJ.

Obtaining the highest value of total energy consumption for stop No 6 is confirmed on the basis of the registered speed in time function for the trolleybus and the bus; it is connected with the inclination of Piłsudski street in Lublin.

Acknowledgements

Research conducted as part of the project "Scholarship for postgraduate members of research teams", Operational Programme Human Capital.

REFERENCES

1. Chłopek Z., Metodyka szacowania zużycia paliwa przez autobusy na liniach komunikacji miejskiej. Technika w lokalnym transporcie zbiorowym, Kielce, 2004.
2. Ciastoń A., Sapań G., Prędkość komunikacyjna pojazdów w miejskim transporcie zbiorowym. Transport miejski i regionalny 10, 2006.
3. Fice M., Zarządzanie rozpływem energii w napędzie hybrydowym. Rozprawa doktorska PŚ, Gliwice 2010.
4. Ostaszewski J., Energochłonność transportu miejskiego. WKiŁ, Warszawa 1984.
5. Šarkan B., Measurement of fuel consumption of a road motor vehicle by outdoor driving testing. Adv. Sci. Technol. Res. J. 7(19), 2013, 70–74.
6. Siłka W., Analiza energetyczna procesu rozpędzania samochodu. Zeszyty Naukowe WSI, Opole 1993.
7. Siłka W., Analiza wpływu parametrów cyklu jezdnego na energochłonność ruchu samochodu. Zeszyt 14 Monografia 2 PAN oddział Kraków 1998.
8. Wołyńczuk P. Energochłonność, porównanie autobusu i trolejbusu. Praca dyplomowa PL, Lublin 2011.

31

THE CONCEPT OF INTEGRATED ENGINEERING AND BUSINESS (EB) EDUCATION SYSTEM

Michał Charlak[1], Marek A. Jakubowski[2]

[1] Fundamentals Technology Faculty, Lublin University of Technology, 38 Nadbystrzycka Str., 20-618 Lublin, Poland, e-mail: m.charlak@pollub.pl

[2] Management Faculty, Lublin University of Technology, 38 Nadbystrzycka Str., 20-618 Lublin, Poland, e-mail: m.jakubowski@pollub.pl

ABSTRACT

In our approach to engineering and business education system an engineer is a man working as creator and user of technical products. We stress that the process of understanding and gaining knowledge of technical reality and creativity of engineers are the essential for EB concept . Next, we describe briefly three perspectives for building the system of innovative product origination as a basis for EB system: 1) designer's perspective; 2) business perspective. 3) consumer perspective.

Keywords: engineering educational system, heuristics, innovation, organization.

INTRODUCTION

Modern engineer must not only be a specialist in engineering (for example: an electrician, mechanic, architect, and so on), but very often he also needs to be a businessman. More and more often an engineer works as a logistics specialist, environment protection expert, a manager, an IT specialist. He works as a constructor, technologist, materials expert and simultaneously as a chief production manager. Very often he must participate in the process of creating the organization and as a work-safety specialist.

It is necessary to create the appropriate place at technical universities to combine multiple disciplines including business, management and ergonomics. We have to choose what knowledge and specific fields of study (perhaps all) students of engineering need to master. The humanities and social sciences must represent sub-disciplines that allow achieving high qualifications to design objects to the various fields of engineering. We propose a new approach – an engineer is a man who is simultaneously a creator and user of technical products. He need modern knowledge and XXI century skills from all disciplines concerning examining various characteristics for own work, for example: physical, psychical, social, health aspects. We name some important fields: physiology, psychology, praxeology, medicine, neurology, anthropometry, and some important sciences related to engineering, such as: management, business, economy, health and work safety management etc. Our modern engineer must be prepared to perform a creative work for innovative organizations within large and continuously and turbulent changing spheres of the needed knowledge and skills.

A process of understanding the process of gaining knowledge is the best model of a process of education at a technical university. This process includes the following four stages: understanding basic facts of reality; the process of understanding, followed by binding acts of understanding into particles of knowledge; enriching the current knowledge with understood elements of knowledge and enlarging the state of understanding; creating knowledge assets. Therefore, the profile of a student educated in the sphere of technical science should include the requirements connected with knowledge of heuristics and methods of creative work, and other human and social sciences [6].

PLACE OF HEURISTICS IN THE PROCESS OF GAINING ENGINEERING KNOWLEDGE

Most often creation is defined as a process based on matching and analogies which bring new and useful products. In many authors' opinions creative process comes for four basic phases:

- preparation – preparatory actions for finding of idea,
- incubation – finders of brain action,
- illumination – actions of idea appearance,
- verification – actions of estimates and modification.

It is possible to identify all of them in the schemes of ergonomic structures of project designs in authors discussing the problem of useing complex knowledge about functioning technically-social system for technique. It is also possible to single out many trends in modern uses of heuristics:

Heuristics as a creative solution of a problem equal to logical, managerial as well as mathematical (e.g. solution of task, building of definition) particularly by experiment, often with the aid of methods of attempts and errors, appealing for analogy, generalization, called as 'applied creation', formulated by A. Góralski as: 'approach oriented on purpose, first of all, drying inspiration from requirements of practices of inventions and serve this purpose'.

Heuristics as a method of instructing, facilitating and inviting a student to discover knowledge in an active and independent manner. Socrates was a precursor of such understanding of heuristics (469–399 b.c.). He developed a method of dynamic dialogue consisting of critical and constructive element. He comprehended teaching not as devolution of knowledge, but as a process of investigating the truth.

One of unmistakable advantages of creative approach for solving project problems is that it does not only develop imagination of a planner but also promotes the inclusion of human factor criterion to a specific technical attitude of mind. It is necessary to remember that there is no algorithm which guarantees the fastest solution to problems. Without creative approach in solving problems the conditions of solving engineering problems will not emerge [6].

THE SYSTEM OF INNOVATIVE PRODUCT ORIGINATION AS A PRIMARY SOURCE OF THE INTEGRATED EB EDUCATION CONCEPT

The development of ideas for an innovative product needs to be subordinated to complex activities of teams of creative people to a systematic procedure (see: Figure 1). The need of recognition is a mental perception which initiates a product development project [9].

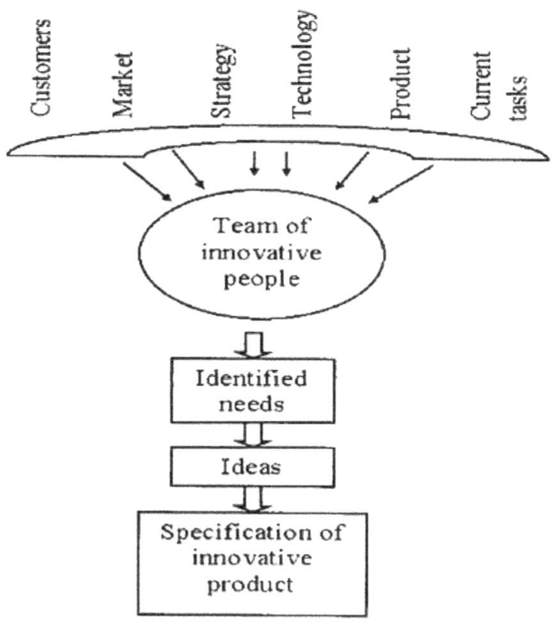

Fig. 1. The system of innovative product development [9]

Companies achieve competitive advantage through acts of innovation. They approach innovation in its broadest sense, including both new technologies and new ways of doing things [7].

We emphasize that innovative research in particular is a relatively young and unconsolidated field. At technical universities we need to prepare your students for the necessity of synthesizing the knowledge that exist in real world. Moreover, this knowledge must be useful for wider community and for various nations and cultures.

Tidd et al. describe four different kinds of innovation in engineering design processes:

- innovation to introduce or improve products,
- innovation to introduce or improve processes,
- innovation to define or re-define positioning of the firm or products,
- innovation to define or re-define the dominant paradigm of the firm [10].

VARIOUS PERSPECTIVES ON INNOVATIVE PRODUCT CHARACTERISTICS, AS BASIS FOR INTEGRATED EB EDUCATION CONCEPT

While the understanding of knowledge which engineers use in practice is rapidly growing, there is still much to be learned from exploring the contribution which related disciplines can offer. Since each discipline has its own way of developing its knowledge base, it is important for engineering science and practice to grasp the underlying manner in which they tackle their work. EB concept must bring together the material about many of these disciplines, which is normally scattered amongst a wide range of sources. It must be presented in a manner which is both accessible and relevant to engineering community for its practice and education. It will help engineers in practice and scientists to identify the ways in which the knowledge they acquire through study or experience can be used constructively to develop the service they offer to their clients.

The concept has the following assumptions:

- It is directly related to engineering practice and education at technical and corporate universities.
- Specialist contributors must explore complex ideas and present them succinctly and clearly
- We need a way in which the theory which is taught and the theory which is used in practice can be brought together to improve engineering practice and technical science.

Our concept will offer insight into some of the fundamental changes which need to occur, especially in Poland as EU member, for engineering attitudes and thinking. It will provide stimulating reading and discussion material for practice engineering, technical universities professors etc. It will be useful for teachers and instructors working in degree and post-registration courses, as well as for practitioners who wish to keep at the forefront of a profession which is undergoing continuing and dramatic change.

According to Hansen and Andreasen one can distinguish three different perspectives on the product idea which we use as a basis for our integrated EB education concept;

- The engineering designer perspective.
- The business perspective.
- The consumer perspective [3].

Following the basis of a very short literature review we describe a subjective selection of some important research works about innovative product characteristics.

The designer perspectives

Asimov writes: "The first step in the study is to demonstrate whether the original need, which was presumed to be valid, does indeed have current existence, or strong evidence of latent existence".

He writes: "We assume that we have been given a primitive statement of needs. By primitive we mean that the statement represents opinion based mainly on causal observations, but unsupported by organized evidence. Such opinions are valuable as starting points when they come from people who have had the opportunity and have the ability to make observations and to temper them with considered judgment". And: "Effective needs we will define as those which have an existence in the market place, by virtue of consumers' willingness and ability to acquire the means for their satisfaction" [1].

According to Asimov the work needs to be strongly integratied with the ideas about their products and the market place, by virtue of consumers and clients.

Harrisberger writes: "A feasibility study is an iterative process – check and revise, then check again – until all possible alternative ideas for achieving the requirements of the design, are explored". And: "The ultimate goal is a product idea that has enough merit to be worth extensive design, development, and testing. However, it is also likely that a feasibility study will verify that the idea would not be practical in terms of cost or utility; that is, 'Sure we can do it – but is it worth the effort?" [4].

Ulrich and Eppinger write: "Needs are largely independent of any particular product we might develop; they are not specific to the concept we eventually choose to pursue". They write: "A product concept is an approximate description of the technology, working principles, and form of the product. It is a concise description of how the product will satisfy the consumer needs. A concept is usually expressed as a sketch or as a rough three-dimensional model and is often accompanied by a brief textual description" [11].

We stress, that Ulrich and Eppinger approach engineering concepts based on strong integration between engineers' ideas and which must satisfy the consumer needs.

The business perspectives

In the USA Cooper described and recommended a state-gate model of the design process. The spirit of stage 1 is to 'spend a little money, gather some information, so that the project can be reevaluated at gate 2 in the light of better information'. Therefore, this first stage is a quick and inexpensive assessment of the technical merits of the project and its market prospects. Preliminary market, technical and financial assessments constitute stage 1.

Cooper writes that an idea occurs when technological possibilities are matched with market needs and expected market demand. Ideas may be generated by the marketplace – a recognition of unsatisfied customer needs, direct request from customers, or a competitive product. Such market-pull ideas represent the source for the majority of new product projects. But technology-push ideas – which are generated by researchers, science or technology, or result from serendipitous discoveries – also play an important role, particularly in radical innovations or breakthrough products [2],

We stress, that in the approach called 'Total design' market needs and demands are very strongly integrated with engineering activity and have priority before the concept of engineering design.

Leifer et. al. define a radical innovation project, which is the one with the potential to produce one or more of the following: an entirely new set of performance features; improvements in known performance features; of five times or greater; a significant (30 percent or greater) reduction in cost.

He writes that in some cases, radical innovation starts with a technical idea or a set of technical ideas. The idea may be born out of natural curiosity of a scientist or engineer, or be stimulated by a challenging problem. The technical idea can take a form of a discovery of novel technology, new insight into an old problem, or a new way of linking the existing technologies. In other cases, radical innovation has its roots in a market need, an industry 'Holy Grail' (a great and unsolved challenge in the company's industry), or the strategic vision of the firm's leadership. Each of these can catalyze the development of technical ideas with a potential to be breakthroughs. Because ideas can come from so many sources, noticing them is difficult. Many are missed for lack of an alert receiver.

In the following text Leifer et. al. clearly describe the concept of strong integration between engineering and business integration. By analogy, this idea is very important for education of engineers in XXI century – century of turbulent market economy.

For a radical idea to move forward, someone must recognise its business potential, in other words, make a connection between a novel technical solution and a compelling market need [5].

The consumer perspective

Creusen focuses on consumers' 'responses to product appearance'. In many purchase situations product appearance determines the consumers 'impression', and based on this impression, consumers reason about other product attributes, e.g. performance and quality. Creusen distinguishes six roles of product appearance: attention drawing, categorization, communication of practical, ergonomic, hedonic, and symbolic product information. The roles are not independent, and consumers evaluate product appearance in a holistic way into accounting a global product appearance, rather than the way the specific functions are designed. Tidd et. al. discuss buying behavior of consumers. Two types of theories describing consumer behavior are presented: utilitarian theories and behavioral approaches. Utilitarian theories assume that a consumer is rational and make purchase decision based on a comparison of the product utility with consumer's requirements. However, such a rational process is seldom seen in practice [10].

Henry Mintzberg was the first one who regards the most crucial part of managerial activity (Table 1) as that concerned with making decisions. He divided the managerial activity into three categories: interpersonal relationship, information processing and decision making.

For Henry Mintzberg the roles that he places in the managerial activity roles are based on different classes of decision, namely: entrepreneur, disturbance handler, resource allocator, negotiator.

As enterpreneurs, managers make decisions about changing what is happening in an organization. They may have to both initiate change and take an active part in deciding exactly what

Table 1. Managerial roles [9]

Interpersonal	Informational	Decisional
Figurehead	Monitor	Entrepreneur
Leader	Disseminator	Disturbance handler
Liaison	Spokesperson	Negotiator
–	–	Resource allocator

is to be done. In principle they are acting voluntarily. This is very different from their role as disturbance handler, where mangers have to make decisions which arise from events beyond their control and unpredicted. The ability to react to events as well as to plan activities is an important managerial skill in Mintzberg's eyes.

The resource allocation role of a manager is central to much of the organizational analysis. Clearly, a manager has to make decisions about the allocation of money, people, equipment, time and so on. Mintzberg points outs that in doing so a manager is actually scheduling time, programming work and authorizing actions. The negotiation role is put in the decisional category by Mintzberg because it is 'resource trading in real time'. A manager has to negotiate with others and, in the process, be able to make decisions about the commitment of organizational resources.

For Mintzberg these roles provide a more adequate description of what managers do than any of various schools of management thought. In these roles it is information that is crucial; the manager is determining the priority of information. Through the interpersonal roles a manger acquires information, and through the decisional roles it is put into use. The scope for each manager to choose a different blend of roles means that management is not reducible to a set of scientific statements and programs. Management is essentially an art and it is necessary for managers to try and learn continuously about their own situations. Self study is vital. According to Mintzberg 'the management school has been more effective at training technocrats to deal with structured problems than managers to deal with unstructured ones [8].

We emphasise that according to Mintzberg's text we have no suitable management theory for teaching our students, he need to try and learn on basis own experience.

Mintzberg presents a way of understanding the design of organizations and suggests that there are seven types. As shown in Table 2, the first five types are differentiated according to which the basic part of the organization forms the key to its operations. In the entrepreneurial organization it is the strategic apex which is key. In a manufacturer, for example, this would be the President of Chief Executive, the Board of directors, and their personal staff. In a machine organization, it is the 'technostructure' which is key, this includes those in planning, finance, operations research and work study and production scheduling. The key part in a professional organization is the 'operating core', those at the working base of the organization. While in a manufacturer this would be the buyers, machine operators, sales people and dispatchers, in a professional organization it might be doctors and nurses (in a hospital) or teaching staff (in a college). It is important for mangers to understand the configuration of their particu-

Table 2. Seven organizational types according to H. Mintzberg [8]

Seven Organizational Types			
Organizational configuration	Prime coordinating mechanism	Key part	Type of decentralization
Entrepreneurial	Direct supervision	Strategic apex	Vertical and horizontal centralization
Machine	Standardization of work processes	Technostructure	Limited horizontal decentralization
Professional	Standardization of skills	Operating core	Horizontal decentralization
Diversified	Standardization of outputs	Middle line	Limited vertical decentralization
Innovative	Mutual adjustment	Support staff	Selected decentralization
Missionary	Standardization of norms	Ideology	Decentralization
Political	None	None	Varies

lar organization in order to ensure that the various parts 'fit together' and are consistent in what they do. But, Mintzberg warns against forgetting that there will always be contradictions among the forces in organizations. Managers should use these contradictions creatively rather than ignoring or trying to suppress them [8].

CONCLUSIONS

1. An engineer is a man interpreted as a creator and user of technical products.
2. Without knowledge of heuristics and knowledge management the condition will not emerge for achievement of progress in engineering education systems.
3. Basis for integrated engineering and business (EB) education system is subordination the complex (engineers and business people) teams to a systematic procedures.

REFERENCES

1. Asimov M., Introduction to design. Prentice – Hall Inc., Englewood, N.J., 1962.
2. Cooper R.G., Winning at new products – accelerating the process from idea to launch. 2nd ed., Addison-Wesley Publ. Co., 1993.
3. Hansen C.T., Andreasen M.M., On the content of a product idea. In: Proceedings of 14th International Conference on Engineering Design (ICED 03), Stockholm 2003.
4. Harrisberger L., Engineersmanship – a philosophy of design. Brooks/Cole Publ. Co., Belmont, Calif., 1966.
5. Leifer R., McDermott C.M., O'Connor G.C., Peters L.S., Rice M.P., Veryzer R.W., Radical innovation: how mature companies can outsmart upstarts, Harvard Business School Press, Boston, Massachusetts 2000,
6. Marcinkowski J.S. et al., Ergonomic education at the technological university. Publishing House of Poznan University of Technology, Poznan, 2012.
7. Porter M., The competitive advantage of nations. McMillan, London 1990.
8. Pugh D.S., Hickson D.J., Writers on organizations. An invaluable introduction to the ideas and arguments of leading authorities on management. Fifth edition, Penguin Books, 1996.
9. Rohatynski R., Innovative product management in Innovation by Design. Ed. R. Rohatynski, Publ. House - Wroclaw University of Technology, Department of Mechanical Engineering, Wroclaw 2011.
10. Tidd J., Bessant J., Pavitt K., Managing innovation. John Wiley and Sons, 2005.
11. Ulrich K.T., Eppinger S.D., Product design and development. 3rd ed., McGraw-Hill Inc., 2004.

ANALYTIC EVALUATION OF RECTILINEARITY OF LOW RIGIDITY SHAFT DURING HARDENING PROCESS

Antoni Świć[1], Oleg Draczew[2], Wiktor Taranenko[3], Sadi Majdalawi

[1] The Institute of Information Technology Systems, Lublin University of Technology, ul. Nadbystrzycka 36, 20-618 Lublin, Poland, e-mail: a.swic@pollub.pl

[2] Togliatti State Technical University, Russia.

[3] Luck State Technical University, Ukraine.

ABSTRACT

The essential influence of the unevenness of temperature distribution while heating in the technological process on dimensions stability of low rigidity elements was shown. The new approach was applied to formulate mathematical models, which describe the elastic and inelastic behaviour of piece using transfer functions and block diagrams, allowing to use frequency method for evaluation of the behaviour of dynamic semi-finished element as the rigid body.

Keywords: non-rigid shaft, shaft axis rectilinearity, hardening.

INTRODUCTION

In the long shafts' manufacturing process, rolled semi-finished products which have a considerable curvature when supplied, are used.

Before undergoing machining process, which allows for obtaining a shaft confirming with the assumed tolerance for curvature of its axis, the semi-finished product must undergo straightening process in order to gain the rectilinearity of 0.5 mm for 1 m of length [3, 5, 7].

The disadvantage of straightening by bending is the emergence of irregular residual stress in the longitudinal section of the shaft's semi-finished product. In the machining process, the stress leads to the emergence of bending moment and to the change in the rectilinearity of the shaft's axis [1, 2].

ANALYTIC EVALUATION OF THE RECTILINEARITY OF SHAFT'S AXIS DURING HARDENING PROCESS

Axial straightening is widely implemented. In the process of straightening of a cylindrical semi-finished product, all fibers reach the extremes of their plasticity, thus any preexisting axial residual stress is removed. During axial straightening all residual stress is removed when permanent deformation equals 0.5–1% [4, 6].

Research conducted on the heat-stress treatment device [2, 5], in case of axial straightening of shafts made of X10CrNi18-8 austenitic steel, is coupled with hardening process (to improve steel's corrosion resistance) but tempering is not required [7]. Austenitic steels possess a high thermal expansion factor and low level of plasticity, thus in order to induce axial stress of straightening after heating up the semi-finished product in the preparation for hardening, the product's extreme ends are fixed in a device made of steel whose thermal expansion factor is lower than the factor of the shaft's product. Cooling of the device is executed with the product fixed to the device. As a consequence of the difference in thermal expansion factors, during cooling process axial force emerges, which brings about axial straightening of shaft's semi-finished product. The aforementioned device consists of a steel 20 pipe with perforations enabling contact with the surrounding environment. Mountings are located at both ends of the device.

Qualitative axial straightening of shaft's semi-finished product is understood as the process where the result of the operation of axial straightening force is that none of the fibers in every cross-section of the product is under stress exceeding the limit of plasticity. During compression and then stretching, on the surface of any cross-section, stress not lower than the limit of plasticity emerges.

As a maximum input bend of the semi-finished product we understand such initial bend which, in the fiber undergoing extreme compression, brings about stress which does not exceed the limit of plasticity in the temperature corresponding to gaining rectilinearity by the curved axis. Otherwise, a compression cold work emerges in the semi-finished product's cross section, which is not removed in the further stretching process-Bauschinger effect. As a consequence, an asymmetric crumple zone emerges which corresponds to asymmetric distribution of residual stress, as regards axial symmetry of the cross-section, influencing shaft's rectilinearity. It was established that the bend of the long shaft semi-finished product's axis closely corresponds to the quadratic parabolic equation:

$$f(z) = y(z) = \frac{4y_{\max}}{l} \cdot z \cdot \left(1 - \frac{z}{l}\right), \tag{1}$$

where: y_{max} – maximum output bend of the semi-finished product, l – distance between centers of the extreme sections of the shaft, z – current bend value.

Maximum stress emerging in the extreme fiber during stretching process of shaft's bent axis to reach rectilinearity is described as:

$$\sigma_{\max} = \frac{y_{\max} \cdot F_{roz}}{W_x}, \tag{2}$$

where: F_{roz} – axial force, W_x – shaft semi-finished product's cross-section's strength.

F_{roz} value is defined in accordance with equation [1] whose transformation gives the following formula:

$$F_{roz} = \frac{(3\pi \cdot d^4 \cdot E)}{20l^2}, \tag{3}$$

where: d – shaft semi-finished product's diameter, E – Young's modulus.

Determining the relationship of l to d via K_0 the following was obtained:

$$y_{\max} \leq \frac{(5K_0^2 \cdot d \cdot \sigma_{02})}{24E}. \tag{4}$$

Due to the fact that maximum compression stress in the compressed fiber emerges during shaft's cooling by temperature $\Delta T°$ corresponding to shrinking to the value of maximum initial bend and also shift of σ_{02} along with temperature drop ($\sigma_{02} = \sigma_{02}(t)$, $E = E(t)$), (4) can be transformed into:

$$y_{\max} \leq \frac{5K_0^2 \cdot d \cdot \sigma_{02}(t - \Delta T°)}{24E(t - \Delta T°)}. \tag{5}$$

While changing temperature by $\Delta T°$, its length will decrease by:

$$\Delta l = \Delta\alpha(t) \cdot l \cdot \Delta T°, \tag{6}$$

where: $\Delta\alpha(t)$ – disparity of thermal expansion factors of device's material and shaft's semi-finished product functionally related to temperature.

Shaft's shortening Δl during cooling ought to allow for achieving rectilinearity of the bent axis, i.e.:

$$\Delta l = \delta_1 + 2\delta_2, \tag{7}$$

where: δ_1 – distance required for the "selection" of shaft's semi-finished product's bend, defined as:

$$\delta_1 = \frac{8 \cdot \delta_{\max}^2}{3l},$$

δ_2 – backlash between the head of the device and the head of the abutment equal to:

$$\delta_2 = \frac{2d_w \cdot \delta_{\max}}{l}$$

where: d_w – diameter of abutments.

The relationship of the nut's diameter to the diameter of the product was designated as C, from (4) $\Delta T°$ was defined:

$$\Delta T° = \frac{1}{\Delta\alpha(t) \cdot l^2}\left(\frac{8}{3}\delta_{\max}^2 + 4C \cdot d \cdot \delta_{\max}\right). \tag{8}$$

Substituting (8) to (5) and solving δ_{max}, maximum initial bend of the product was determined as regards its geometry and physical characteristics of the material.

In particular, in case of shaft's semi-finished product made of X10CrNi18-8 steel and the device made of steel 20, they are defined as the following:

$$\Delta\alpha = 1{,}43 \cdot T^{-0{,}205}, \quad E = 2688 \cdot T^{-0{,}075}, \quad \sigma_T = 84{,}1 \cdot T^{-0{,}314}.$$

When comparing values of the initial qualitative bend with the data regarding real semi-finished products made of X10CrNi18-8 steel with the diameter of 20–80 mm and 30/80 length/ diameter ratio, it was determined that semi-finished products whose initial bend is larger than initial qualitative bend exist. Therefore, before conducting long shaft's straightening, shaft's ought to be selected according to their initial bend. In case of X10CrNi18-8 steel shafts, semi-finished products of 1050 mm length and 30 mm diameter, their initial qualitative bend ought to amount to no more than 3.4 mm. However, semi-finished products whose bend equals 6mm and measurements correspond to those above exist.

Semi-finished product's bend which can be made rectilinear by heat-stress processing, provided that such straightening is applied to the product whose initial bend does not exceed the maximum qualitative bend, is defined as the following: when cooling, the shortening of the product ought to compensate for to sum of elongations

$$\Delta l = \delta_1 + 2\delta_2 + \delta_3 + \delta_4, \tag{9}$$

where: δ_3 – shortening of the device as a result of implementing a load from the shaft's shortening semi-finished product which equals:

$$\delta_3 = \frac{\sigma_{02} \cdot l}{E_{prz} \cdot K_s},$$

where: σ_T – the limit of semi-finished product's plasticity at 20 °C, E_{prz} – elastic modulus of the device's material at 20 °C, K_s – device's rigidity vs. semi-finished product's rigidity ratio, δ_4 – elongation which ought to be applied to the product so that extreme fibers undergoing compression stress gain stress not lower than the limit of plasticity of the shaft's material at stretching equal to:

$$\delta_4 = \frac{(\sigma_k + \sigma_{02}) \cdot l}{E_{20^0}},$$

where: σ_k – stress in the extreme fiber during semi-finished product's maximum bend occurring during stretching to gain rectilinearity.

In accordance with (4) in order to establish σ_k:

$$\sigma_k = \frac{24d \cdot E_t \cdot \delta_{max}}{5l^2},$$

where: E_t – elastic modulus of the shaft's material in heating temperature.

Solving (9) as regards δ_{max} the following was obtained:

$$\delta_{max} = -d\left(0.9\frac{E_t}{E_{20}^0} + \frac{3}{4}C\right) + d\frac{1}{2}\sqrt{\left(1.8\frac{E_t}{E_{20}^0} + 1.5C\right)^2 + 1.5K_0^2\frac{\sigma_{02}}{E_{20}^0}\left(\frac{\Delta\alpha \cdot t \cdot E_{20}^0}{\sigma_T} - \frac{E_{20}^0}{E_{CT}K_S} - 1\right)}, \tag{10}$$

where: T – heating temperature, $\Delta\alpha$ – disparity of thermal expansion factors of device's material and shaft's semi-finished product in temperature range of 20 °C to heating temperature.

In order to improve precision and stability of the part's geometric shape of low rigidity, a method coupling straightening and heat treatment was devised. The utilization of the method allows for axial straightening of shaft's semi-finished products during their heat treatment, and improvement of the initial bend of the product as a result of fibers stretching in cross-section up to the limit of plasticity and forming symmetrical residual stress in the product.

SUMMARY

Universal equations describing the dynamic system of heat-stress treatment of axiosymmetric parts of low rigidity including changes in their geometric parameters were obtained and analyzed. These relationships allow for the evaluation of secondary defects of elasticity.

A considerable influence of irregularity in the distribution of heating temperature in the technological process on forming spatial stability of low rigidity parts was shown. Application of apriori information does not provide basis for calculations with the required precision due to a considerable spread of physical and mechanical properties and geometric parameters of semi-finished products.

Relations of input and output parameters of the controlled subject – a dynamic system, are defined with the use of linear differential equations with constant quadratic and higher coefficients.

Studies in axial straightening of shafts coupled with hardening process (implemented in order to improve corrosion resistance of steel) were carried out in a device for heat-stress treatment device. Suitable relations describing the process were developed.

REFERENCES

1. Dračev O.I.: Tehnologiâ izgotovleniâ maložestkih osesimmetričnyh detalej. Sankt-Peterburg: Izdatel'-stvo Politehnika, 2005, p. 289.
2. Draczow O., Taranenko W.: Technologia modułowa wytwarzania osiowosymetrycznych wałów długowymiarowych. Zesz. Nauk. Politechniki Rzeszowskiej. Mechanika, 230(67), 2006: 47–50.
3. Hałas W., Taranenko G., Taranenko W., Świć A.: Modelowanie poszczególnych operacji procesu technologicznego wytwarzania wałów. Modelowanie inżynierskie, 6(37), 2009: 115–130.
4. Ratchev S., Liu S., Huang W., Becker A.A. Milling error prediction and compensation in machining of low-rigidity parts. International Journal of Machine Tools and Manufacture, 15(44), 2004: 1629-1641.
5. Świć A. Technologia obróbki wałów o małej sztywności. Wydawnictwo Politechniki Lubelskiej, Lublin 2009, p. 144.
6. Taranenko W., Świć A. Urządzenia sterujące dokładnością obróbki części maszyn o małej sztywności. Wydawnictwo Politechniki Lubelskiej, Lublin 2006, p. 186.
7. Taranenko G., Taranenko V., Szabelski J., Swic A.: Indentification and automation of shaft machining in elastic-deformable condition. Information systems architecture and technology, Model Based Decisions, Part IV, Wroclaw Uniwersity of Technolodgy, Wroclaw 2008, p. 203–223.

33

CALCULATION OF THE SHAPE OF TOOL CUTTING EDGE FOR PRODUCTION OF INVOLUTE WORM

Ján Kráľ Jr.[1], Ján Kráľ[2]

[1] Faculty of Mechanical Engineering, Centrum of Informatics, Technical University of Košice, Slovakia, e-mail: kral.jan@tuke.sk

[2] Faculty of Mechanical Engineering, Department of Technologies and Materials, Technical University of Košice, Slovakia. e-mail: jan.kral@tuke.sk

ABSTRACT

The article deals with theoretical analysis of the production of worm gears and also the calculation of the tool cutting edges for their production. Based on the design parameters of worm and worm wheel, it is possible to calculate the coordinates of the shape of the tool cutting edge. The exact calculation of the coordinates or parameters requires mathematical formulations and applying mathematical relationships we can transform the design of worm screw profile to the profile of tool. Worms of different types can be produced in various ways, e.g. with a lathe and the plane to set a tool is either axial or normal. In this article we consider their production on the lathe and tool setting in the normal plane. Axial profile of worm is a curve which is the intersection of worm surface and the plane passing through its axis, and normal profile is perpendicular in a selected point to worm profile. This normal profile in more accurately way characterizes the shape of the worm profile, therefore, the calculation of the shape of the tool cutting edge is designed with a tool set to the normal plane. Based on the analyses, the new computation equations were obtained, flowchart compiled and, according to a given regulation, the computer aided calculation of the coordinates can be implemented.

Key words: shape, edge, algorithm, worm gear, operations.

PREFACE

Worm gears can be produced in different ways, depending on the size of the transmission, number of gearing courses, pitch angle of worm profile, number of units produced, intended application, etc. The worm gears can be made as cylindrical worms with globoid gears, globoid worm with globoid worm gear and globoid worm with cylindrical worm gearing [2].

There are a few types of cylindrical worms with globoid worm gear wheels and they can be divided according to the shape of the side curve of the worm tooth in frontal plane. The shape of the side of the tooth is given by the location and shape of the tool applied to produce the worm gears. In the manufacture of worm gears, tools for their production may have shapes of cut edges, either straight or curved.

Cylindrical worms with straight tooth flank can be spiral, general, involute and concave ones. The screw is spiral when the side curve of the tooth in frontal plane is Archimedes spiral. The worm is spiral if the curve of the side of the tooth in frontal plane is Archimedean spiral.

General worms are those with linear tool cutting edges (generating line of the side of the tooth) located in the normal plane passing through the center of tooth gap during manufacturing.

Involute worms are those with an involute curve of the side of tooth in the frontal plane. Cylindrical worms that are manufactured by tools with curved cutting edges are worms with concave (convex) side of the tooth profiles.

Worms can be roughly machined by milling cutters of disc, shank pin, hobbing types, rack-shaped cutters, slotting heads, single-bit cutters and on a lathe by shaped turning tools. Finishing operations are mainly performed on grinding machines using grinding tools of shrank, straight, straight cup or disk types [1].

The shape of the side surface of the worm depends on the shape of the tool profile and its settings due to the worm to be machined. According to the shape of the side curve of the tooth of worm in frontal plane we distinguish worm gears with toothing of worm:

- spiral worms, where the shape of the side curve is Archimedean spiral,
- general worms, where the shape of the side curve is a general curve,
- involute worms where side curve is an involute,
- concave worms, the shape of the side curve is the part of the circle (concaved on the side of the tooth) [4].

BASIC DEFINITIONS FOR SURFACES OF WORM

The term "worm motion of point (curve, ...) in the space" means the motion composed of rotation with constant angular velocity around the fixed axis and translation with constant velocity in the direction of this axis. This axis can be called the axis of a worm motion.

Considering the arbitrary curve in the space, under term "the worm area determined by this curve", we understand the surface generated by worm motion of this curve around the considered axis. This axis can be denoted as the axis of the worm surface.

Under term "axial profile of worm surface" we understand the curve that is the intersection of this surface and the plane passing through its axis.

The term "the front profile of worm surface" means the curve that is the intersection of this surface and the plane perpendicular to this axis.

Under term "the thread of worm surface" we understand a curve which is the result of a worm motion of particular point on worm surface.

Under term "the normal profile of worm surface" we understand a curve that is the intersection of this surface and the plane perpendicular to the thread. This definition is ambiguous. In particular cases, it is necessary to define which thread is considered. Worm surface can be defined as a surface generated by worm motion of the considered profile.

EQUATIONS OF WORM SURFACE

To derive the equations of worm surface, consider a right-handed rectangular coordinate system (0, x, y, z) connected to this surface the way that the axis of the worm surface is identical to the axis z, the axial profile is lying in the plane y = 0 and the frontal section in the plane z = 0.

Considering the curve by its worm motion the corresponding worm surface is generated and can be expressed by parametric equations:

$$\begin{aligned} x &= x(u) \\ y &= y(u) \\ z &= z(u) \quad u \in \langle u_1, u_2 \rangle \end{aligned} \tag{1}$$

where: u – parameter.

Worm surface determined by the curve can be expressed by parametric equations:

$$\begin{aligned} x &= x(u).\cos\varphi - y(u).\sin\varphi \\ y &= y(u).\sin\varphi + y(u).\cos\varphi \\ z &= z(u) + p.\varphi \quad u \in \langle u_1, u_2 \rangle \ \varphi \in \langle \varphi_1, \varphi_2 \rangle \end{aligned} \tag{2}$$

Parameter φ is the angle of rotation around the z axis. The constant p is called the slope parameter, i.e. size of translation corresponding to rotation about one radian [6]. When we denote t as the distance between corresponding generated points at worm motion about angle $\varphi = 2\pi$ then:

$$t = 2.\pi.p \quad \text{or} \quad p = \frac{t}{2.\pi} \tag{3}$$

Equations of the worm surface determined by its axial profile. Consider that the axial profile of the worm surface is given (i.e. its section of plane y=0) and the parametric equations of the profile are:

$$\begin{aligned} x &= r \\ y &= 0 \\ z &= f(r) \quad r \in \langle r_p, r_h \rangle \end{aligned} \tag{4}$$

where: r – parameter that defines the distance of point of profile from the axis z.

Parameter r is a given function that defines the profile from r_p to r_h, it is so-called foot or head radius [4].

According to Figure 1 worm surface can be expressed by equations:

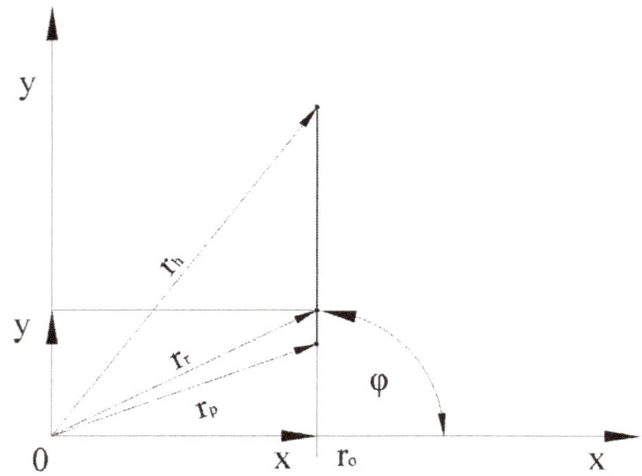

Fig. 1. Generating line of involute

$$x = r.\cos\varphi$$
$$y = r.\sin\varphi \qquad (5)$$
$$z = f(r) + p.\varphi\,;\ r \in \langle r_p, r_h \rangle;\ \varphi \in \langle \varphi_1, \varphi_2 \rangle.$$

FRONTAL PROFILE OF WORM SURFACE DETERMINED BY ITS AXIAL PROFILE

Consider a worm surface determined by its axial profile (4), i.e. surface determined by the equations (5). Its frontal profile is its section of the plane $z = 0$ (Fig. 2).

Each point of the frontal profile is given by parameters:

r – distance from the starting point,

δ – rotation around the z axis, where r is either a function of δ (i.e., $r = r(\delta)$, or δ is a function of r (ie, $\delta = \delta(r)$), as appropriate.

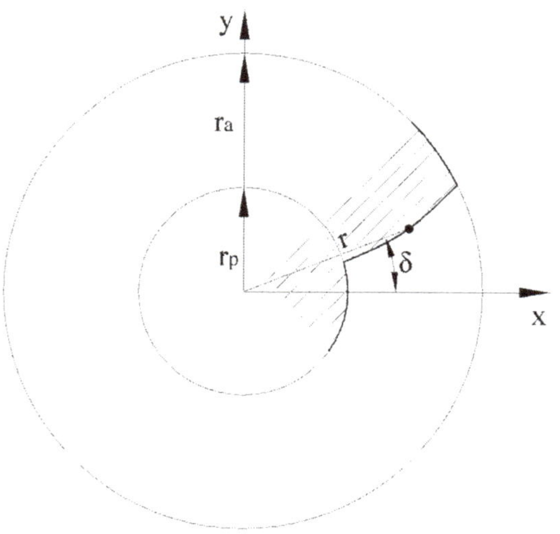

Fig. 2. Involute profile in frontal plane

Specific point of front profile is then given by relations (see polar coordinates). Concrete point of the front profile is given by equations:

$$x = r.\cos\delta$$
$$y = r.\sin\delta \qquad (6)$$
$$z = 0$$

Equations of worm surface given by its front profile according to (2) will be:

$$x = r.\cos\delta\cos\varphi - r.\cos\delta\sin\varphi = r.\cos(\delta + \varphi)$$
$$y = r.\cos\delta\sin\varphi + r.\sin\delta\cos\varphi = r.\sin(\delta + \varphi) \qquad (7)$$
$$z = p$$

According to (4) the axial profile must be:

$$r.\cos(\delta + \varphi) = \mathbf{r}$$
$$r.\sin(\delta + \varphi) = 0 \qquad (8)$$
$$p.\varphi = f(r)$$

From (8) we obtain:

$$\delta + \varphi = 0 \ \text{ or } \delta = -\varphi \qquad (9)$$

and

$$\varphi = \frac{f(r)}{p} \qquad (10)$$

From relations (9) and (10) we obtain:

$$\delta = \frac{f(r)}{p} \qquad (11)$$

or

$$r = f_{-1}(-\delta_p)$$

where: f_{-1} is inverse function to function f.

THE NORMAL PROFILE OF THE WORM SURFACE DETERMINED BY ITS AXIAL PROFILE

Consider a worm surface determined by its axial profile (Fig. 1) i.e. surface defined by equations (5). Under term "the normal plane" we understand a plane passing through the x axis and perpendicular to the thread, which is wound on the pitch cylinder of radius r_r in the point $(r_r, 0, 0)$. This thread is generated by worm motion of this point and has parametric equations [3]:

$$x = r_r.\cos\varphi$$
$$y = r_r.\sin\varphi \qquad (12)$$
$$z = p.\varphi$$

The point $(r_r, 0, 0)$ corresponds to the value of parameter $\varphi = 0$. Normal vector $\vec{n}$ of this plane is, therefore, identical withthe contact vector $\vec{t}$ of this thread at point $(r_r, 0, 0)$ and thus:

$$\vec{n} = \vec{t} = \left.\frac{dx}{d\varphi}\right|_{\varphi = 0} \vec{\imath} + \left.\frac{dx}{d\varphi}\right|_{\varphi = 0} \vec{\jmath} + \left.\frac{dx}{d\varphi}\right|_{\varphi} \vec{k} \qquad (13)$$

By substitution of derivations (12) we obtain:

$$\vec{n} = \vec{t} = r_r\vec{j} + p\vec{k} \tag{14}$$

Subsequently, considering the unit normal vector, the normal plane can be expressed:

$$ay + bz = 0 ,$$

where:

$$a = \frac{r_r}{\sqrt{r_r^2 + p^2}}$$

$$b = \frac{r_p}{\sqrt{r_r^2 + p^2}} \tag{15}$$

considering:

$$a^2 + b^2 = 1 \tag{16}$$

Normal profile of worm surface will be the curve which is the intersection of the surface (5) and the plane (15). We can mathematically express it with equations (5) and (15) excluding the parameter φ. By substitution (5) into (15) we obtain:

$$a.r.\sin\varphi + b(f(r) + p.\varphi) = 0 \tag{17}$$

For each fixed value of parameter $r \in \langle r_p, r_h \rangle$, the (17) is a transcendent equation for the unknown parameter $\varphi = \varphi(r)$. This equation can be solved only approximately, e.g. using Newton's method. If we calculate approximate angle φ for a given $r \in \langle r_p, r_h \rangle$, the corresponding point of normal profile, lying on a cylinder of radius r is given by relations (5). If we introduce the tool coordinate system $(0_n, x_n, y_n)$ in the corresponding normal plane the way that and axis y_n completes this system as rectangular and also clockwise; then, the point of normal profile will have the coordinates $[x_n, y_n]$, as follows:

$$x_n = x$$

$$y_n = a.z - b.y \tag{18}$$

where: (x, y, z) are calculated coordinates of point of normal profile in the coordinate system (0, x, y, z).

Flowchart for calculation of the shape of tool cutting edge for production of the worm for normal profile of involute worm gearing

SUMMARY

In this paper the formulas for calculation of the shape of tool cutting edges for production of worms are given, assuming the tool setting in the normal plane. The proposed equations allow the automated calculation in any available programming language. The coordinates of the points of tool cutting edge shape are calculated and based on them the tools will be sharpened. Considering

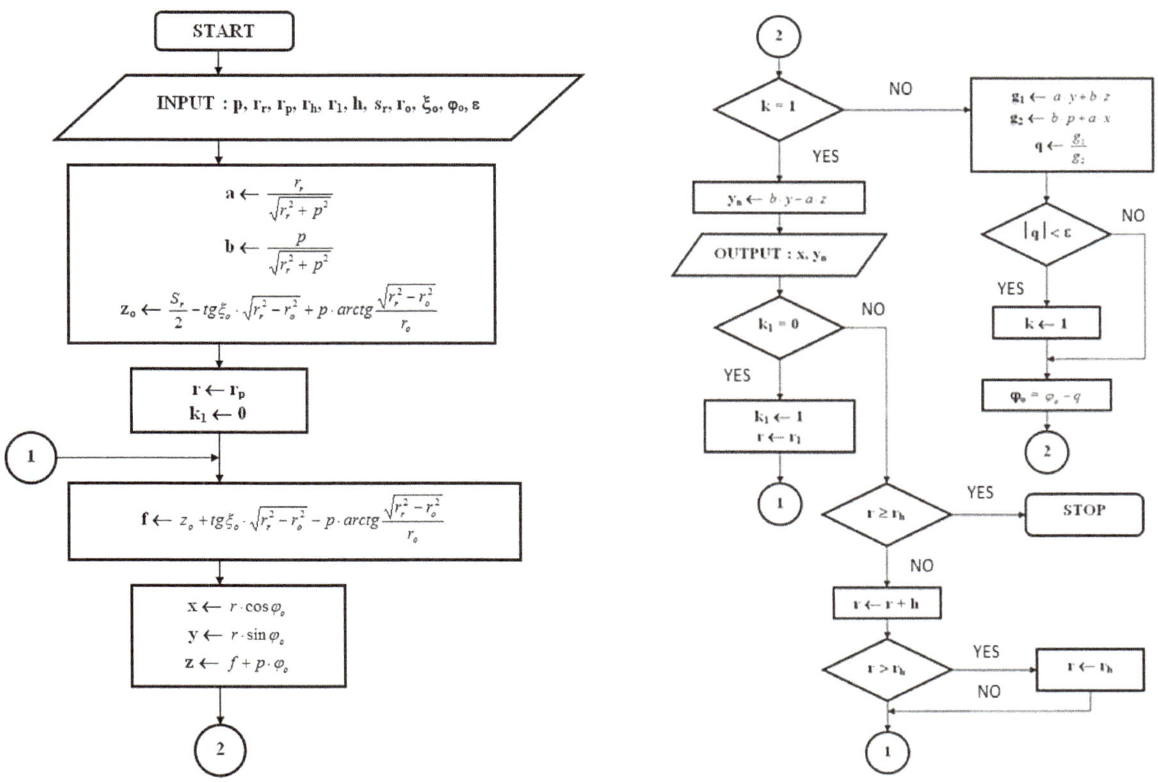

Fig. 3. Flowchart for normal profile

the possible technological deviation of tool profile in the production, it is recommended the calculation are rounded to more decimal places for production for more accurate profiles. The shape of the tool cutting edge is designed for tool setting in the normal plane.

REFERENCE

1. Janáč A. at al. 2004. Technológia obrábania. STU, sig: 11330, Bratislava.
2. Kráľ J. 1994. Výpočet, konštrukcia a výroba rezných nástrojov pre závitovkové prevody pomocou výpočtovej techniky. Habilitačná práca. TU Košice, p. 102.
3. Kráľ J. 1988. Výskum určovania dotykových kriviek a záberových pomerov pri spoluzábere závitovkových súkolesí: Návrh a realizácia riadiaceho systému pre nástrojovú brúsku KON 250 CNC. TU, SjF, Košice, p. 22.
4. Kráľ J., Řehoř, J., Spišák E., Kráľ J. Jr. 2009. Technologické a informačné činitele obrábania. Košice.
5. Kráľ J. at al. 2007. Programovanie NC strojov. 1. vyd. TU, SjF,. Košice, p. 181.
6. Řasa J. at al. 1986. Výpočetní metódy v konstrukci řezných nástrojů. STNL, Praha.

34

COMPARATIVE STUDY OF SELECTED LUBRICATING PROPERTIES OF BIOFUELS

Leszek Gardyński[1], Jolanta Kałdonek[2]

[1] Department of Material Engineering, Lublin University of Technology, 36 Nadbystrzycka Str., 20-618 Lublin, Poland, e-mail: l.gardynski@pollub.pl

[2] Department of Material Engineering, Lublin University of Technology, 36 Nadbystrzycka Str., 20-618 Lublin, Poland, e-mail: jolanta.kaldonek@gmail.com

ABSTRACT

The paper presents the research results of lubricity of selected vegetable oils, rapeseed oil methyl esters, and esters with addition of oleic acid. Higher wear of samples during lubrication by rapeseed oil methyl esters in comparison to the tested vegetable oils was obtained. The addition of oleic acid to esters resulted in the improvement of their lubricating properties.

Keywords: vegetable oils, rapeseed oil methyl esters, lubricating properties.

INTRODUCTION

Ability to limit access to the ON (shortage or depletion of oil in the world), as well as political issues and environmental hazards associated with the extraction and processing of oil, inspired the search for alternative fuels of vegetable origin to power diesel engines. To supply diesel engines (CI) most commonly used are biofuels the following: vegetable oils, fatty acid alkyl esters and bioethanol [1].

The world produces mostly soybean and palm oil, and rapeseed oil, along with sunflower oil, is only on the third-fourth place. [2]

The work carried out on the use of vegetable oils as fuels depends on the geographic region. In temperate climates soybean and sunflower oil are tested, in the northern temperate zone other oils are tested: rapeseed, linseed and corn oil, and in warm climates: palm, peanut, cotton, sesame and coconut oil [1].

Vegetable oils are the glycerol esters and fatty acids of chemical nature containing from 14 to 22 carbon atoms in the molecule. Fatty acids are composed of carbon chains of different lengths and different number of double bonds between the carbon atoms [3].

In the studies described in [4], it was observed that the lubricity of pure fatty acid compounds increases to some extent with the length of the carbon chain. However, greater improvement in lubricity was achieved with the increased number of double bonds than in the case of increasing the chain length. Triglycerides of oleic acid also showed a better lubricity than the corresponding methyl ester, in contrast to literature reports [5]. The carboxyl group is likely to be most effective in improving the lubricity [4]. Electrons of double bonds at the end of the carbon chain are also very effective in improving lubricity [6]. The sequence of oxygenated groups to improve lubricity according to [4] is as follows: COOH> CHO> OH> $COOCH_3$> C=O> C-O-C. Improved lubricity caused by COOH and OH groups correlates with the known observation of ionic interaction of the metal substrate with the lubricant molecules caused by hydrogen bonds and the Debye orientation forces, which are much stronger than the interaction based on the van der Waals forces [7].

The lubricating properties of alkyl esters are lower, because it does not give the ionic interaction with the metal, because of the absence of free OH groups. In the tests described in [4] is pointed out, that the lubricity decreases with decreasing number of OH groups, and that the addition of free fatty acids for fuels with low lubricity improves lubricity, which is consistent with literature reports [8, 9, 10]. Geller and Goodrum [11] in 2004 formed the initial correlation between the degree of unsaturation and lubricity. In a number of investigated fatty acid methyl ester: stearic acid ester (C18: 0), oleic acid ester (C18: 1), linoleic acid ester (C18: 2) and linolenic acid ester (C18: 3), with the increase of the degree of unsaturation, the increase of lubricity was observed. However, the research [12] shows that with the increase of the degree of unsaturation of fatty acids in sunflower oil, the lubricating properties of the oil decreased.

This paper presents the results of lubricity of selected vegetable oils, rapeseed oil fatty acid methyl esters and esters with oleic acid added for potential use as a fuel or fuel additives for petroleum products.

MATERIALS AND METHODS

The study was conducted on a test stand presented in detail in [13]. The method of testing was to simultaneously wear three samples (*Φ* diameter about 5 mm) by means of a rotating counter-sample in a form of a flat ring, in the lubrication conditions with the test agent at the temperature of 333K, by friction of about $2 \cdot 10^4$ m, at a pressure of 29.43 kN. Samples and counter-samples were made of steel 100Cr6 (H15SG), which is a very common material used for the production of injection equipment. The test agent during the test circulated in a closed loop, in the amount of 30 dm^3, which was thermally filtered and stabilized. The measure of lubricating properties is the mass and surface loss of samples. The study determined the coefficient of friction.

Refined vegetable oils: sunflower, soybean, rice, peanut, corn and rapeseed (characteristics in Table 1), as well as fatty acid methyl esters of rapeseed oil Trzebinia and esters produced at Lublin University of Technology were used in the research. The rapeseed oil methyl esters Trzebinia with the addition of free fatty acid in the form of oleic acid were also tested.

RESULTS

First of all, the vegetable oils were subjected to the tests. The result charts concerning the lubricating properties of these oils are presented below.

The graphs in Figure 1 indicates that the soybean has got the best lubrication properties, measured by the degree of wear of the steel among the tested vegetable oils. The biggest loss of mass and the largest surface of trace of cooperation was observed for the sunflower oil, which means the worst lubricating properties of the oil.

Then rapeseed oil methyl esters Trzebinia were tested to define their characteristics, which are shown in Table 2. The study was conducted in the laboratory of the University of Life Sciences in Lublin. Results of the lubricating properties of pure rapeseed oil methyl esters are summarized in Table 3.

The research shows that rapeseed oil methyl esters are characterized by poorer lubrication properties than the tested vegetable oils. A similar degree of wear of test samples in conditions of ester lubrication, regardless of manufacturer, was obtained.

The third group subjected to the tests consisted of rapeseed oil methyl esters Trzebinia with addition of oleic acid, for which the results are shown in Figure 2.

Rapeseed methyl esters with the addition of oleic acid have got better lubricating properties than pure esters. The addition of oleic acid in the

Table 1. The fatty acid content in vegetable oils tested, % of mass [1]

Oil types	C18: 0 stearic acid	C18: 1 oleic acid	C18: 2 linoleic acid	C18: 3 linolenic acid
Rapeseed oil	1.2–2.0	60–73	19–22	9–10
Soybean oil	2.1–6	22–34	53–60	2–10 8 [2]
Sunflower oil	2.9–4	17–18 22 [2]	71–74 66 [2]	0.5
Corn oil	1.5–4	30–50 28 [2]	34–56 58 [2]	1.0
Peanut oil	2.0–2.4	48–48.5	32–34	0.9–1.0
Rice oil [14]	1.9	42.5	39.1	1.1

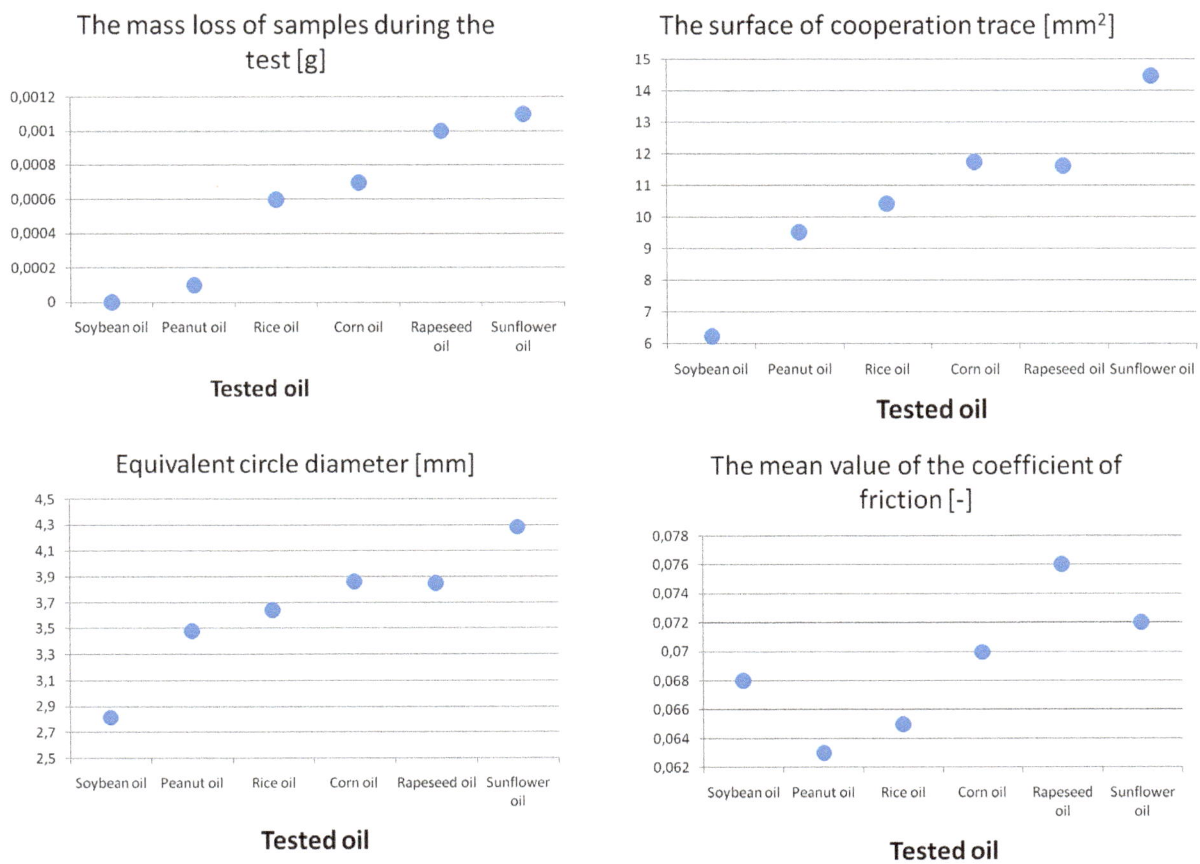

Fig. 1. Summary of the results of the lubricating properties of vegetable oils tested

Table 2. Characteristics of rapeseed oil methyl esters Trzebinia (Biodiesel B-FAME). Research date 23.03.2012

1.	Research parameter	Research method	Requirements according to standards	Result % (m/m)	Expanded uncertainty
2.	The content of fatty acid methyl esters	Procedure CLA-GC-17-2011 based on the PN-EN 14103:2004 standard	min. 96.5	91.95	15.34
3.	The content of linolenic acid methyl ester	Procedure CLA-GC-17-2011 based on the PN-EN 14103:2004 standard	max. 12	9.7	1.56
4.	Methyl alcohol content	Procedure CLA-GC-15-2011 based on the PN-EN 14110:2004 standard	max. 0.2	0.004	0.0005
5.	Total glycerol content	Procedure CLA-GC-13-2011 based on the PN-EN 14105:2004	max. 0.25	0	0

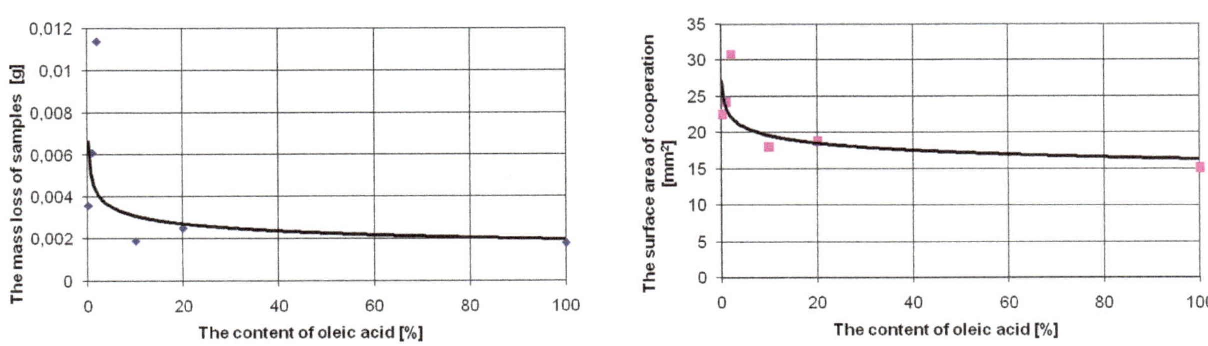

Fig. 2. Summary of the results of the lubricating properties of rapeseed oil methyl esters with addition of oleic acid according to [15]

Table 3. The results of lubricating properties of rapeseed oil methyl esters

The test agent	Loss [g]	The surface of trace of cooperation [mm^2]	Equivalent circle diameter [mm]	Final pressure [MPa]	The coefficient of friction [-]
ESTER I (Trzebinia)	0.0049	23.02	5.41	129	0.082
ESTER II (Lublin University of Technology 1)	0.0049	25.78	5.73	114	0.089
ESTER III (Lublin University of Technology 2)	0.0036	22.47	5.35	131	0.101

amount of several percent provides improved lubricating properties of rapeseed oil fatty acid methyl esters.

ANALYSIS OF RESULTS

Best lubricating properties among the tested vegetable oils were observed in the case of soybean oil, for which there has been almost no mass loss of the samples. The explanation of the observed effect of the best lubricating properties can be based on the correlation between the degree of unsaturation and lubricity, because soybean oil contains the highest number of unsaturated fatty acid esters among all of the tested oils.

It was found out that regardless of the manufacturer, the increased wear of samples during lubrication by pure fatty acid methyl esters of rapeseed oil was reported.

Free fatty acids [2] are present in the vegetable oils in small amounts (0.1–2%) that are acids containing COOH group, which according to the group sequences improving the lubricity [4] is the strongest one. The presence of the group COOCH3 determines the lubrication properties of esters, and so the group of further positions in the sequence. The obtained results of vegetable oils and pure rapeseed oil methyl esters are compatible with the above-mentioned sequence of oxygenated groups, which indicates that a small amount of free fatty acids provides lubricity of vegetable oils. An addition of free fatty acid (oleic acid) to methyl esters of rapeseed oil also had a positive effect on the lubricating properties and confirmed the accuracy of the sequence. Wherein, it is reasonable to add it in the amount of several percent (similar to the content in vegetable oils). Over 10% increase in the content of this addition will virtually have no effect on the result. Free fatty acids are, therefore, the materials to improve lubricity, but they are believed to be contamination and are limited by the standard.

CONCLUSIONS

On the basis of the conducted research, we can draw the following conclusions:

- Among the tested vegetable oils, the soybean oil is characterized by the best lubrication properties, which contains the highest number of unsaturated fatty acid esters.
- Increased wear of the samples in the lubrication conditions of rapeseed oil methyl esters in comparison to vegetable oils was obtained, which is consistent with the sequence of oxygenated groups improving lubricity.
- The addition of oleic acid to methyl esters of rapeseed oil ester increased lubricating properties of esters, confirming the validity of the said sequence.

REFERENCES

1. Baczewski K., Kałdoński T. Paliwa do silników o zapłonie samoczynnym. Wydawnictwa Komunikacji i Łączności. Warszawa 2008.
2. Szlachta Z. Zasilanie silników wysokoprężnych paliwami rzepakowymi. Wydawnictwa Komunikacji i Łączności. Warszawa 2002.
3. Górski W., Ostaszewski W., Wiślicki B. Krajowe oleje roślinne surowcem dla paliw silnikowych i olejów smarowych. Paliwa, Oleje i Smary w Eksploatacji, 2001, 90: 5-11.
4. Knothe G., Steidley K.R. Lubricity of Components of Biodiesel and Petrodiesel. The Origin of Biodiesel Lubricity. Energy & Fuels, 2005, 19: 1192-1200.
5. Van Gerpen J.H., Soylu S., Tat M.E. Evaluation of the Lubricity of Soybean Oil-Based Additives in Diesel Fuel. [In:] Proceedings of the 1999 ASAE/CSAE-SCGR Annual International Meeting, 1999, Paper No. 996134.
6. Barbour R.H., Rickeard D.J., Elliott N.G. Understanding Diesel Lubricity. SAE Tech. Pap. Ser. 2000, 2000-01-1918.
7. Liang H., Totten G.E., Webster G.M. Lubrication and Tribology Fundamentals. [In:] Totten G.E.,

Westbrook S.R., Shah R.J. (Eds.) Fuels and Lubricants Handbook, ASTM International: West Conshohocken, PA, 2003, p. 909-961.

8. Kajdas C., Majzner M. Boundary Lubrication of Low-Sulphur Diesel Fuel in the Presence of Fatty Acids. Lubr. Sci., 2001, 14: 83-108.
9. Kajdas C., Majzner M. The Influence of Fatty Acids and Fatty Acids Mixtures on the Lubricity of Low-Sulfur Diesel Fuels. SAE Tech. Pap. Ser. 2001, 2001-01-1929.
10. Kenesey E., Ecker A. Oxygen Compounds for Improvement of the Lubricity in Fuels. (In Ger.) Tribol. Schmierungstech., 2003, 50: 21-26.
11. Geller D.P., Goodrum J.W. Effects of specific fatty acid methyl esters on diesel fuel lubricity. Fuels, 2004, 83: 2351-2356.
12. Fox N.J., Tyrer B., Stachowiak G.W. Boundary Lubrication Performance of Free Fatty Acids in Sunflower Oil. Tribol. Lett., 2004, 16: 275-281.
13. Gardyński L. Stanowisko do badania odporności materiału elementów aparatury paliwowej na zużycie w warunkach smarowania. [In:] Materiały konferencji Silniki Spalinowe w Zastosowaniach Wojskowych SILWOJ'2005. Akademia Marynarki Wojennej, WAT, Warszawa 2005, p. 93-100.
14. http://sliwkirobaczywki.blogspot.com/2013/04/olej-ryzowy-wprowadzenie.html
15. Gardyński L., Hys L., Kiernicki Z., Sawa J. The effect of biofuel acid number on some operational and wear properties of DI diesel. [In:] Procedeengs of 6th International Colloquium Fuels, Stuttgart/Ostfildern, Germany 2007, p. 641-646.

EXPERIMENTAL INVESTIGATION OF A VARIABLE GEOMETRY DUCTED PROPELLER

Tomasz Muszyński[1], Piotr Strzelczyk[2]

[1] The State School of Higher Education in Chełm, Pocztowa 54, 22-100 Chełm, Poland.

[2] Faculty of Mechanical Engineering and Aeronautics, Rzeszow University of Technology, Al. Powstańców Warszawy 8, 35-959 Rzeszów, Poland, e-mail: piotstrz@prz.edu.pl

ABSTRACT

The paper presents preliminary results of experimental research on the variable geometry of a ducted propeller. The purpose of this work is explore the area of application of the ducted propellers of variable geometry. The paper contains the description of a test station and a model, initial tests' results of some selected geometries intakes and exit vents in comparison with an open propeller.

Keywords: ducted propeller, variable geometry, flow velocities.

INTRODUCTION

According to the research on the development of ducted propellers in research institutes and leading companies in aviation industry it was possible to use them for two ranges of velocity. The first velocity range is of over 200 m/s (0.6 Ma); fan jets have almost entirely overwhelmed the market of communication and military subsonic aircrafts. There are some designs of power plants with free propellers designed for such velocities, such as Piagio Avanti, An 70 with turbofan, Tu 95.

However, due to a number of limitations of this type of propulsion, they are in small numbers – i.e. a fraction of percent among all turbofan planes. Another range of speeds for ducted propellers is 0–40 m/s. Ducted propeller is frequently used in airships and hovercrafts. Presently there are not any large scale manned aircrafts flying under 0.6 Ma. The shroud for small and large velocities are designed on the basis of different criteria, thus, their nozzles are of different shape. The decelerating nozzles used for turbofans slow down the incoming stream and Ma number of blade tips what decreases efficiency as a result of compressibility. Moreover, a ducted propeller blade works the same way as the blade with endplate – extension is more effective, outer part of the blade can be much more loaded, larger thrust and power can be obtained from the propeller surface with significantly lower level of noise. Much smaller diameter of ducted propeller lowers the speed of blade tip. For lower speed, accelerating nozzle is used. Increased air velocity in the leading edge is accompanied by decreased pressure, what results in additional thrust. Thanks to additional thrust from the ring, ducted propeller generates higher thrust for the same power.

Numerous research that proved the increase of ducted propeller in comparison to free propeller have been conducted. Sources [1, 2, 3] state that the increase of static thrust for the ducted propeller, depending on design, is from a few percent for constant pitch propellers, up to 39% for controllable-pitch propellers. In present, constant geometry shrouds are used. Constant geometry of the shroud must be designed in the same way as constant pitch propeller as a compromise between take-off and in-flight conditions – for large and small propeller advance ratio.

PRESENTATION OF RESEARCH PROJECT

Alike in propellers a change of blade setting angle angle seems a natural solution, a change of shroud profile with increasing velocity seems to be natural for shrouds. Nevertheless the task is much more complicated than in case of propeller setting angle. It was assumed that changeable leading and trailing edges of the shroud cannot be used, as such geometry is not possible to obtain in currently known technologies. The first, and the most absorbing stage, as it turned out later, was to develop a feasible construction. The use of shape memory alloys was not taken into consideration, as they are still in the testing phase. A number of variable geometries of construction variants were considered for intakes and outlets of aircrafts with thrust vectoring and without it, and also the models existing in the nature. The design of variable outlet geometry was more time-consuming, as the solutions from supersonic aircrafts were impossible to use for lower velocities. Another problem was the control of geometry as such. The use of solutions typical for large velocities was too complicated, expensive and, in our case, useless, because for smaller velocities the strengths in the tunnel are much lower. The selection of geometric measurements of the model was made on the basis of statistics and the results from available research and reports. On the basis of statistics and the assumed conditions, the main measurements of the shroud were selected, i.e. the ration length / case diameter 1.1:1 and relative thickness of case profile – below 10%. Inner diameter was 306 mm, outer diameter in the propeller surface was 386 mm, the length of case profile chord was 420 mm (for zero dilation angle of intake and outlet). Other geometric parameters of the model were also analysed, e.g. the size of spacing between propeller's edge and the case wall. As on the basis on articles [4] and [6] it was concluded that the size of slit has a significant influence on propeller's efficiency, whereas in other articles, concerning the exploitation of hovercrafts, it was concluded that the slit cannot be too small because it generates a risk of damaging the propeller. Finally, the slit was accepted as 1% of propeller' blade radius. Initially, the case consisted of 128 elements; after numerous attempts it was simplified thanks to the use of biomimetics – overlapping and pressing "feathers" inspired by birds' tails. Finally, 64 elements were used. The system of case cover control consists of 32 ball joints connected with two independent rings – one in the intake and one in the outlet. The turn of the ring causes the change of connectors angles against the crosswise plane, which generates a change in dilation angle of case intake and outlet. The design and construction of the case geometry was the most laborious task so far. A number of attempts were made with paper, composites, aluminium for different geometries, with the use of different solutions for hinges and geometry control systems.

Both thrust and moment are measured with beam strain gauges. The drive was fixed to the inside part of the bearing. The possibility of turn was limited by weight balance. The measurement of pressure difference on airscrew disc was realised with a total head probe. The measurement of dynamic pressure with Prandl' probe. Pressure sensors were connected to a digital manometer. The measurement of rotational speed was done with a transoptor, temperature and humidity were measured with a portable mini weather-station.

a) b)

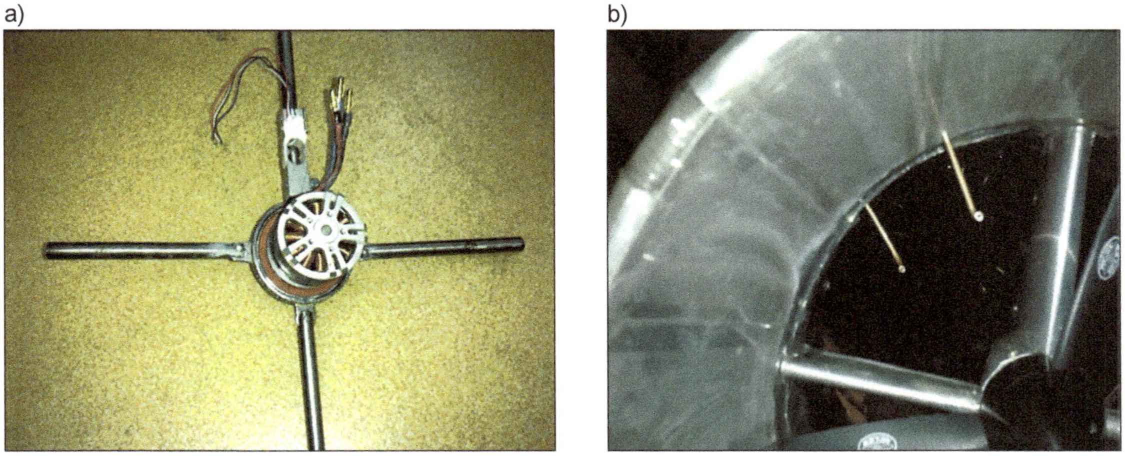

Fig. 1. a) drive fastening with strain gauge for moment measuring, b) measurement of total pressure jump at propeller disc

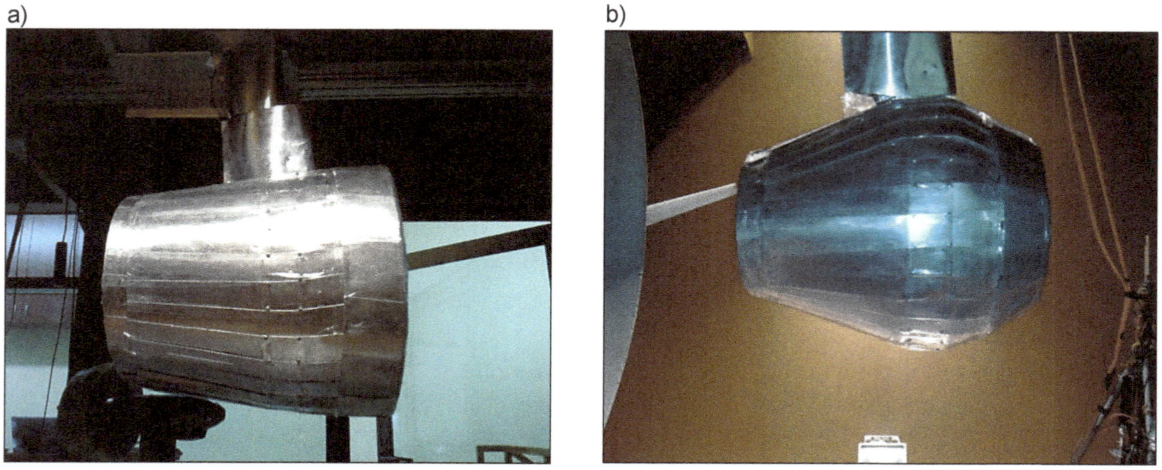

Fig. 2. The photographs present a ready shroud: a) shape appropriate for small flow velocities, b) shape appropriate for large flow velocities over 40 m/s

Initial characteristic measurements of the geometry were assumed as a ratio of intake surface to the surface of propeller actuator disc to the ratio of outlet to the surface of propeller working disc. The geometry marking 1.69/1.09 says that it is the geometry of inlet surface equal to 169% of the propeller disc and the outlet surface equal 109% of the propeller disc.

For the purpose of the experiment a computer measurement system was used in a module version operating in DasyLab environment. This allowed for conducting the experiment according to the requirements of modern metrology. The frequency of probing was 10 000 Hz, and the size of sample was 4096 points for each channel. Initial pilot measurements were made, and then 33 most promising configurations were selected. The range of intake regulations ranged from 238 mm to 405 mm, what makes from 60 to 177% of the propeller disc. The measurements were made within 10 days in 2 months (February – May), when pressure, humidity and temperature changed. The aim of the experimental research was to examine the influence of shroud geometry changes on aerodynamic characteristics of the power unit and to increase the usability of ducted propeller by the velocities that have been uneconomical so far or this type of drives without variable geometry. The investigation was conducted in aerodynamic tunnel TA 1000 in the Department of Fluid Mechanics and Aerodynamics in Rzeszów University of Technology.

RESEARCH RESULTS

The information was obtained about the aerodynamic characteristics of individual geometries, e.g. change of thrust coefficient for different outlet geometries for constant intake diameter and for reverse situation. The largest static thrust was obtained for the largest possible opening of average outlet dilation (geometry 1.75/1.16), what is in line with the theory of ducted propeller theory [1, 2, 7]. The increase of flow velocity in the front part of the tunnel, which is an airfoil, by the work of the propeller and advance velocity cause significant velocity increases in the leading edge, thus significant underpresure that gives a resultant pointed forward which generates an additional thrust. Slight broadening of the part of duct behind the propeller's disc is used to lower the velocity of outgoing air from the shroud, what makes the energy loss significantly lower. For geometry 0.67/0.63 the largest velocity of thrust zeroing was obtained, larger than the velocity thrust zeroing of the propeller itself.

The results of the experiments are still at the stage of processing, however, initial conclusions can be already drawn. For a given power unit, for individual geometries the relation between the ability to generate static thrust and thrust for large velocities is reversely proportional: the larger thrust increase, the larger resistance and lower total thrust of the drive for larger velocities with geometries generating the largest thrust. By changing the shroud geometry, its efficiency can be increased significantly.

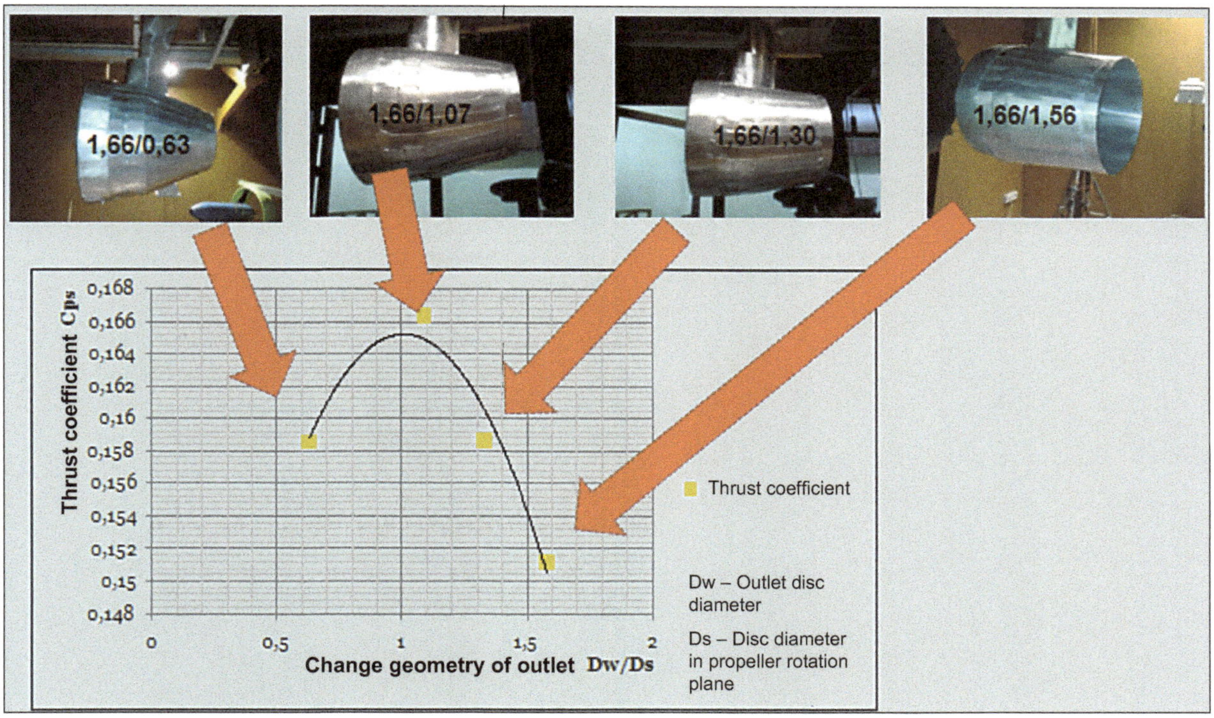

Fig. 3. The value of thrust coefficient for the changing outlet geometry of constant inlet diameter of 166% of the propeller disc for 5 m/s

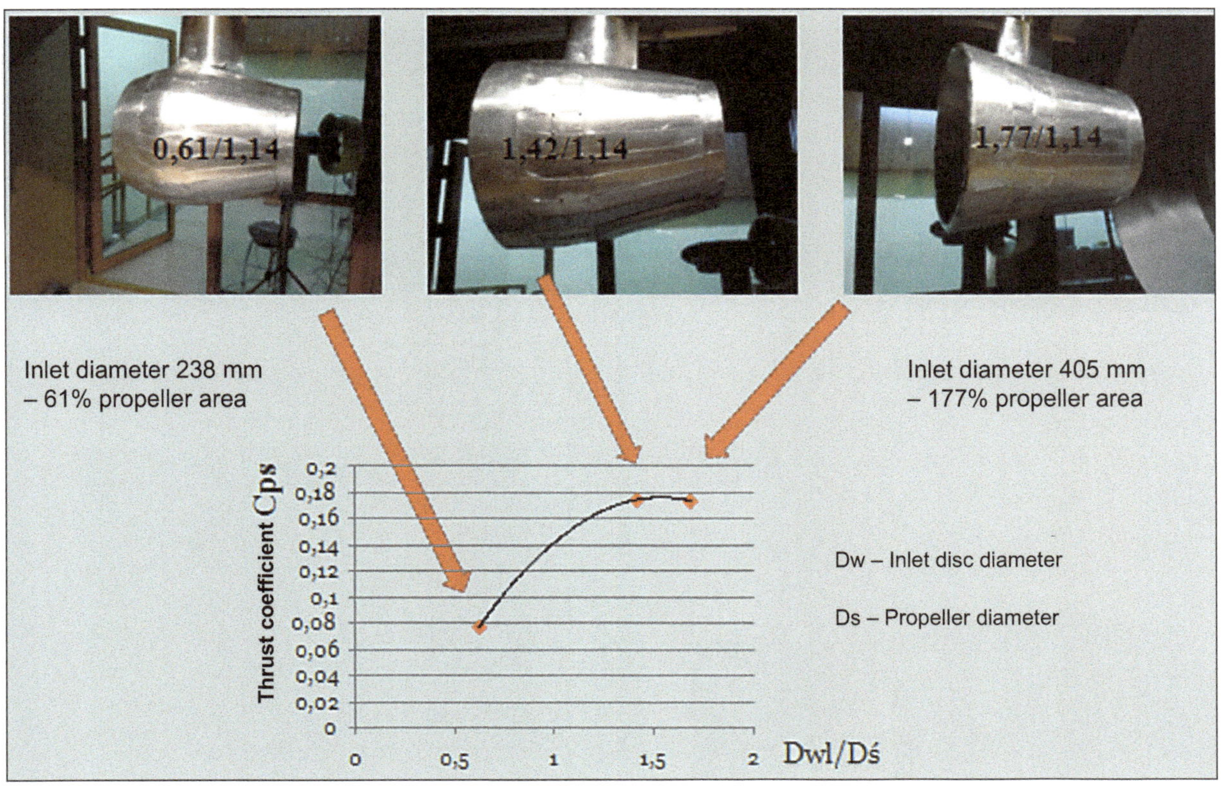

Fig. 4. The value of thrust coefficient for the changing inlet geometry of constant outlet diameter of 325 mm for 5 m/s

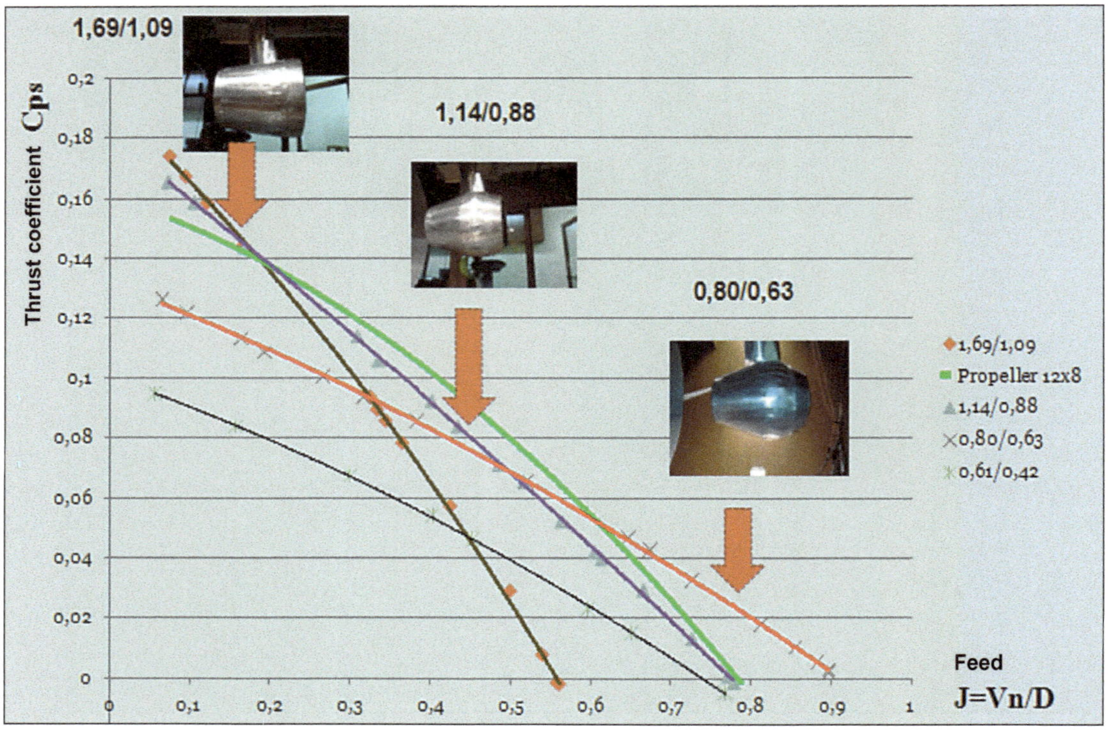

Fig. 5. The list of thrust coefficients and feed for selected geometries

a) b)

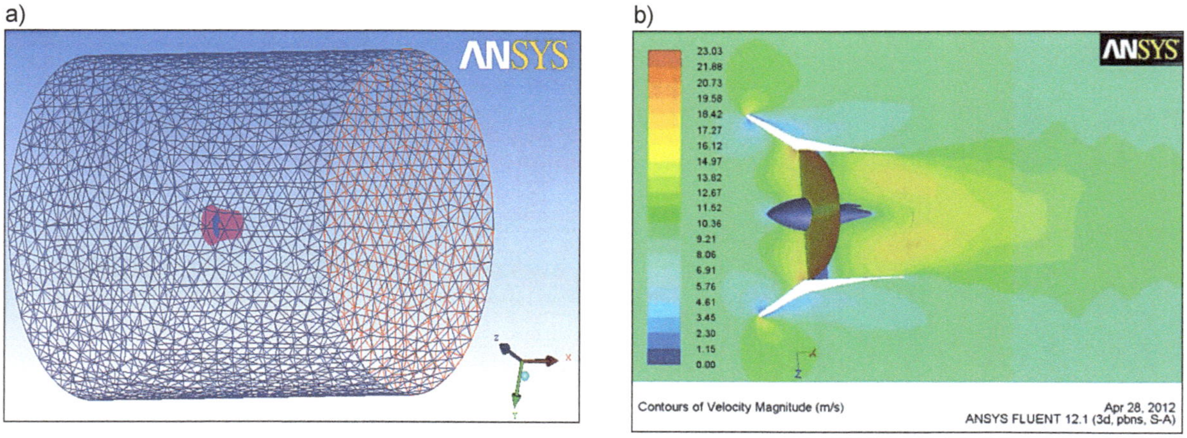

Fig. 6. a) Calculation domain with the model, b) distribution of velocity for geometry 1.69/1.05

a) b)

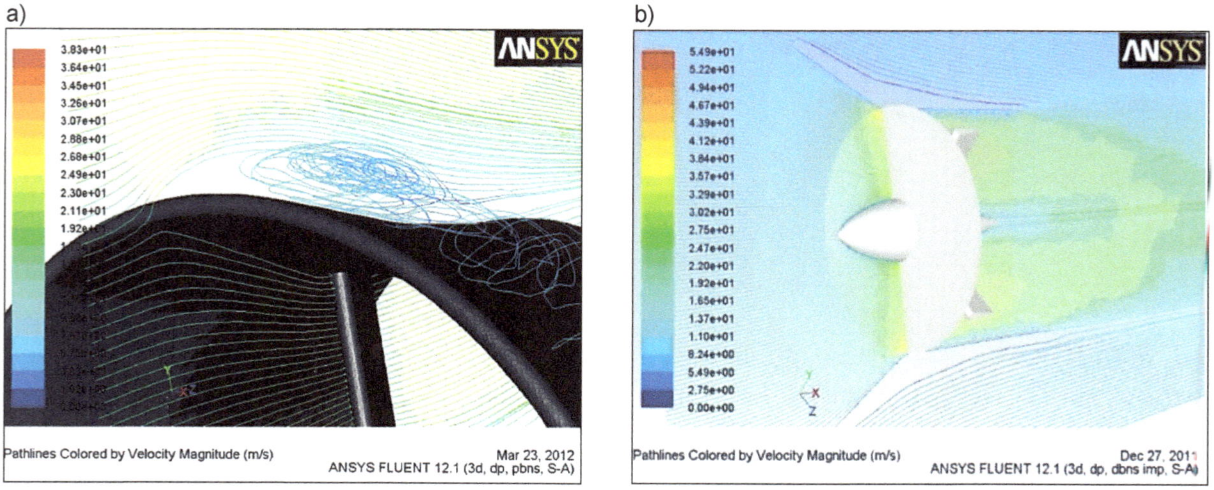

Fig. 7. Distribution of velocity with the flow lines of shroud 1.69/1.05 for 5 m/s for cases: a) without a propeller, b) with the propeller – pressure step

Numerical research of flows were made with the method of finite volumes with a commercial bundle Ansys/Fluent. The continuation of numerical research in one of Polish academic centres is planned. The calculations are not at initial stage. In order to define optimum calculation conditions the comparative analyses of for the used non-structural, structural and hybrid grid, the model of flow and the method of mapping the wall surface and turbulence. In order to generate calculation grids ICEM CFD programme was used. In most simulations concerning the body a grid of PRISM type "wall layer" was used. In the area of the domain, a grid of TETRA type was used. The connection between the wall layer and body surface was made of PYRAMID elements. Numerical examinations will be conducted for a number of model configurations, in the first stage, for the shroud only – without the propeller; only after the results comparable to experimental data a propeller will be added with FAN function. It allows reconstructing the flow generated by the work of a given element, without giving its detailed geometry. The condition is defined by the moment added to the surface, proportional to the given pressure step.

In the visualisations of flow field with the flow lines the disorders in outer surface of shroud surface characteristic eddies and irregularities for the flow without the propeller can be seen.

CONCLUSIONS

Ducted propellers have many advantages, yet their basic disadvantage is their lower efficiency in comparison to free propellers. Our research shows that the change of shroud geometry has a significant influence on the change of aerodynamic characteristic of the drive and can make the quickest way to improve efficiency/productivity of the drives in certain ranges of velocities (propeller advancement). The development of advanced design methods allows generating new and more complicated tools. Advanced design systems using FSI analysis and the works on memory shape alloys can have a significant influence on the development of drive systems. The range of using such solutions is still enlarging – small and medium haul aircraft market and GA planes. Also the military UAV constructions of vertical take-off and landing use ducted propeller equally often due to the possibility to load a small propeller with larger power and higher efficiency for small velocities of flight. Despite such a high level of complexity in such constructions, it seems that full-scale attempts are not so difficult and should be conducted as soon as possible, due to potential benefits.

REFERENCES

1. Lewandowski R.: Możliwości napędu śmigłami obudowanymi. Technika Lotnicza i Astronatyczna nr 1, Warszawa 1983.
2. Neal & Slack W.E.: NRCC Aeron. Rep. LR-445 „Static and low forward Speer tests on several six-foot diameter tractor and pusher ducted propellers", Ottawa 1965.
3. Szafran K.: The Laboratory Experience of Fan Propeller Unit Power for Unmanned Airship. Warsaw Institute of Aviation, Proceedings of Conference Scientific Aspects of UAV's, May 07–09, Warszawa 2008.
4. Akturk A., Camci C.: Tip clearance investigation of a ducted fan used in VTOL UAVS. Proceedings of ASME Turbo Expo Turbine Technical Conference, June 6-10, Vancouver, Kanada 2011.
5. Roberts S.C.: Trecom Technical Report 64-41. The Marvel projects, MSU, USA 1964.
6. Taylor T.: Experimental investigation of the effects of some shroud design variables on the static thrust characteristics. LTL, Washington, USA 1958.
7. Strzelczyk P.: Wybrane zagadnienia aerodynamiki śmigieł. Wydawnictwo Politechniki Rzeszowskiej, Rzeszów 2009.

COMBINED SYSTEMS OF ENERGY GENERATION – A CHARACTERISATION AND CLASSIFICATION

Jan Gilewski[1], Jerzy Montusiewicz[2]

[1] Elektroprojekt S.A., Lublin, Poland, e-mail: bjgilewski@o2.pl

[2] Institute of Computer Science, Lublin University of Technology, Nadbysrtrzycka 36b, 20-618 Lublin, Poland, e-mail: j.montusiewicz@pollub.pl

ABSTRACT

The study presents issues concerning technical solutions of combined systems of energy generation which can be used primarily in low-level power plants, installed in various types of public utility sites. A detailed description is given of selected ways of powering combined energy generation systems, presenting conceptual outlines of their operation and information on their advantages, disadvantages and applications. The following systems are introduced: gas-steam, back-pressure steam turbine, extraction-condensing steam turbine, gas turbine, gas microturbine, Stirling engine, fuel cells and internal combustion piston engine. Moreover, the study addresses economic aspects of energy generation in combined systems, discussing different methodologies of cost calculation, including the one used by the European Union. The article also gives a detailed review of piston engine combined-system aggregates available in the Polish market. Type series of associated systems designed for low-power appliances are shown, produced by Polish and foreign companies such as Viessmann, Centrum Elektroniki Stosowanej CES, H. Cegielski – Poznań, KWE Technika Energetyczna, TEDOM Poland or the EPS System.

Keywords: Combined Heat and Power (CHP), combined energy generation, cogeneration systems.

INTRODUCTION

The basic energy carriers such as electricity, heat and cold are produced, both in Poland and in the world, mainly as a result of thermal processes using the chemical energy contained in solid, liquid or gaseous fuels. Such energy technologies should fulfil the following technical-ecological and economic specifications [1]:

- be characterised by the greatest efficiency in energy processing,
- possess the most advantageous indices of economic profitability, i.e. have a short time of investment return and ensure a high profit value,
- have the least conceivable negative influence on the environment, i.e. emit a minimum of toxic dusts, gases, noise and waste, using in the process as little water as possible.

The realisation of the above-mentioned goals depends both on the kind of technology of electricity and heat production that we use (e.g. application of combined or separate systems) and the type of fuel used (e.g. solid, liquid or gaseous).

Combined systems of energy generation using gas module engines are regarded as distributed generation systems. There is no firmly grounded and generally accepted terminology in this regard. According to the CIGRE 37-23 Working Group (WG 37-23) [2] distributed generation refers to power sources from 50 to 100 MW whose construction is not centrally planned. Such sources are not subject to central disposition of power, although in most cases they are part of the distribution network. In Poland the power limitation of distributed generation systems to 150–200 MW results from their influence on the course of operation of the distribution network of no more than 110 kV. In practice there are examples of combined systems which are never part of the distribution network and are entirely autonomous systems. This is the situation on boats and ships [3], where the entire energy produced is divided into driving, electric and heating energy.

The notion of combined systems is broader than that of cogeneration systems. The latter produce two types of energy, e.g. electricity and heat, whereas the former can have more energy factors.

Considering the information prepared by the CIGRE 37-23 Working Group [2], the definition of combined systems could be generalised. Thus they are small systems of distributed generation of nominal power of up to 150 MW, or systems attached directly to distribution networks, or ones localised directly in the consumer's electric energy network behind the metering and billing system, which produce electricity from original energy (e.g. chemical one) in association with the generation of heat or cold.

The basic task of combined systems is to meet the energy demand of both community and industrial consumers. Such systems, known as CHP ones (Combined Heat and Power), belong to the most efficient technologies of distributed power engineering and bring numerous technological and ecological advantages. The electricity produced in such systems can be used in a building in its entirety or be partly sold to other consumers. On the basis of catalogue data analysis it is worth mentioning that the general efficiency of a combined system may reach 90%, while that of a power plant only between 30 and 40%.

Users of combined systems include:

- public utility sites (hotels, office blocks, shopping centres),
- sports centres (sports halls, swimming pools and skating rings),
- educational centres (universities, schools),
- industrial plants of diverse specialty,
- health service sites (hospitals, out-patient wards, sanatoriums),
- community sites (clusters of single- and multi-family housing, waste treatment plants), etc.

TECHNICAL ASPECTS OF COMBINED SYSTEMS

A description of different technical solutions of combined systems is given, among other places, in Kiciński and Lampart [4], as well as in reference [5, 6, 7, 8]. The authors present selected solutions for those systems which can be used in CHP-type arrangements.

Today there are many different kinds of technological solutions of combined systems:

- gas-steam systems,
- with back-pressure steam turbines,
- with extraction-condensing steam turbines,
- with gas turbines,
- with gas microturbines,
- with Stirling engines,
- with fuel cells,
- with internal combustion piston engines,
- and others (meeting the criterion of simultaneous generation of electricity and heat and/or cold).

Gas-steam systems

Combined gas-steam systems serve the generation of heat and electricity in a system of a gas machine (gas turbine) and a steam machine (steam turbine). In gas-steam systems there is an interrelation of the thermodynamic circuits of a gas turbine and a steam turbine via the exhaust waste heat utilised in the heat recovery boiler. Due to this arrangement a hybrid system of increased efficiency is created, combining a high-temperature circuit realised in a gas turbine with a low-temperature steam circuit [9]. A diagram of the working of a cogeneration gas-steam heat and power plant is shown in Figure 1.

Two types of gas-steam heat and power plants can be distinguished:

- systems working in an open circuit – the closure of the gas fuel circuit takes place via the surrounding atmosphere (air);

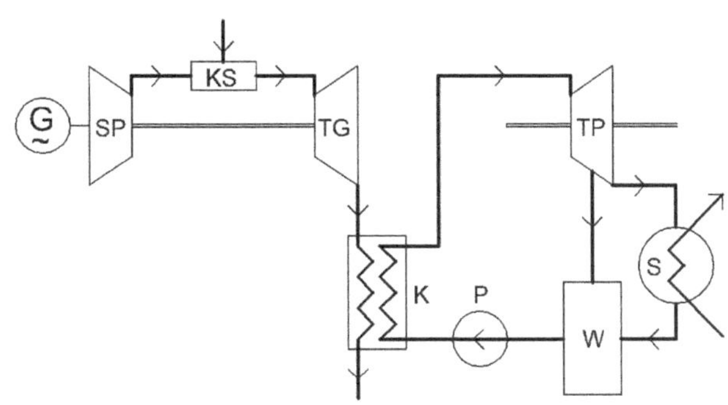

Fig. 1. Outline of a cogeneration gas-steam heat and power plant (with a gas turbine and a heat recovery boiler); G – power generator, SP – air compressor, KS – combustion chamber, TG – gas turbine, K – recovery boiler (steam exchanger), TP – steam turbine, S – condenser, W – heat box (reservoir), P – circulation pomp

- systems working in a closed circuit – in such a system a constant amount of working factor (gas exhaust) keeps circulating.

In view of their high investment costs, in Poland combined gas-steam systems are only used in industrial installations and low-level power industry electrical engineering, in which they reach capacity between 20 and 200 MW. Cogeneration systems with gas-steam turbines combine the advantages of gas and steam systems, which causes an increase in the difference of the exit process temperatures (steam) compared to the entrance process (gas exhaust), which in turn raises the thermal efficiency of the whole system.

Systems with a backpressure steam turbine

A cogeneration system with a back-pressure steam turbine is an example of a cogeneration system of power industry electrical engineering. Such a system has a back-pressure turbine working in a close circuit, supplied by decompressed overheated steam produced in a boiler. Having come through the turbine the steam passes to the exchanger, where it gives off its excess heat to heat up the water in the heating network. A diagram of a cogeneration system with a back-pressure turbine is shown in Figure 2.

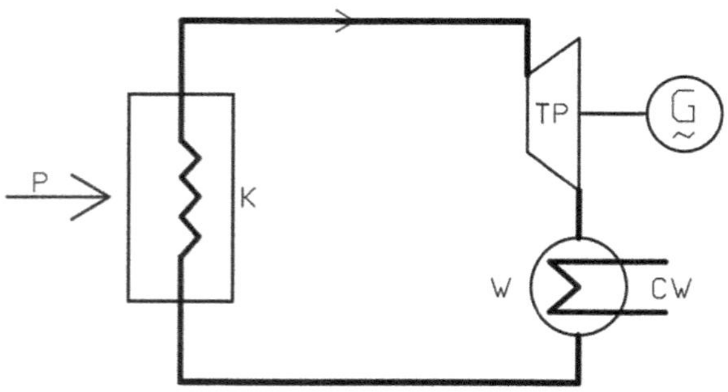

Fig. 2. Outline of a cogeneration heat and power plant with a back-pressure steam turbine; P – fuel, K – boiler, TP – steam turbine, G – power generator, W – condenser, CW – hot water [4]

Such a system is characterised by a simplicity of structure and a relatively low demand for water. A disadvantage is the high dependency between the electricity produced and the demand for heat and the exploitation of only high pressures in the production of electricity. Backpressure turbines as cogeneration systems are characterised by high total efficiency. The steam obtained is used for technological and heating purposes. They are mainly applied in (power) industrial installations.

Systems with an extraction-condensing steam turbine

A cogeneration system with an extraction-condensing steam turbine works similarly to one with a back-pressure steam turbine, but differs by the presence of several heat extractions (several degrees of steam extraction) from the turbine as one goes towards the steam entrance. Such a solution ensures the exploitation of a wider scope of pressures and temperatures in electricity generation.

Just like cogeneration systems with a backpressure turbine, systems with an extraction-condensing steam turbine are characterised by a high efficiency of the whole system. They are also mainly applied in (power) industrial installations in view of the high powers installed and the investment costs. A outline of a system with an extraction-condensing turbine is shown in Figure 3.

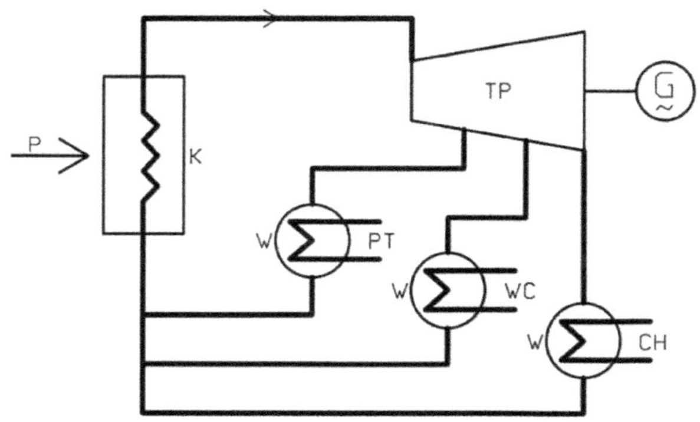

Fig. 3. Outline of a cogeneration heat and power plant with an extraction-condensing steam turbine; P – fuel, K – boiler, TP – steam turbine, G – AC generator, W – heat exchanger, PT – technological steam, WC – hot water, CH – cooling water [4]

Systems with a gas turbine

The basic elements of a cogeneration system with a gas turbine are: a compressor, a generator turbine (on a common shaft) and a combustion chamber between them. The driving energy is obtained from the exhaust of the gas burning in the combustion chamber. The exhaust drive a turbine, which in turn drives the generator and compressor. At the turbine exit we get the exhaust, which makes its way out (an open system) or pass through the heat exchanger (recovery boiler), from which heat is recovered (a closed system) (Fig. 4.)

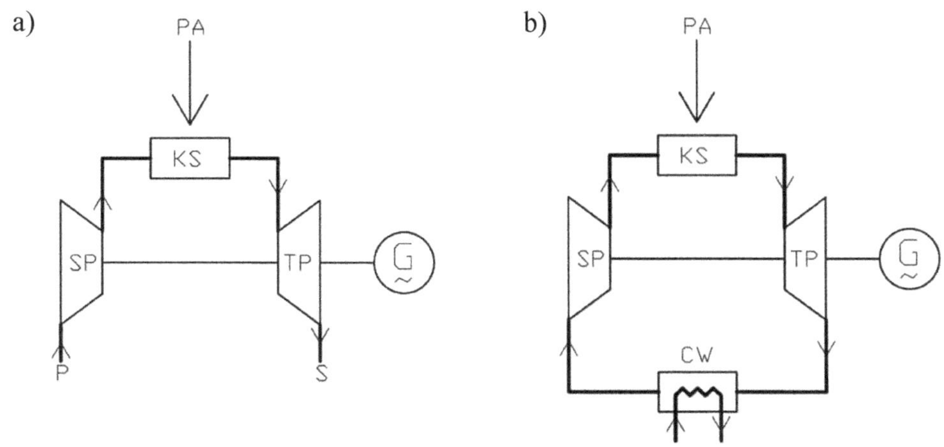

Fig. 4. Outline of a cogeneration heat and power plant with a gas turbine: a) open system, b) closed system; PA – fuel, KS – combustion chamber, TP – gas turbine, G – AC generator, SP – compressor, CW – low-temperature heat exchanger [1]

Systems with a gas microturbine

Systems with gas microturbines are gas turbosystems of small power. Their construction is similar to classical gas turbines except for slight differences. However, their capacity does not surpass a few hundred kilowatts.

Just as in a classical gas turbine, a gas microturbine is made up of [1]:

- mechanically and thermally connected one-step radial turbines and compressors;
- a regenerative exchanger between the compressor and combustion chamber, increasing the efficiency of the whole system;
- a high speed generator, together with a power-electronic converter system, allowing to adjust the parameters of the energy produced to the demands of the electric energy network.

The advantages of gas microturbines include:

- small overall dimensions,
- high efficiency of the whole system,
- reliability of operation and ease of use,
- low emission of pollutants,
- ability to supply different fuels (gaseous, liquid),
- ability to work in a grid or insular system.

Examples of applications:

- emergency power supply of public utilities such as offices, hospitals, etc.,
- as a source in distributed energy engineering.

Systems with a Stirling engine

Cogeneration systems with a Stirling engine use reciprocating heat engines with external combustion, which means that the working element – which is the piston moving in the cylinder – is supplied by energy from an external source. Since energy is passed to the system from outside it, it becomes possible to supply such an engine from an arbitrary source of heat. A source of heat can be the process of burning energetic fuel, but also concentrated solar radiation, geothermal energy, etc. At the basis of the operation of a Stirling engine lies continuous heating and cooling, that is continuous supply of potential difference of heat to the appropriate zones of the piston in the cylinder. A engine works in a closed circuit with any working gas, together with heat regeneration while keeping the gas volume constant (in Carnot's cycle). Despite the fact that the first copies of those engines were developed already in the 19th c., such systems found no general use, but are still within the interest of constructors and scientists.

Systems with fuel cells

Fuel cells are galvanic cells generating electric power from chemical energy on the basis of chemical fuel constantly supplied from outside. The mass flow of the fuel is correlated with the current cell load. In contrast to classical galvanic cells, electrodes take no direct part in electrochemical processes, only steer those processes, therefore their lifespan is theoretically limitless. The lifetime of the electrodes is limited by their structural stability, not chemical one.

In a fuel cell there occurs a direct transformation of chemical energy into electrical one without the mediation of heat and limits resulting from the application of Carnot's circuit, due to which a high efficiency of the whole process is achieved [10]. Just as ordinary galvanic cells, it contains two electrodes and electrolyte. The reacting substances (fuel and oxidant) are supplied non-stop. Apart from electrical energy, a product of a fuel cell's work is heat. The temperature achieved reaches around 200 °C (low-temperature cells) or 1000 °C (high-temperature cells) [11]. At present, both in the USA and in Europe, there are initiatives to commercialise the production of fuel cells as solutions of distributed cogeneration. Their capacity reaches some 200 kW. The greatest fuel cell systems achieve the capacity of 11 MW [12] and are used in power industry.

Systems with internal combustion piston engines

Cogeneration systems with piston engines are characterised by a wide range of rated power, modularity, low capital and investment costs and a short construction and installation time. They also feature a fast start, flexible keep-up with the load, high efficiency at partial load and high reliability. They can be used in objects of wide utility as emergency or peak-time sources, but they can also be put to sub-peak or even basic service, providing power, heat or even cold. Their fundamental flaws are pollution emitted into the atmosphere and noise emission.

Current production includes low-emission engines with spark ignition, powered by natural gas or biogas with a low degree of compression in relation to engines with compression ignition. Such engines are characterised by 60–80% of the power of diesel engines, which ultimately contributes to higher investment costs [11]. Yet when powered by natural or biogas, they have lower pressure in the cylinders and less bearing load, which as a consequence causes a lengthening of exploitation time. Figure 5 shows a outline of a cogeneration system using a piston combustion engine.

The basic cogeneration module using piston engines consists of:

- a system generating electricity (generator) and heat;
- electric safety system;
- auxiliary drives switchgear;
- installation of automatic oil refill;
- noise mufflers at exhaust and air exits;
- master control cabinet enabling monitoring and visualisation of the operating parameters;
- electricity network synchronisation system;
- emergency cooling system;
- soundproof enclosure.

A piston engine powered by natural gas, propane or biogas is placed on a common shaft with an asynchronous generator. The work of such a kit enables the production of electricity and heat. As a result of gas combustion the piston engine produces heat as an additional product of energy conversion. Silnik tłokowy zasilany gazem ziemnym, propanem lub biogazem posadowiony jest na wspólnym wale z prądnicą asynchroniczną. The work of such a kit enables the production of electricity and heat. As a result of gas combustion the piston engine produces heat as an additional product of energy conversion. This heat is received from the body of the engine and the exhaust expelled, then recovered by the heat exchanger circuits, added up and by a water or glycol system transmitted to the receptor. To achieve the assumed temperature of the water or glycol at the output of the module and for a stable op-

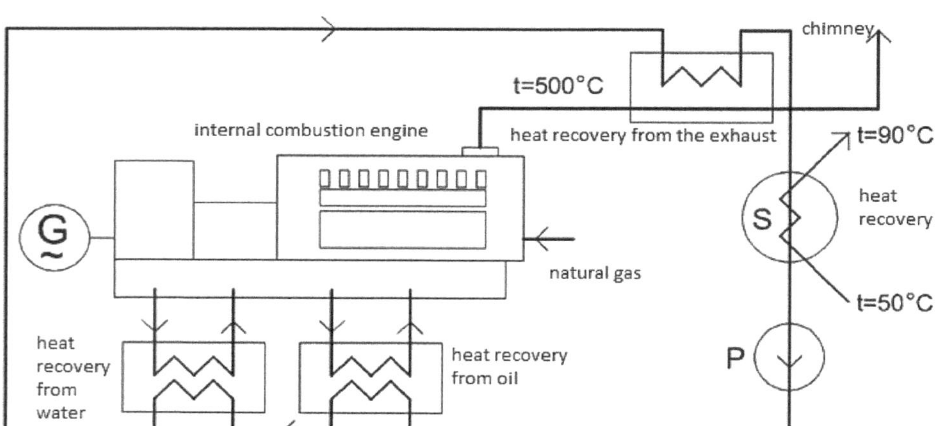

Fig. 5. Outline of a cogeneration circuit using an internal combustion piston engine: P – circulation pump, S – heat exchanger, G – electricity generator [13]

eration of the cogeneration system an auxiliary drives switchgear is used. This switchgear controls the valves of the emergency cooling system and continuously monitors the coolant:

- in the case of too hot water at the entrance to the cogeneration system, it redirects part of it to an additional cooling system;
- in the case of too cold water a by-pass is activated, warming it up to a designated temperature.

For CHP modules water parameters are assumed at 70 °C at input and 90 °C at outlet [9]. High temperature heat is obtained just from the flue gas at a temperature of 400–600 °C, while the heat obtained from cooling the engine block does not exceed 95 °C [14].

ECONOMIC ASPECTS OF COMBINED SYSTEMS

Calculating the cost of electricity and heat generation in combined systems in order to assess their economic competitiveness can be done by designating the unit cost of production, taking into account the entire period of its existence. The elements of the economic calculation of the combined technology of power generation are:

- investment costs;
- operating costs, understood as system servicing, current supervision, etc.;
- fuel costs;
- the costs of decommissioning at end of life.

In the literature one can find several methods to determine the unit cost of production of electricity and heat, including how this is to be understood in combined systems [11, 12, 15]. In [16] the authors consider the explores the effects of micro CHP systems for energy companies.

One of the criteria for selecting the most efficient design of electricity and heat generation is unit cost. According to the authors of paper [10] discounted long-term unit cost of production in the proposed systems of electricity generation in combined technology can be calculated from equation (1):

$$k_j = \frac{\sum_{t=0}^{n} I_t a_t + \sum_{t=0}^{n} K_t a_t - WM_N a_n}{\sum_{t=0}^{n} A_t a_t} \quad (1)$$

where: a_t – discounting factor $a_t = \frac{1}{(1+p)^t}$;

a_n – discounting factor of the last n-th year of analysis, $a_n = \frac{1}{(1+p)^n}$.

p – discounting rate, or the period of analysis including the period of the object's construction (b years) and operation (N, e.g. 30 years);

I_t – investment outlay in year t;

K_t – current operating costs (maintenance, renovation and fuel) in year t;

WM_N – object value in year N (final, as yet unamortised, value of fixed capital);

A_t – energy generated in year t;

„*0*" – base (zero) year, or the year of the first expense or the year of economic analysis.

A similar methodology for calculating the cost of production of electricity and heat in diffuse sources was applied by the European Commission in 2008 with the publication of a collection of documents under the title: „Second Strategic Energy Review – An EU energy security and solidarity action plan". One of the documents it includes is: "Energy Sources, Production Cost and Performance of Technologies for Power Generation, Heating and Transport". The method presented can calculate the equivalent unit cost of production of electricity and heat – see formula (2), [15]:

$$COE = \frac{SCI \cdot (1+IDC) \cdot CRF \cdot}{8760 \cdot LF} + \frac{FOM}{8760 \cdot LF} + VOM + FC + CC + CTS \quad (2)$$

$$COE = \frac{SCI \cdot CRF}{8760 \cdot LF} + \frac{FOM}{8760 \cdot LF} + VOM + FC$$

where: 8760 – ratio equivalent to the number of hours per year;

COE – equivalent unit cost of electricity generation, € / (MW × h);

COH – equivalent unit cost of heat generation, € / toe;

toe – tonne of oil equivalent – energy equivalent to one metric tonne of crude oil with a calorific value equal to 10,000 kcal / kg;

SCI – individual investment in manufactured facility, € / MW or € / toe;

IDC – interest on capital expenditures during the construction of the facility;

CRF – equity installment (return on equity);

LF – annual utilisation of the capacity of the facility;

FOM – equivalent (for the time of operation) annual fixed operating costs, € / MW or € / toe;

VOM – equivalent unit variable operating costs , € / (MW × h) or € / toe;
FC – equivalent unit fuel costs, € / (MW × h) or € / toe;
CC – equivalent unit CO_2 emission costs, € / (MW × h);
CTS – equivalent unit costs of transport and storage of captured CO_2 € / (MW × h) (in the case of energy sources equipped with facilities for CO_2 interception and storage).

COGENERATION SYSTEMS USING PISTON ENGINES

Analysis of the parameters of the produced gas cogeneration units with piston engines reveals that:

- the largest amount of heat comes from the cooling system of the engine block and the circulation of lubricating oil cooling system;
- additional heat is recovered by cooling the exhaust gas;
- the amount of heat recovered from the engine cooling system, relative to the chemical energy of fuel, represents about 30%;
- the amount of electric power produced relative to the heat is distributed approximately half-in-half, with a slight advantage of the electric power over the thermal power;
- recovered heat is most commonly used for domestic hot water.

On the basis of analyses of selected manufacturers' data sheets it can be stated that systems associated with compression are currently available in a wide power range from about 400 kW to about 5000 kW of electrical power and heat. Therefore, such systems are used primarily where less heat output is needed. They also have limited capabilities in terms of steam generation, but have a higher round-trip efficiency (close to 90%) compared to combined systems based on a gas turbine.

Currently, the domestic market has several major suppliers of gas aggregate systems combined with piston engines, who offer units built with components from different manufacturers. Table 1 shows a list of selected suppliers and their products belonging to so-called small-scale energy appliances (5 KW to 5 MW [4]).

The features of the above systems include [19]:

- control system package;
- handling and visualisation of the functional processes and control by a PC and a touch terminal;
- the use of additional control functions (such as adjusting the gas pressure or heat generation);
- interconnection, supervision and control of the multiport systems via Ethernet;
- possibility to connect to a master control system;
- support of various communication protocols (e.g. Ethernet, Profibus DP, 3964R, Modbus RTU);

Table 1. List of firms producing type series of associated systems with combustion engines

Company	Modules offered
VIESSMANN Sp. z o.o [17]	Vitobloc 200 EM-18/36 … EM-401/549 (from 18 kWe/36 kWt to 401 kWe/549 kWt)
CES Sp. z o.o. [18, 19]	MTU Seria 400 (from 119 kWe/198 kWt to 420 kWe/504 kWt) MTU Seria 4000 (from 772 kWe/872 kWt to 1948 kWe/2156 kWt) MWM Seria TCG 2016 (from 400 kWe/428 kWt to 800 kWe/856 kWt) MWM Seria TCG 2020 (from 1200 kWe/1189 kWt to 2000 kWe/1977 kWt) MWM Seria TCG 2032 (from 3333 kWe/3206 kWt to 4300 kWe/4164 kWt)
H. Cegielski – Poznań S.A. [20]	module with motor WAUKESHA APG 1000 – 1002 kWe module with motor MWM TCG 2016 V12C – 600 kWe module with motor MWM TCG 2016 V16C – 800 kWe
KWE Technika Energetyczna Sp. z o.o. [21]	Jenbacher type 2 (J208) – 200-300 kWe Jenbacher type 3 (J312, J316, J320) – 500-1000 kWe Jenbacher type 4 (J412, J416, J420) – 1000-1500 kWe Jenbacher type 6 (J612, J616, J620, J624) – 2000-4500 kWe Jenbacher type 9 (J920) – 9,5 kWe
TETOM POLAND Sp. z o.o. [22]	Micro T7, T30, T50 (from 7 kWe/17,2 kWt to 48 kWe/91,0 kWt) Cento M50, T80, T100, T120, T160, T180, T200 (from 50 kWe/79,0 kWt to 200 kWe/239,0 kWt) Quanto D400, D580, D770, D1200, D1600, D2000 (from 400 kWe/456,0 kWt to 2000 kWe/2155,0 kWt)
EPS SYSTEM [23]	60 GA (41 kWe/72,8 kWt) 85 GA (57,8 kWe/99,5 kWt) 150 GA (111 kWe/170 kWt) 2500 GH (1151 kWe/1388 kWt) 4100 GH (1501 kWe/1840 kWt) 5300 GH (1927 kWe/2324 kWt)

- record of the history of operations and events in the database (from the period of 6 months);
- optional remote diagnostics system via ISDN;
- option to notify the operator via text messages / e-mail (reports, alarms).

An example of a large public utility site using combined generation with a 1.6 MW piston engine is the complex of buildings of Prince Charles Hospital in Brisbane, Australia. The system generates electricity and hot utility water and technological steam [24]. Another interesting example is the designing of combined generation aggregates for already existing office blocks in Sydney [25].

CONCLUSIONS

The article presents an overview of cogeneration systems used in distributed power generation as well as in power industry. Based on the analysis of existing design solutions one can make their basic division into those used in the power industry and in distributed networks. Due to more attention currently paid to environmental issues the boundaries of classical divisions are becoming blurred. At present application systems in power industry are mainly associated with high power ratings, while in distributed power – with low ratings (less than 1 MW, though some move this limit to the value of 5 MW [4]). Installed systems using gas and steam turbines are primarily used in power industry and are characterised by high power ratings – above 1 MW.

Cogeneration systems used in distributed power generation are selected taking into account the characteristics of the object supplied. In the case of power industry, systems supply the national electricity system, which has a relatively rigidly defined parameters and it is not possible to determine the parameters of individual loads received.

In Poland, public utilities are most commonly powered by cogeneration systems using piston engines.

The authors deal with the topic of the article in the context of the multicriterial optimalisation of the selection of combined generation systems with regard to the needs of powering a public utility site such as a hospital. Such systems can also be used for powering other utility sites, using – apart from electricity – technological heat, steam and technological cooling. Such a solution, apart from economic effects, can raise the power generation reliability ratio. Using several units of lower power can also raise the power generation flexibility on a yearly basis, depending on the energy needs of the site powered. Thus the designed associated technologies of energy generation should be characterised by the highest possible capacity, modular structure, scalability and the possibility of flexible enlargement. Further, they should display the most advantageous indices of economic return, low environmental impact through low emission of dust, gas, sewage and noise, as well as low water consumption.

Associated systems used in dispersed energy production are selected with regard to the load characteristics of the site powered. For an optimal choice, different criteria should be used and the whole process should be a multi-stage one. In the case of professional energy production the systems supply the national electricity generation system, characterised by relatively inflexible parameters and the lack of the possibility of specification of reception parameters. Power generation in public utility sites In Poland most often uses combined generation systems with piston engines in the power range of several kilowatts to several megawatts.

REFERENCES

1. Skorek J., Kalina J., Technologie i efektywność ekonomiczna generacji rozproszonej w układach gazowych. Seminarium cykliczne „Elektroenergetyka w procesie przemian": – Generacja rozproszona. Gliwice, Politechnika Śląska 2002.
2. Impact of increasing contribution of dispersed generation on the power system. Working Group 37.23. CIGRE, Paris, February 1999.
3. Ciesielski S., Sroka K., Optymalna eksploatacja systemu energetyczno-napędowego statku morskiego. [W:] Kulikowski R., Sosnowski J.S. (red.), Badania systemowe, tom 2, Instytut Badan Systemowych PAN, Omnitech Press, Warszawa 1990.
4. Kiciński J., Lampart P., Cogeneration in a large and small scale, Kwartalnik Energetyków Acta Energetica, 2/2009, 21–28.
5. Thermal Energy Equipment: Cogeneration, Energy Efficiency Guide for Industry in Asia, 2006, www.energyefficiencyasia.org
6. http://www.facilitiesnet.com/powercommunication/article/How-Cogeneration-Systems-Can-Save-Energy-Costs--10288
7. Mikielewicz J., Micro Heat and Power Plants Working in Organic Rankine Cycle, Polish J. of Environ. Stud. Vol. 19, No. 3, 2010, 499–505.

8. By Lindsay Audin, Combined Heat And Power Systems, Building Operating Management, December 2008, http://www.facilitiesnet.com/powercommunication/ article/How-Cogeneration-Systems-Can-Save-Energy-Costs--10288
9. Kotowicz J., Stan i perspektywy rozwoju układów gazowo-parowych. Współczesne problemy rozwoju technologii gazu. Gliwice 2012.
10. Płatek W., Buczak A., Skojarzone wytwarzanie energii elektrycznej i ciepła w oparciu o paliwa gazowe – agregaty kogeneracyjne. Aspekt techniczny i ekonomiczny III FORUM ENERGETYKI, Białystok, 20 kwietnia 2004.
11. Paska J.: Wytwarzanie rozproszone energii elektrycznej z ciepła. Oficyna Wydawnicza Politechniki Warszawskiej, Warszawa 2010.
12. Paska J., Wytwarzanie energii elektrycznej Oficyna Wydawnicza Politechniki Warszawskiej, Warszawa 2005.
13. Kamiński T., Zagadnienia inżynierskie i ekonomiczne związane z produkcją energii w układach kogeneracyjnych Prezentacja Gazoprojekt, Politechnika Wrocławska, Wrocław 2011.
14. Figat K., Kogeneracja – Optymalizacja doboru technologii szansą rozwoju przedsiębiorstwa ciepłowniczego Czasopismo „Instal", 2011, Tom nr 10, pp. 14–19, Ośrodek Informacji „Technika instalacyjna w budownictwie"
15. Energy Sources, Production Cost and Performance of Technologies for Power Generation, Heating and Transport. COM, 2008.
16. Harrison J., Kolin S., Hastevik S.: Micro CHP implications for energy companies. COSPP, issue 2, March – April 2000.
17. Combined generation for heat and power production, Catalogue materials, Viessmann Sp. z o.o., 2012
18. Combined energy generation, Company catalogue card, Centrum Elektroniki Stosowanej CES Sp. z o.o., 2013.
19. Combined Production of Electricity and Heat from Natural Gas. Prospectus promotional. Center for Applied Electronics, CES Sp. z o.o., 2013.
20. Review of combined energy generation. Catalogue materials, H. Cegielski – Poznań S.A., 2012.
21. Combined generation aggregates with JE Jenbacher gas engines, Catalogue materials, KWE Technika Energetyczna Sp. z o.o., 2012.
22. Tedom CHP Units, Catalogue materials, TEDOM Poland Sp. z o.o., 2013
23. Gas combined energy generation aggregates, Catalogue materials, EPS System, 2013.
24. The Prince Charles Hospital cogeneration project, EcoGeneration — May/June 2012, http://ecogeneration.com.au/news/the_prince_charles_hospital_cogeneration_project/075483/
25. Retrofit of Sydney Building Includes Cogeneration Plant, October 29th, 2009:
26. http://cogentenergy.com.au/2009/10/article-retrofit-of-sydney-building-includes-cogeneration-plant/

RMS – SYSTEM OF THE FUTURE OR NEW TREND IN SCIENCE?

Arkadiusz Gola[1], Weronika Konczal[2]

[1] Department of Enterprise Organization, Faculty of Management, Lublin University of Technology, 38 Nadbystrzycka Str., 20-618 Lublin, Poland, e-mail: a.gola@pollub.pl

[2] Management of Production Students' Scientific Association, Lublin University of Technology, 38 Nadbystrzycka Str., 20-618 Lublin, Poland, e-mail: aka41091@interia.eu

ABSTRACT

Manufacturing companies in the 21st century will face unpredictable, high-frequency market changes, driven by global competition. The article presents a new concept of manufacturing systems – which address these challenges - known as Reconfigurable Manufacturing Systems (RMS). The concept began in 1999 at University of Michigan and is widely developed in many scientific publications. The paper first analyses main features of RSP. The next sections are devoted to analyze basic assumptions of RMS design and technical measures of RMS. Finally, the answer for the question if the RMS has the perspective of wide industrial implementations is discussed.

Keywords: reconfigurable manufacturing system, RMS, reconfigurability, flexibility.

INTRODUCTION

Frequent changes in the market, driven by global competition make production companies face new challenges and problems whose solutions must be cost efficient in cost and productivity [3, 8]. Aggressive competition forces adaptation to these requirements by new solutions that meet searching expectations of buyers and government regulation and legislation. High frequency of launching new products, changing production volumes and demand fluctuations are the most frequent problems that 21°C. companies face. To do this and to keep the pace of technological, economic and social changes, which have a significant impact on the progress in the organization of production, the company must react in a cost-effective, efficient and fast way [9]. A trend that emerged in modern production systems solutions shown in Table 1.

As shown in Table 1 striving for improvements in the organization of production is progressing in the direction in which the existing solutions are no longer sufficient and adequate. There is a need, which is characterized by J. Honczarenko, who wrote: "*Produce in a flexible, but in the simplest possible way and at the best price*" [4]. The basis of savings is to reduce the time needed to design, build or reconfigure production system to a minimum until its start-up and a high quality of production will be achieved. Today, lead-time becomes a bottleneck, the area reduc-

Table 1. The modern trend of development and changes of the production strategy of enterprises (own study on the basis of [4])

Present condition	Trend of development
• Priority productivity at the lowest possible level of the costs of production.	• Priority smooth flow of materials.
• The total dependency on access to a range of machinery and equipment.	• Dependency on the range of products.
• Control according to production plans.	• Control according to customer orders.
• Production of economically viable parties.	• Production takes into account the needs of customers.

ing the speed of production, from which earned income and market success of the product depends on to large extent.

As a response to emerging issues of efficiency and full use of technical resources in the production system, in 1999 at University of Michigan (United States), the concept of reconfigurable manufacturing systems (RMS) implemented was [6, 10]. RMS combines the advantages of existing, conventional production systems. The objectives set by engineers at the University of Michigan was:

- maximal reduction of the time which is needed to launch the production of a new product in the system,
- minimization of the costs associated with the design of the production system adapted to the production of a new product,
- eliminate the costs associated with excessive levels of system flexibility.

The purpose of this objective has set a framework for a new system and fully allowed to define this new paradigm. Namely, the RMS Reconfigurable Manufacturing System means the production system designed for functionality and customization for production capacity, for the tasks, arising from the changing needs of the market, by changing the structure of the system, both in terms of production equipment, and software. The system is defined as the evolutionary continuation of the existing solutions and an attempt to achieve the set objectives. Research conducted since 1999, as well as initial tests in the world, give satisfactory results, but further steps and improvements are needed.

CHARACTERISTICS OF RMS

The continuing evolution of science and technology in a natural way has led to the development of a new production system that is RMS. As already mentioned, it is the answer, and rather the exploration for recent years trend, developed according to market needs and revealing shortcomings of conventional production systems, particularly automated production lines and flexible manufacturing systems. As shown in Figure 1, RMS combines the advantages of both major systems (Dedicated Manufacturing Line – DML and Flexible Manufacturing System – FMS) into one cohesive production system.

Automatic lines (such as Dedicated Manufacturing Lines – DML) are characterized by high throughput and production of one type of products. This solution is very effective when demand for a particular offer is high and stable. Then, the rhythm and scheduled flow make them introduce to the market a large amount of one type of products that nourish it, while the level of demand is the same. In such a favorable situation DML, as referred, are cost effective, efficient and cheap (Figure 2). The problem arises when the offer ceases to have buyers or when demand falls to a level at which capacity ceases to be in full use. This situation is very common in the market, then the DML also face the need to reconfigure, which is a long and costly process. In addition, at the beginning and at the end of product life cycle demand for the product is not high enough, which makes automatic line working below its full capacity. Hence the problem of scalability of this system arises. The scalability cannot be attained

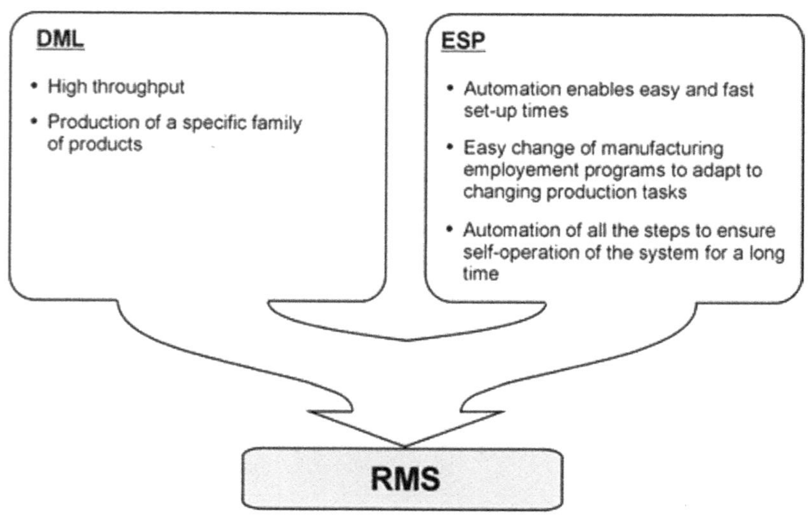

Fig. 1. Advantages of conventional manufacturing systems combined by the RMS

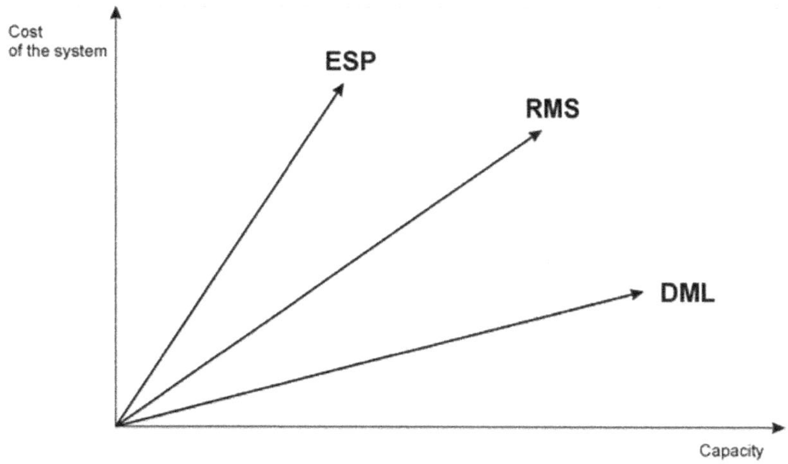

Fig. 2. The costs of RMS, FMS, DML compared with a production capacity (own study based on [3])

due to the machines suitable for the production of one type of products.

Scalability is achieved by the FMS system which is able to produce various kinds of products with variable and adjustable frequency. So it uses the work of multi-functional machines which, at the time of purchase, are suitable for diverse production and equipped with a set of spare parts and modules. However, due to general-purpose machine tools for FMS, its throughput is much lower, and the cost of plant and machinery does not always aligned to its profits. High costs cause many companies not to choose FMS, and even make the system be unaffordable for them. Therefore, it is relatively rare as a solution used in production.

The combination of the two systems, their main advantages, with non-performances elimination clarified certain characteristics to be met by the new system. Thus, the RMS is characterized by these six factors: diagnosability, scalability, convertibility, rationality, modularity, integrity [1, 2]:

- **Modularity** – means that the technical solution is designed and built with the prepared individual parts or assemblies, "*which as modules, [are] often different from each other with embodiments, reached by a combination of various general features of the system*" [4]. Modularity allows easily change of the structure, conservation of modules or their improvement. This solution reduces the cost and adapts the system to the market demand for product.
- **Integrity** – is the ability to quickly and efficiently integrate the modules into a cohesive system. It is possible by combining mechanical, computer and control connections of individual elements. The integrity at the machine-level considers the possibility of such a combination, which aims at shaping the machine according to the current need. The integrity of the system means development of the system, so that, the machinery, transport equipment and other items through appropriate drivers to operate efficiently as a whole.
- **Rationality** – means a strict focus on the flexibility, adapted to market needs. It is a key element distinguishing RMS from FMS, which often generates unnecessary costs through the part of system's flexibility. RMS uses different tools in the same machine, which reduces costs and increases productivity. As it is shown in Figure 1 RMS is designed for a particular narrow family of products manufacturing, which distinguishes it from flexible systems.
- **Convertibility** – the ability to adapt quickly the functionality of the system, its equipment and drivers, in order to adapt to new market needs. The rate of changes in the system is a prerequisite for its efficiency and profitability, for this reason retooling methods should be adjusted to the needs: conventional and modern, including quick machine calibration.
- **Scalability** – means adaptation of production capacity level according to the needs. This change is made in two ways: first by changing the ability of the individual elements of production, either by full or partial reconfiguration of the system. Scalability having a close reference to variability, can also be considered as a machine or system scalability.
- **Diagnosability** – this is system's ability to read, detect and diagnose (the state of the ma-

chine and poor quality production). It also includes taking immediate corrective action. Diagnosing recognized in terms of poor quality of manufactured parts is critical. This is due to the fact that frequent change of the system or individual modules, forces equipment calibration and quality of manufactured products verification. The time required to perform these activities will largely determine usefulness and profitability.

The design of the system described with the above features, makes such a system fully adapted to the changing market needs and demand fluctuations, and its modular design makes it efficient and cheap. The aforementioned characteristics define a new, different from the FMS and DML, class of Reconfigurable Manufacturing Systems (RMS).

DESIGN AND CONSTRUCTION AND TECHNICAL MEASURES OF FMS

Before the problem of design and commissioning RMS will be discussed, as well as using technical resources in modular structure of individual machines and entire system, it should be noted that the proposed system will not always be the best solution. If a producer wants to apply, he should start with the analysis connected with desirability of application of RMS to a specific product. It is evident that the demand for some products is constant over long periods of time, then the implementation of the production can be the best by using DML. It is not necessary choosing RMS, or FMS. Therefore, the first essential step in the design of the system is the analysis of the purpose of implementation [7].

After decision of using RMS, the next step is to design a reconfigurable system. Its configuration, defined as a set of machines (including controllers, means of transport and connections between them) is to comply with the six, already mentioned, characteristics: modularity, integrity, rationality, convertibility, scalability, and diagnosability. The machine consists of modular machining centers and modular machine. To design a system which is to meet reconfigurable conditions, the input must be gathered (what is shown in Figure 3), and identified all the failures connected with the completeness and proper configuration of the existing system (a whole as well as individual modular units and the modules themselves).

Therefore, system designing requires, not only the selection of appropriate components of modules, but also continuous the analysis of their optimality and the possibility of improvement. Design of the new RMS system, as well as reconfiguration of an already existing, can be extracted, in order to adapt to market needs.

The problem of monitoring and control in the designed system is a separate issue. The physical connection of the modules is not enough (although is problematic due to the efficient connection of the modules of specific interfaces, mechanical (its construction, drives, hydraulics, electrical and electronic) for their efficient work. Control of complex systems requires precise algorithms of their control and combine them into a single database. In addition, large number of modules, generates a large number of different system configurations and connection units, from serial to parallel through mixed ones. For example, five machines have 16 different con-

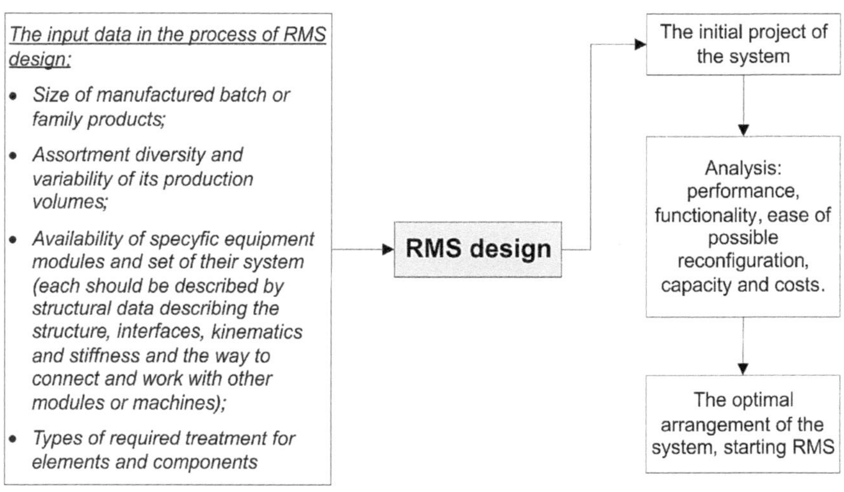

Fig. 3. RMS design stages (*own study based on* [9])

figurations, that is, for n-machine 2^(n +1) configurations. An efficient system should be able to choose an optimal configuration and control it so that all the production is both efficient and qualitatively adequate. Master RMS control system integrates into a coherent whole, therefore, the work of individual modules on machines connected in a structured system whose devices are designed for their measurement, control, and transportation of materials, products and intermediates. Thus, the problem is based on open controlling (open-architecture), which is characterized by several important features [5]:

1. Openness in the choice of hardware – thanks to modularity various components can be changed without the need to replace the entire set. This solution is far cheaper than standard solutions.
2. Openness of CNC operating systems for compatibility – allowing import of the existing standard software.
3. Openness of input – output configuration interfaces.
4. Openness for manufacturers' machines – by the opportunity to display individual wishes and settings.

Only the development of a standard for manufacturers of modules and equipment allows to develop RMS paradigm and make the production much cheaper for efficient control and combining standardized components from different companies.

The condition of modularity in the design of the system includes both the system as a whole coherent and logical structure, as well as the construction of various machinery and equipment [11]. Only the design of the system allows to reduce costs and fast start-up production of new products.

TECHNICAL MEASURES OF RMS

The key element which determines manufacturing system reconfigurability is the technical equipment – and especially machine tools. There are two types of machine tools included to manufacturing subsystem of RMS:

- Reconfigurable machines,
- Modular-machining centres.

Reconfigurable machines are modular devices that can be easily adapted to individual needs, by changing, adding or removal of individual functional modules. Such machines are created by consumers postulate of the principles of lean – machining for lean – production, so the installation of base units in these machines is easy for operators, does not require highly skilled personnel. For example, tasks of machining different elements are applied to a modern lathe (apart from the typical turning operations) processing elements of different shape, such as smooth and threaded holes, windows, gouges, grooves. In this way, almost all the element processing, including turning, drilling and milling, is performed by a single device, which greatly reduces both the time of manufacture (no storage between positions, unnecessary transport between devices) as well as costs.

Modular-machining centers are defined as CNC machines, whose structure provides the ability to run a large number of different machining treatments, using different tools in order to get the item for the most trimmed part (if possible-completely).

Modular design brings mutual benefits for producers and buyers of equipment listed in Table 2.

As shown above, modular design of the system and machine construction provides tangible benefits that make the trend in science, which is the

Table 2. The benefits for producers and buyers, resulting from the use of a modular machine tools (own study based on [4])

Producer	Buyer
• Investment in commercial products and technological - construction documentation is done only once (the next contract renewal of existing offers).	• The possibility to make changes in a using system, its development, and removal.
• Additional work on the design documentation are provided only for individual customers.	• Quick replacement and repair of defective parts. (Availability of the elements on the market).
• It is to prepare production schedules, production is stabilized.	• Short term of supply of whole systems and individual modules.
• Production of the modules can be done regardless of the order in batches of production that minimizes costs.	• Facilitating and acceleration the design work (using CAD), by having many modules and components database.
• The division of batch production scheduling allows subsequent modules storage and assembly.	• Ability to create custom modules and solutions from the range.

RMS, begins to grow rapidly in the direction of an isolated system which is efficient and cost effective market response to the needs of customers.

Machining centers and reconfigurable machine tools are not separate elements, but form a coherent whole in the RMS through wired and wireless connections, which are effectively controlled by the master control system, together with monitoring devices and transport.

CONDITIONS OF RMS APPLICATION IN MODERN COMPANIES

Reconfigurable production is the latest development trend in the wider field of integrated production systems. Here are examples of its applications, which, can be termed as the RMS, but they are significant steps forward in field of reconfigurability [3]:

- 1976 "Molins System 24" – flexible and integrated system developed by Mr. Williamson, in order to ensure greater productivity. Machining centers were combined into an automated, touched system to machining elements prepared on pallets.
- 1972 Aurebach – plant system of production is M250/02 CNC. It was equipped with two three-axis machining centers, but the service was performed manually from a central operator station.
- 1977 – Trial Tskuba factory in Japan, containing modular machines and assembly robots to produce the kind of party with assistive devices, which improve the flow of pieces by the production process.

At the end of the 1970s FMS's began to develop, which halted the development of a reconfigurable systems. Since mid-1990's RMS trend has been growing-systems that are able to react to market changes quickly, with providing desired functionality and capacity at any time. Several documented projects, describing the problem, however, in a general way, have been completed. The European Union has drawn up a report on the European system of machine tools, which states that if the machines are designed and built in a modular way then they can specialized in different modules, rather than be complete systems. System integrators will build a complete system of modules. This strategy, as the report states, requires a distribution of machine tools for autonomous functional units, which are based on a plug - and-play basis and can be used according to specific, contract needs. Several projects have been completed in this way. These include:

- Modular Synthesis of Advanced Machine Tools (MOSYN) – the project realized by the University of Hannover, focusing its research on the specification of machine configuration according to different customers' needs.
- Special Research Program 467 – a project of the University of Stuttgart, in which one of the objectives was the development of machining systems, which, thanks to reconfigurations could increase its functionality, as well as the ability of the production. This project was to combine short-term adaptation to the changing conditions of machine tools production and fluctuations in demand.
- 1996 – Engineering Research Center of the University of Michigan (in cooperation with the National Science Foundation and 25 manufacturing companies) implement the concept of RMS, which are investigated and developed to this day, for example by yielding patent for reconfigurable machine tools.

The above examples indicate, not only the scientific trend, that theory concerning contemporary research on the development of production systems and technological for manufacturing of machinery which has set, but also the direction of specific actions, which is undertaken in order to create an optimal system: adapted to the changing environment – reconfigurable manufacturing system.

CONCLUSION

Strong evolutionary trend of the development of production systems cause the solutions which help companies keep up with market changes in a quick and cost-efficient way are still searched for. Therefore, a start-up of system is a key issue for the purpose of staying competitive. Today the time from a product idea to a full capacity production must be cost-effective and reduced to a minimum, with constant search for new ideas, proposals and solutions (Figure 4).

Previous proposals for production systems has begun to show their lack of adaptability to the changing environment. Development trend, based upon FMS and DML, develops in a key direction, which is modularity: the entire system, individual, machinery and components. Modular-

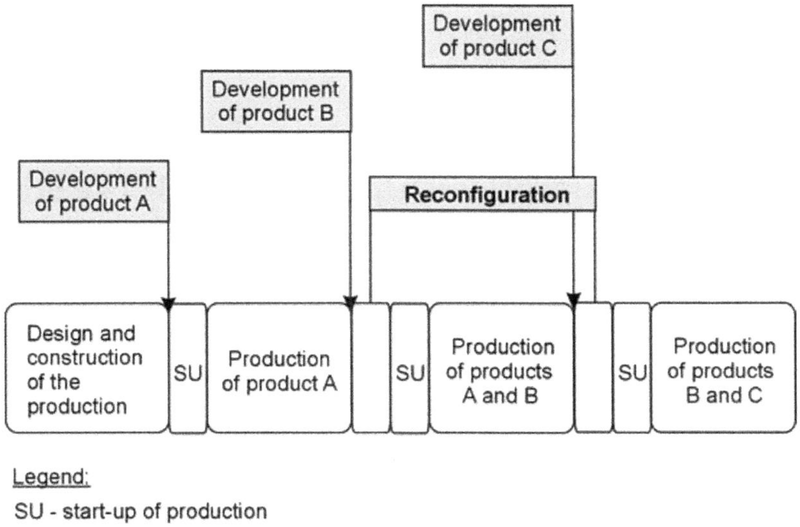

Fig. 4. The product life cycle on the market today (own study based on [6])

ity is a challenge for today's designers, as well as for the manufacturers in the field of engineering of manufacturing, because it means normalization and standardization of solutions, such as base units and pieces of equipment. It is a wide-ranging problem involving functional elements in a compact system. This problem develops on the underlying substrate links: mechanical, electrical, electronic and control: algorithms of signal flowing through reconfigurable modular system of libraries – still dilated databases. Will RMS become the new solution, fully accepted and implemented in modern enterprises? Reconfigurable system, according to the principles, meets all the expectations of flexible, low-cost and easy to use production. It is admitted, that RMS is only the first step, created by the assumptions, whose shortcomings cannot be seen yet. No companies, based on the system, confirms the difficulties in the implementation of this paradigm. Previous research, produce positive results also raise the idea that, what today is still learning, based on the examples, in the future can become a reality – a natural next step in the evolution of production systems. All things considered Reconfigurable Manufacturing System is a trend and direction in which manufacturing industry growing rapidly today.

REFERENCES

1. Bi Z., Lang S., Shen W. Reconfigurable Manufacturing Systems: The State of the Art, International Journal of Production Research, 2007, 46(4): 967-992.
2. ElMaraghy H.A.. Flexible and Reconfigurable Manufacturing Systems Paradigms, International Journal of Flexible Manufacturing Systems, 2006, 17(4): 261-276.
3. Gola A., Świć A. Directions of Manufacturing Systems' Evolution from the Flexibility Level Point of View. [In:] R. Knosala (ed.) Innovations in Management and Production Engineering, Oficyna Wyd. Polskiego Towarzystwa Zarządzania Produkcją, Opole 2012, p. 226-238.
4. Honczarenko J. Elastyczna automatyzacja wytwarzania obrabiarki i systemy obróbkowe". WNT, Warszawa 2000.
5. Joergensen S.N., Nielsken K., Joergensen K.A. Reconfigurable Manufacturing Systems as an Application of Mass Customisation. International Journal of Industrial Engineering and Management, 2010, 1(3): 111-119.
6. Koren Y., Heisel U., Jovane F., Moriwaki T., Pritschow G., Ulsoy A.G, Van Brussel H. Reconfigurable Manufacturing Systems. A Keynote Paper. Annals of the CIRP, 1999, 48(2): 527-540.
7. Koren Y., Shpitalni M. Design of Reconfigurable Manufacturing Systems, Journal of Manufacturing Systems, 2010, 29(4): 130-141.
8. Koren Y. The Global Manufacturing Revolution: Product-Process-Business Integration and Reconfigurable Systems. John Wiley & Sons, 2010.
9. Malhotra V., Raj T., Arora A. Reconfigurable Manufacturing System: An Overwiew. International Journal of Machine Intelligence, 2009, 1(2): 38-46.
10. Mehrabi M.G., Ulsoy A.G., Koren Y. Reconfigurable Manufacturing Systems: Key to Future Manufacturing. Journal of Intelligent Manufacturing, 2000, 11: 403-419.
11. Shabaka A.I., ElMaraghy H.A. A model for generating optimal process plans in RMS. Interantional Journal of Computer Integrated Manufacturing, 2008, 21(2): 180-194.

PREPARATION OF CONSTRUCTION PRODUCTION OF METAL SHEET FOR MEANS OF TRANSPORT

Piotr Penkała[1], Beata Płowaś[2]

[1] Institute of Technological Systems of Information, Faculty of Mechanical Engineering, Nadbystrzycka 36, 20-781 Lublin, Poland, e-mail: p.penkala@pollub.pl

[2] Institute of Technical Sciences and Aviation, The State School of Higher Education in Chełm, Pocztowa 54, 22-100 Chełm, Poland,, e-mal: bplowas@pwsz.chelm.pl

ABSTRACT

The design of sheet metal parts, pressed, used in the automotive industry is very complicated. Many factors influence the final shape of the part. Contemporary designer does not need to have the knowledge needed to understand the essence of its all requirements that are placed on parts of the body. It is only important that they are aware of their existence and know who in the company can help them in their fulfilment of the construction. Nowadays, only the constructor creates a CAD model geometry, which is assumed to provide the functionality. The rest of the aspects such as the provision of adequate stiffness, manufacturability, assembly features, vibration analysis, etc., are the arena of other specialists. This is the essence of constructing simultaneous, where many cell companies often work on the same element, giving it a set of features impossible to obtain by one expert on everything. Therefore, the role of the designer is often limited to being only a CAD system operator.

Keywords: simultaneous construction, pressed parts, CAD, curvilinear surfaces.

INTRODUCTION

Constant increase of consumers' requirements entail an increase of competition between car producers, what generates increasingly high expectations towards product development, not only to pressed parts, but also the whole vehicle [1]. The development is aimed at modernity, functionality, effectiveness, productivity, quality and uniqueness. Considering these aspects from the point of view of highest profitability (what is presently most important), one can additionally state that good style and image of the product are more important than all other features and are the most important factors which decides about clients' choice [1].

In case of constructing the elements of car body, four most important objectives are planned to be fulfilled: functionality, minimum weight, high passive safety and low costs of the enterprise [2]. Additionally, one should consider not only about the features of an individual sheet metal part, but also about its interactions with neighbouring parts. Simultaneous design should be conducted in such a way to make all the neighbouring parts play their roles to be assembled and have positive look at lowest possible cost [3]. Therefore, during the development at least such aspects as ensuring the possibility to connect one sheet metal element with another, i.e. ensuring appropriate amount of space for welding tongs (at least 5mm between the tongs and another element of the car body). Additionally, in case of pressure welding the width of connecting area must be at least 12 mm [4]. Constructor must take consider such aspects as tolerance, special properties of the materials the elements are made of, number of operations necessary to make the element, price of material, ensuring low mass, high resistance to humidity and corrosion. In case of the elements of platen wall and the floor it is important that the sheet metal element have enough places with minimum curvature of 150 mm and the diameter of 30 mm for welding the bolts for fixing devices, hydraulic wires and fuel pipes. One should also consider occasional loads resulting from e.g. lifting the car with a hydraulic ramp.

Fulfilling all these requirements frequently lead to conflicts between the experts from different fields. Therefore, in case of so complicated processes, simultaneous development is necessary so that the construction is made by all stakeholders within the institution at the same time. In the simulation process the exchange of information and cooperation at the level of individual employees and departments is crucial. Good example of simultaneous development is a situation when one of bolts is impossible to weld because there is a collision between the welding head and another part of the body. In order to solve this problem, a meeting of all specialists responsible for the issue from a given field is organised and they search for the cheapest solution to the problem. In this case there is a number of possibilities. Additional welding dock can be added in the assembly line, which fixes the bolt before the colliding element. Unfortunately, this generates additional coast in the station and increases the production time. The bolt can be moved to a different place, however, then it is required to move. In the new place there need to be enough space for it and the generated heat cannot have negative impact on other devices. Additionally, the assembly of the whole system must be analyzed (electronic device plus a handle) in a new position.

Each of the specialists responsible for individual areas must evaluate how expensive and complicated the change be. Then the costs of modifications are estimated and the cheapest and easiest option is selected. The situation shows how complicated the process of the development of pressed parts is and how the simultaneous construction works. In a traditional process, in case of head collision with the car body, new head, with the size that allows collision-less bolt fixing, should be designed,.

The advantages of this type of designing include: regular exchange of data, the necessity to accept changes by everyone, high level of flexibility in changes, better results, higher efficiency of employees, lower costs of development. The scheme of simultaneous process of development is presented in Figure 1.

THE USE OF CATIA SYSTEM IN DESIGNING CAR BODY

The construction of the outer surface, called the "strack" starts from the import of sketches in four projections to module "Sketch Tracer" in Catia. Then, in the *spline* function the sketches are outlined and making a digital interpretation of curves creating the main conditions of the car body (Fig. 2). Considering the fact that in case of outer construction of the car body only one part (left of right) the curves for only one side are developed. All curves are made on planes XY, ZX, Zy or the parallel planes. Plane XY is parallel to the road surface, plane ZX is the plane of

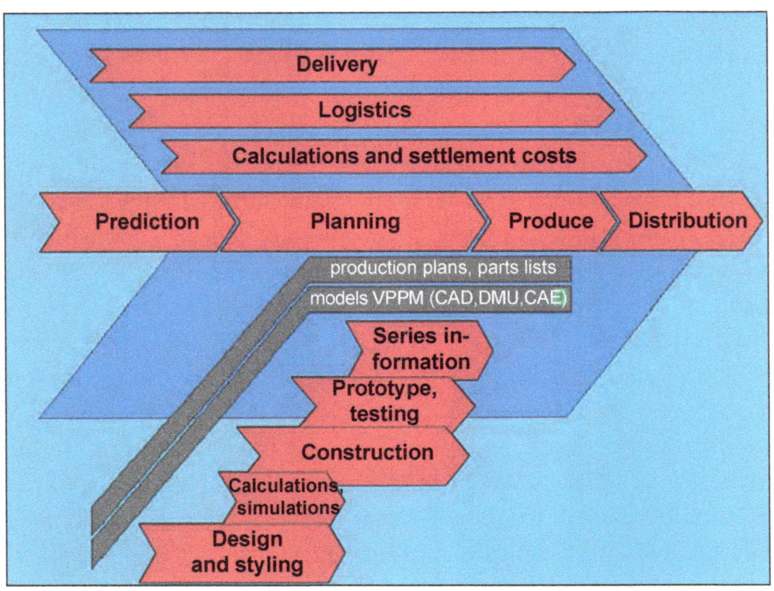

Fig. 1. Simultaneous development process [5]

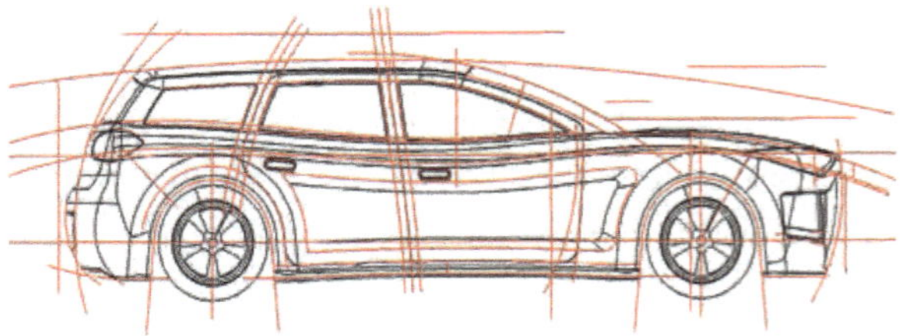

Fig. 2. Beizer curves in side projection

longitudinal symmetry, which is at the same time parallel to the direction of movement. Plane ZY is perpendicular to the ZX plane and turns into the front axis of the vehicle. The point of crossing all the planes lies exactly at the frontal axis of the car and makes a zero point for all the curves.

Considering the fact that one side of the car body is designed, all curves which lie on the YZ plane perpendicular to ZX plane, so that there was not a discontinuation of surface on the reflection according to this plane. This rule is reflected in Figure 3, where the bonding curve is tangent at the point on ZX surface to the auxiliary straight to this perpendicular straight.

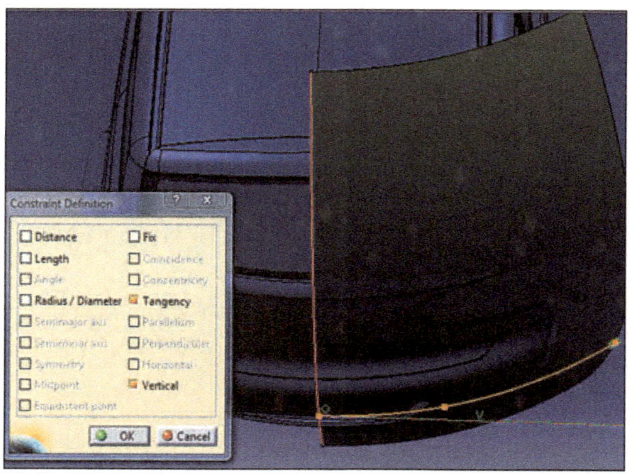

Fig. 3. Perpendicularity of bonding curve to ZX plane

The obtained curves are used to build elementary surfaces, which make the outer surface of the car body. During this operation it is important that the whole outer surface was broken into elementary surfaces this can be described with crossing curves. The rule is illustrated in Figure 4.

Each elementary fragment should be curvilinear. It is a basic rule of constructing such type of construction. straight-linear look concave, therefore, only invisible B and C class elements are used. In order to ensure curvi-linearity a tool called "Sweep" in Catia should be used. In order to use this tool, two curves, lead curve and profile are necessary. In the dialog window (Fig. 5), in "Profile" and "Guide curve" bookmarks the curves must be selected.

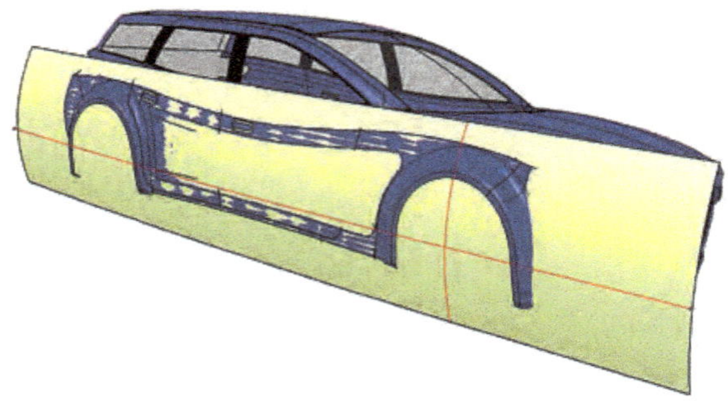

Fig. 4. Elementary side surface of the car body, described with two bonding curves

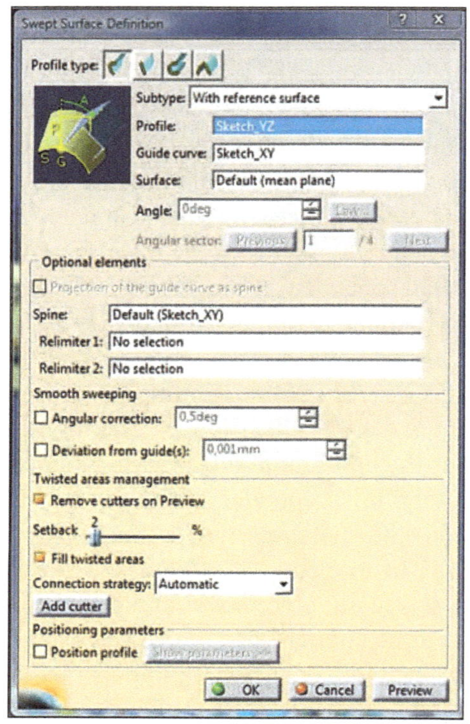

Fig. 5. Dialog window of „sweep" tool

External surface of the car body consists of a number of pieces, therefore, it is necessary to order a historic tree of the model and to place the surface and geometry necessary to build in a separate geometric "set", which should have such a name that enables the constructor to find appropriate surface in case of changes (Fig. 6). The names that correspond to the function of the surface should be used, e.g. side surface, roof, boot lid, etc. Individual pieces (Fig. 7) are connected with such functions as "fillet" and "trim". The effect of such connection should be inserted under two "sets" where the connected surfaces are present, and its name should contain the information about the pieces it connects (Fig. 8).

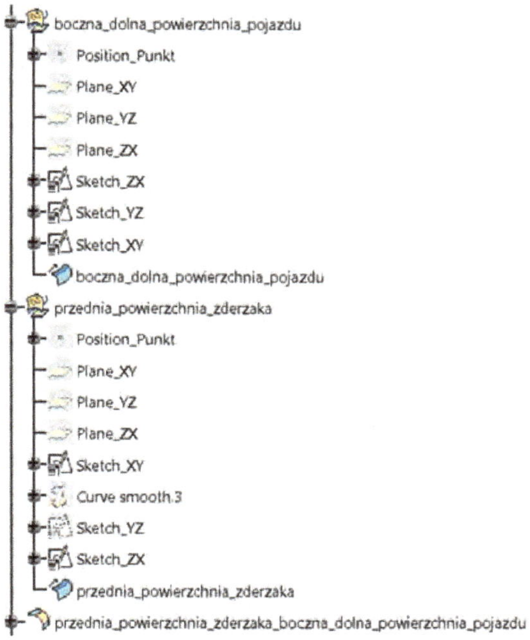

Fig. 6. Ordering the history of the model

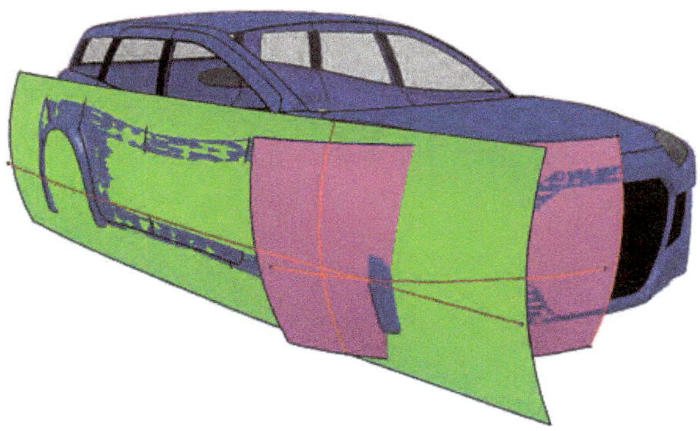

Fig. 7. Connecting the surface pieces

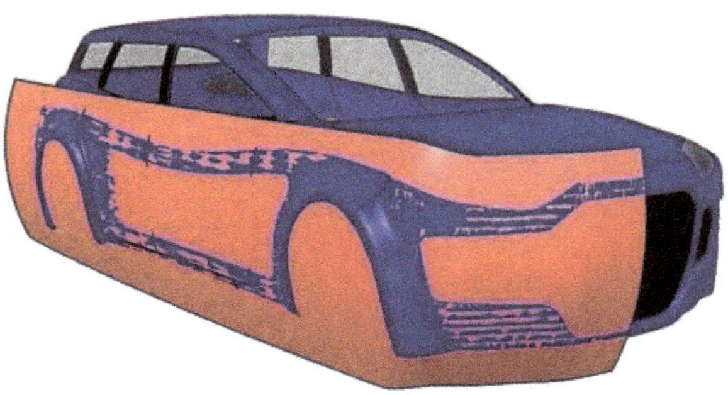

Fig. 8. The effect of connection

Each geometric set should be appropriately constructed so that possible changes in geometry were easy to do. Each car concern has its methodology to construct a history tree in Catia. The methodology defines the look of each geometric "set", but also defines the location of surfaces imported from neighbouring elements, cutting operations, drilling, cuts and solid operations. The methodology frequently exerts certain methods of constructing by providing the constructor with the so called "Model start", which is a base file that all the operations should start with. The file contains an empty history of the mode, which is filled with constructor's own operations, located in the slots determined by the start model. With such an approach, each model has the same form, regardless of the constructor that operated on it. This allows quick navigation in model history and modification of geometries by different users.

DIVISION OF THE OUTER SURFACE INTO COMPONENTS

The division of the outer surface is determined bz numerous, frequently contradictory factors. In the discussed example the division was made on the basis of design and technological assumptions for individual parts. Then, each element was checked for its manufacturability, i.e. if the depth of pressing are not too deep and if the shapes angles of formation are not too small [6].

Considering the fact that the present article is of only theoretical character, the width of spaces was 4 mm. In industrial practices a conflict of interests appears at this point. According to car designers this interval should not be present at all; however, from technical point of view it is not possible. Thus, designers press other members of the team for the smallest possible value. On the other hand, constructors cannot allow for such an approach due to the assembly tolerances in the car bodies, which make +/- 2 mm. This means that the used 4 mm space would disappear completely in case of overlapping the maximum tolerances of two neighbouring elements.

The largest possible value of the space between car body panels is in the interest of the specialists on assembly. Welding machines that position the elements against one another have also their tolerances, moreover the pressed element is also characterised by a tolerance of +/- 0.5 mm. Therefore, considering the increasing pressure for aesthetic look, there is also a pressure for the development of pressing technologies and the assembly of pressed elements in order to limit the tolerance spaces, therefore using smaller weldings between car body panels.

In case of numerous body versions, it important that the construction was divided in such a manner that the versions have the highest possible number of mutual elements [1]. In case of closed body, the whole car body up to B column is identical for each version. As for the floor, it is also identical (most often) up to C column.

In modern constructions, where virtual crash tests are possible, the distribution of deformations in standard NCAP test at the speed of 50 km/h is taken into consideration at the design stage. If a compromise between the designer and constructor is possible, it is done in such a way that limits the number of elements that are destroyed in the collision [1]. The course of car body division is important from the point of view of the designer, e.g. it gives the car silhouette the dynamism or elegance, and despite numerous factors in favour of its change it is not done, because the emotions the look generates are more important as they decide whether the model will be warmly welcomed in the market or not. This frequently entails the use of expensive technologies and significant increase of construction development costs, however, the expenses generate large returns if the model sales well.

Ergonomics is also an important factor that influences the car body division line. The programme that simulates human interaction with the vehicle is called Ramsis. It is used to test if the position of sitting is comfortable. The results of the simulation influence changes in door shape. Certain rule can be noticed. B column in four-door cars is often curved backwards, as in the discussed construction. Whereas, it is curved forward or upright in coupe-type of cars. This results from the posture of a person who enters the car [7].

ASSEMBLY FEATURES OF A PART: WING

Assembly analysis is a broad topic, and in case of the discussed model it was made primarily on the basis of the deformation angle in the depth of pressing. The analysis is presented on the example of a wing. For this purpose "Draft analysis" tool was used.

In order to determine initial direction of forming, edge visualisation should be switched on, and then view the model in such a way, so that all the edges are visible. Next, „plane definition" tool must be applied (Fig. 9) and „equation" option and "Parallel to screen" must be selected. In this way a plane parallel to the screen and perpendicular to the formation direction is obtained. The final step is to make a perpendicular line, with "Line" tool, which defines the direction of the die set.

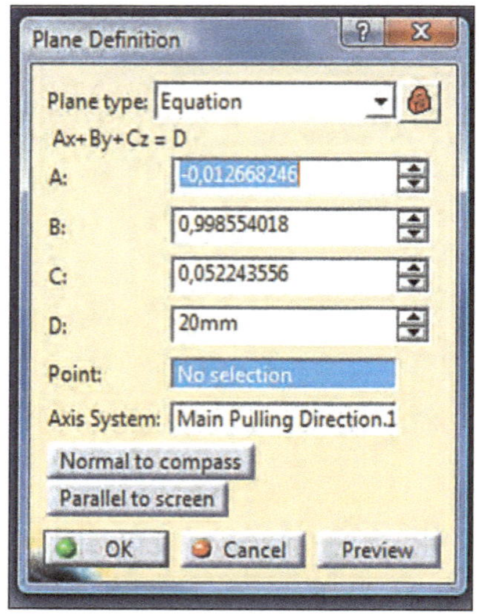

Fig. 9. Dialog window of „Plane Definition" tool

In order to check the distribution of forming angle for each surface part the visualisation option „Shading with material" should be selected and "Draft Analysis" tool should be run and the options as in Fig. 10 should be followed. From the area called "Direction" the icon of a compass should be dragged onto the previously generated line of formation direction in such as way that – 0Z axis was parallel to the line. Then the area for analysis must be marked with the mouse. In the "mode" window the enlarged colour scale should be selected and each colour should be attributed to the individual angle. In this way a set of colours showing the distribution of forming angle is presented.

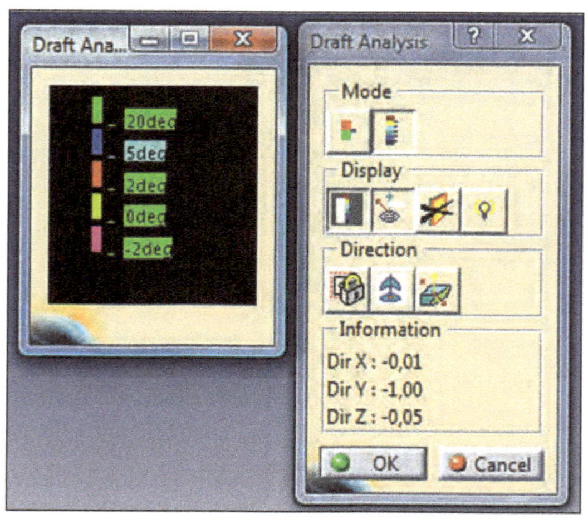

Fig. 10. Dialog window of „Draft Analysis" tool

In case of the analysed surface, minimum angle of 5° was assumed and marked with blue colour. Each part of the surface characterised by the angle between 2° and 5° is described with red colour. The angles between 2° and 180° are visualised with black colour.

Fig. 11. The analysis of formation angle in a front wing

The edge of cutting divides the sheet metal element into two area: the former cooperates with the die and the latter with the matrix. the forming angle of both the matrix and the die should be larger than the assumed minimum angle. Otherwise, the part will be unworkable or it will have to be formed in a number of operations, what increases the production costs.

In the discussed example the upper part of the wing (Fig. 11, marked with red) in the area of window joint and A column, is characterised by the angle within the range 2 – 5°. Similar situation is observed in the are of connecting the wing with the floor sheet. The value of angle that is lower than the assumed value may cause the obstruction of the pressed element in the tool. Therefore, because the critical sphere covers a small area, the risk of interference is low.

In Catia, there is no tool that allows for analysing the depth of pressing. Initial verification of manufacturability in this respect is based on technologist's experience and the final one requires a prototype of the tool and the pressed element. Because the discussed wing is characterised by relatively simple geometry, one can assume that it does not include the areas threatened by ruffling or breaking the metal sheet.

CONCLUSION

Individual stages of construction are connected to one another in such a way to ensure complete associativeness of the model. Most geometries were based on the surface copied from the "Strak..." or on the resultant operations based on them. In case of changes in car design, it is enough to change the old surface into the new one and update the model. If a need arises to change e.g. the door plating, it is enough to make changes in one sketch and the whole related geometry, i.e. the places of holes, shape of stamping, etc., change automatically during the update. Such a construction of models saves time in a situation of inevitable changes.

The present article shows a methodology of developing CAD models in order to ensure the highest level of associativeness and manufacturability. The methods are used by leading car concerns. In the development process, changes are inevitable for constructors. Therefore, attempts are made to save time on changes in CAD models' geometry. In case of logical, associative construction, a change in one parameter responsible for a function of the constructed part is enough to make changes in all the part after the update.

REFERENCES

1. Anselm D.: Die PKW-Karosserie. Vogel Business Media; Auflage: 2, Aufl., 2001.
2. Rusiński E.: Zasady projektowania konstrukcji nośnych pojazdów samochodowych. Oficyna wydawnicza Politechniki Wrocławskiej, Wrocław 2002.
3. Braess H.-H., Seiffert U.: Vieweg Handbuch Kraftfahrzeugtechnik. GWV Fachverlage, 2003.
4. http://www-home.htwg-konstanz.de, styczeń 2011.
5. http://www.rucker.pl/index.php?id=4&id2=4_1, styczeń 2011.
6. Pippert H.: Karosserietechnik. Vogel Verlag Und Druck, 1998.
7. Zieliński A.: Konstrukcja nadwozi samochodów osobowych i pochodnych. WKiŁ, Warszawa 2004.

39

IDENTIFICATION AND CORRECTION OF COORDINATE MEASURING MACHINE GEOMETRICAL ERRORS USING LASERTRACER SYSTEMS

Adam Gąska[1], Maciej Gruza[1], Piotr Gąska[1], Michał Karpiuk[1], Jerzy Sładek[1]

[1] Laboratory of Coordinate Metrology, Mechanical Department, Cracow University of Technology, Al. Jana Pawła II 37, 31-834 Kraków, Poland, e-mail: agaska@mech.pk.edu.pl; gruzam@interia.pl; pjgaska@gmail.com; pkkarpiukmichal@gmail.com; sladek@mech.pk.edu.pl

ABSTRACT

LaserTracer (LT) systems are the most sophisticated and accurate laser tracking devices. They are mainly used for correction of geometrical errors of machine tools and coordinate measuring machines. This process is about four times faster than standard methods based on usage of laser interferometers. The methodology of LaserTracer usage to correction of geometrical errors, including presentation of this system, multilateration method and software that was used are described in details in this paper.

Keywords: LaserTracer, Coordinate Measuring Machine, geometrical error.

INTRODUCTION

One of the main components that affect the accuracy of the measurement are errors of the kinematic system of the machine on which the measurement were carried out. First models of kinematic errors for coordinate measuring machines (CMMs) were created and implemented in practice in 1970's [1], however, first attempts of eliminating the machine tools geometric errors were made at the second half of XIX century [2, 3]. Nowadays, in the era of costs minimization the majority of measuring and machining devices are equipped with geometric errors software correction systems because it is more profitable to produce parts (which built kinematic system of machine) that are more distant from ideal geometry and then compensate geometric errors influenced by these faults, rather than to produce expensive parts with very narrow shape and dimension tolerances.

There are few different models of CMM errors i.e. full rigid body, reduced rigid body which determines geometrical errors described by different number of geometrical components. More often models are supplemented with elastic errors of the machine. The most common model met in the coordinate metrology consists of 21 geometric error components, which include translation, rotation and squareness errors. All of them were presented in Figure 1.

Errors presented (Figure 1) significantly affect the indication error *e* of the measuring machine. Their impact can be written mathematically by the equation [5, 6]:

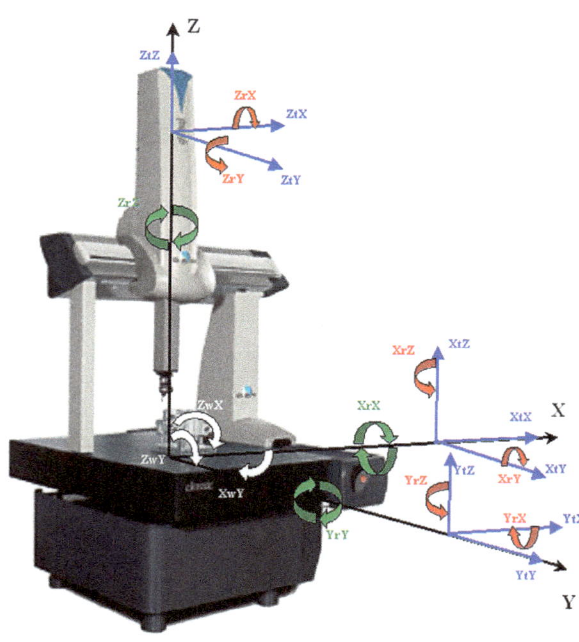

Fig. 1. Geometric errors of CMM [4]

$$e = k \cdot M \qquad (1)$$

where: $e = [e_x, e_y, e_z]$ – indication error of the machine,

$k = [ywz, xwz, xwy, ytx, ypy, ytz, yrx, yry, yrz, xpx, xty, xtz, xrx, xry, xrz, ztx, zty, zpz, zrx, zry, zrz]$ – vector containing 21 geometric errors components,

$$M = \begin{bmatrix} -z & 0 & -y & 1 & 0 & 0 & 0 & z+z_t & -y_t & 1 & 0 & 0 & 0 & z+z_t & -y_t & 1 & 0 & 0 & 0 & z_t & -y_t \\ -z & 0 & 0 & 0 & 1 & 0 & -z-z_t & 0 & x+x_t & 0 & 1 & 0 & -z-z_t & 0 & x_t & 0 & 1 & 0 & -z_t & 0 & x_t \\ 0 & 0 & 0 & 0 & 0 & 1 & y_t & 0 & -x-x_t & 0 & 0 & 1 & y_t & -x_t & 0 & 0 & 0 & 1 & y_t & -x_t & 0 \end{bmatrix}^T$$

– weighted matrix which maps the impact of each element of the k vector on the x, y, z components of indication error.

This equation is a base for correction of geometric errors of CMM using CAA (Computer Aided Accuracy) matrix. The matrix is formed by measuring geometric errors in the evenly spaced reference points in measuring area of machine. After the data obtained during the machine tests is being uploaded to the controller, it is possible to correct on-line, particular errors at any point of the measuring volume using equation (1).

As it was mentioned above, machine errors are measured experimentally only in reference points. In order to make it possible to know the values of errors at any point of measuring area, the interpolation methods have to be used. Measurements can be made using a variety of methods and devices. They will be presented in next clause.

CLASSICAL METHODS FOR GEOMETRIC ERRORS IDENTIFICATION

Classical measurement of geometric errors of CMM involves using a laser interferometer (Figure 2). These devices are characterized by very good metrological features. Measurements carried out with these devices are performed at feed rates up to 1 m/s [7] with nanometer resolution (1 nm). Single-frequency laser accommodates the electronics used for interpolation, increasing stability and counting interference fringes. The laser frequency is calibrated using a reference laser. Additionally due to the influence of environmental factors which affect on the operation of the laser [8], there are systems for compensating the wavelength depending on the conditions under which the tests are carried out, for example, temperature, pressure and humidity. With the use of laser interferometer the following components of geometric errors WMP can be determined: positioning errors of particular machine axes, perpendicularity of axes errors, straightness errors and rotation errors. For determination of each component special optical systems are needed. During researches, reference points are designated in which all measurable geometric error components are determined. Information about them is then sent to the machine controller, for example, in a tabular format. They include information about various errors for each machine axis and also about the applied measurement interval.

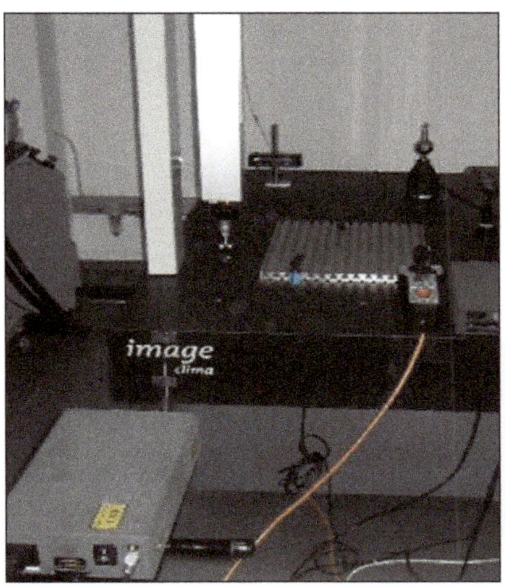

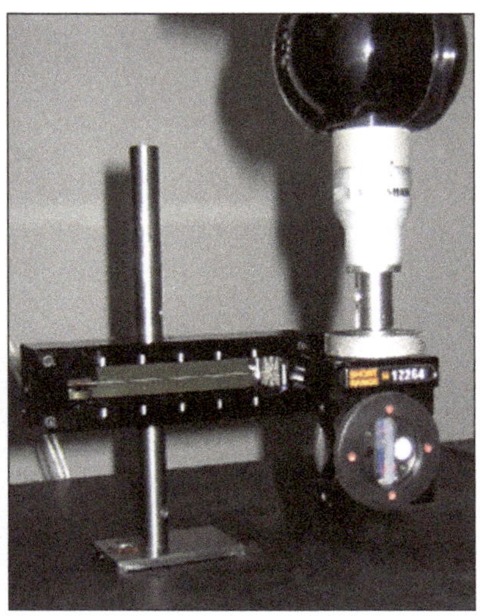

Fig. 2. Exemplary measurement of straightness errors using laser interferometer

Geometric errors can be also expressed using standards of length. Mainly 2D standards, such as a ball or hole plates are used here. One of the methods that uses this methodology is Novel method that was developed in a German laboratory PTB (Physikalisch Technische Bundesanstalt – Braunschweig). In this method, measurements of a two-dimensional standard are made to obtain information about the kinematic errors components, including the information about CMMs geometric errors. Although used standard is a two-dimensional plate standard it allows identification of errors in the entire three-dimensional measuring space by the successive arrangement of standard. The method raises a number of requirements that must be met during the tests. They refer mainly to the arrangement of standard position, the ambient temperature (20 °C), and variations of standard temperature during the measurements in one position. To estimate geometrical errors, standard plate must be measured at least six times in four predefined positions. The calibrated object is measured at three levels of the measuring area, each time using two oppositely oriented measuring tips, in the two different distances from CMM drive system guides. Every time the plate is set in parallel direction to the corresponding plane of machine coordinate system (Figure 3). It should be remembered that standard components (balls or holes) should be measured at the same points, which they were measured during the calibration. The measurements are carried out twice. The second measurement is performed in reverse order so it becomes possible to assess the reproducibility of the results. With this method the systematic changes in the dimensions of the standard in a specific direction (so called drift) can be eliminated [9].

As a result of the measurements, value and direction of a geometric errors vectors in all positions of the plate are obtained. The analysis of results involves comparing the coordinates of the holes or balls centers derived from the measurements to nominal coordinates which are known form earlier calibration. Thr information obtained from this analysis are the basis for the determination of 21 components of CMM geometric errors. With Novel method it is possible to define: the position error in each axis, straightness errors, squareness errors and rotation errors. The errors are defined as follows:

- position error is represented by the distance between points and calibrated lines,
- straightness error determined for each point is interpreted as the distance of the point from the best fit line,
- perpendicularity error is the deviation of angle between two nominally perpendicular best fit lines,
- rotation errors are divided into rotation of the axis errors and errors generated when turning in the direction of this axis.

Next method, which had been gaining a lot of interest lately, is determining the geometric errors using laser tracking systems. These machines determine the coordinates of measuring point on the basis of distance from the retro-reflector to interferometer, taking into account tracking system position angles read from angular encoders. The first attempts with tracking devices aiming in identification of geometric errors were made in the 1970's, but because of low accuracy of the trackers at the time, the results were not satisfactory. Now, when some tracking systems are able to achieve uncertainties of length measurements less than 1 micron, it is possible to successfully determine the geometric errors using this devices. The idea of determining the geometric errors is analogical to the Novel method. The biggest dif-

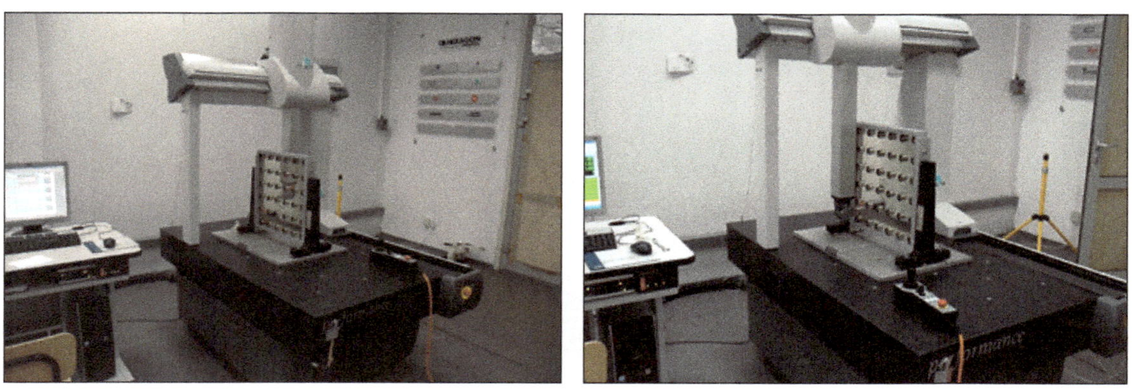

Fig. 3. Measurements of a hole plate in one of the positions recommended in Novel method [9]

ference is that instead of centers of reference elements at 2D standard, as a basis to the computation, the coordinates of points at which the retroreflector mounted on the CMM probe head (or instead of it) stopped during measuring sequence are taken.

IDENTIFICATION OF GEOMETRIC ERRORS USING LASERTRACER SYSTEM

LaserTracer (LT) system is one of the laser tracking devices mentioned in the previous section. LaserTracer (Figure 2) is an interferometric device, which allows tracking movements of the reflector mounted on the probe head of measuring machine or on tool holder of machining tool [10]. The accuracy of this device is given by:

$$U = 0.2 + 0.3 * L/1000 \mu m \qquad (1)$$

where: L – is measured length given in mm.

So high accuracy can be achieved thanks to novel solution in its construction based on reference sphere, which form errors do not exceed 30 nm. This sphere lays in the center of rotation of LT rotary system.

Application of LaserTracer for determination and correction of geometric errors of CMM is based on measuring the distance between a retroreflector mounted at the machine probe head and the LaserTracer. With laser tracking mechanism LT follows the reflector while the machine moves along a prespecified grid of points. LaserTracer works together with metrological software Trac-cal developed by Etalon company to facilitate experiment set-up (planning the mapped volume, setting the measuring path, establishing connections between devices) and processing the results. This software allows also to generate a geometric error correction matrix in a format suitable for machines offered by different manufacturers.

The described method needs to use multilateration technique as LaserTracer is able to measure only the distance (from itself to center of reflector) while determination of geometric errors requires knowledge of points coordinates in measuring volume of the machine. Multilateration is the method that uses only distance measurements from several different positions in order to determine the position of the localized object. This method was primarily used in GPS satellite navigation systems. It has also been used for many years in the so-called Internal GPS measurement systems to measure large objects, and more recently also used to correct the accuracy of the measuring machines and to create a coordinate measuring systems with a very large range.

Next part of the paper shows the exemplary usage of LaserTracer combined with Trac-cal software for geometric error identification done on Leitz PMM 12106 machine located in accredited Laboratory of Coordinate Metrology (LCM) at Cracow University of Technology.

Planning experiment with Trac-cal software

The first thing that has to be done is planning the experiment, including measuring paths. Figure 4 shows the definition of machine parameters and its measuring volume.

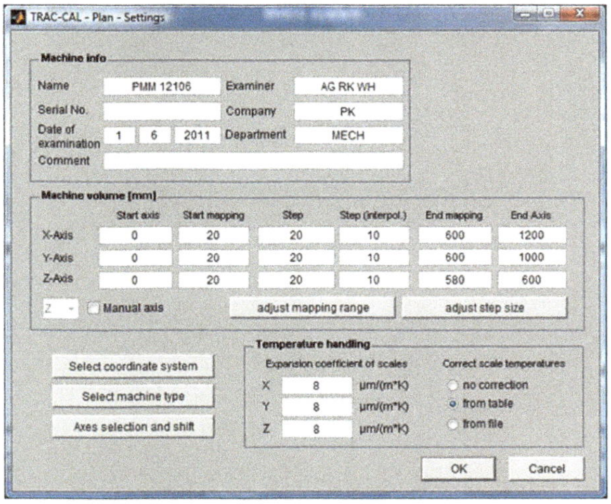

Fig. 4. Configuration of experiment

After that, the generation of measuring paths takes place. In this step, the shape of measuring paths, positions of LaserTracer in machine volume and offsets at which the retroreflector is mounted have to be determined (Figure 5).

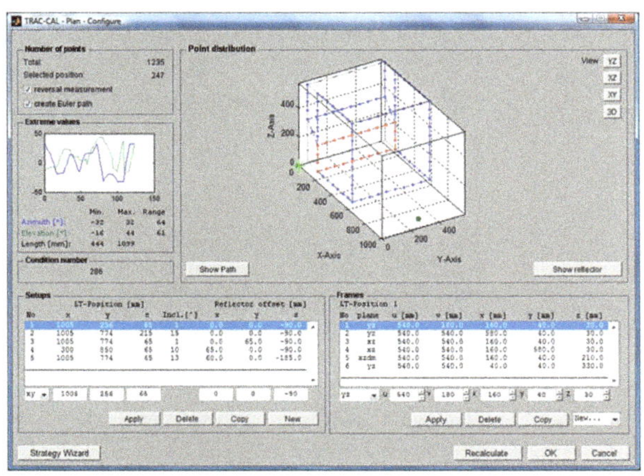

Fig. 5. Planning of the measuring sequences and LT positions

PERFORMING THE MEASUREMENTS

When the experiment is already planned, the required measurements in all specified Laser-Tracer positions have to be conducted. At least four positions of LT should be planned, but more positions could facilitate determination of all geometric error components. It is also important to change the offsets at which the retroreflector is mounted. The offset specifies the distance of reflector center to reference point at the machine probe head. If, for example, all of the sequences from all positions would be measured with reflector pointing in direction *Z* (when *Z* is a vertical axis of the machine) then determination of some rotation errors will be impossible. Figure 6 presents measurements done at LCM.

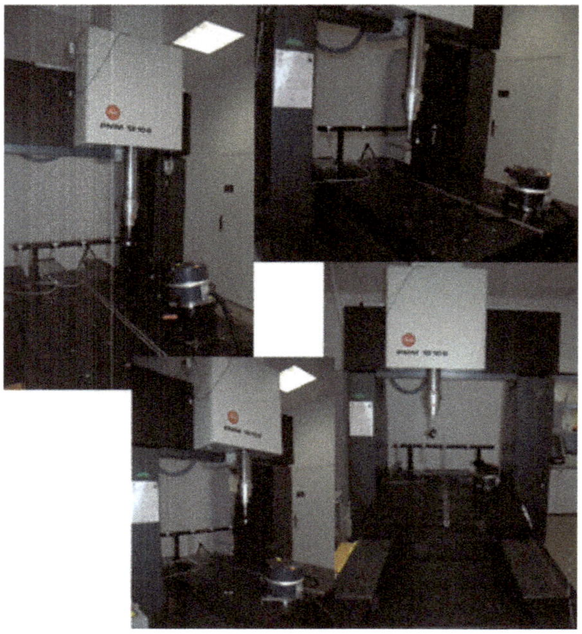

Fig. 6. LaserTracer performing the measurements at the PMM machine

RESULTS

Results could be computed after all planned measurements are finished. Trac-cal software solves the system of equation describing the multilateration method, determines the coordinates of points at which center of reflector stopped during measuring sequence and using methodology similar to that known from Novel method determines components of geometric errors. Results could be presented both in graphical and textual way (Figure 7).

Described software gives also opportunity to create automatic reports from performed inspections and generate compensation matrices for different types of machines, depending on producer and machine controller. This matrix can be then uploaded into the controller and used to compensate the identified vectors.

CONCLUSION

Increasing number of usages of laser tracking systems like LaserTracers to geometrical error compensation and results shown in many papers, including this one, proves that these kind of systems could be successfully used in that field. They can be also regarded as an helpful improvement reducing time and costs of identification and compensation of CMMs and machine tools geometrical errors. User friendly software and comparably small dimensions of LT (whole unit including its driver and cables fits into one case) makes it even more comfortable device for geometrical errors investigations.

Results obtained using LT system are comparable to those obtained using other methods

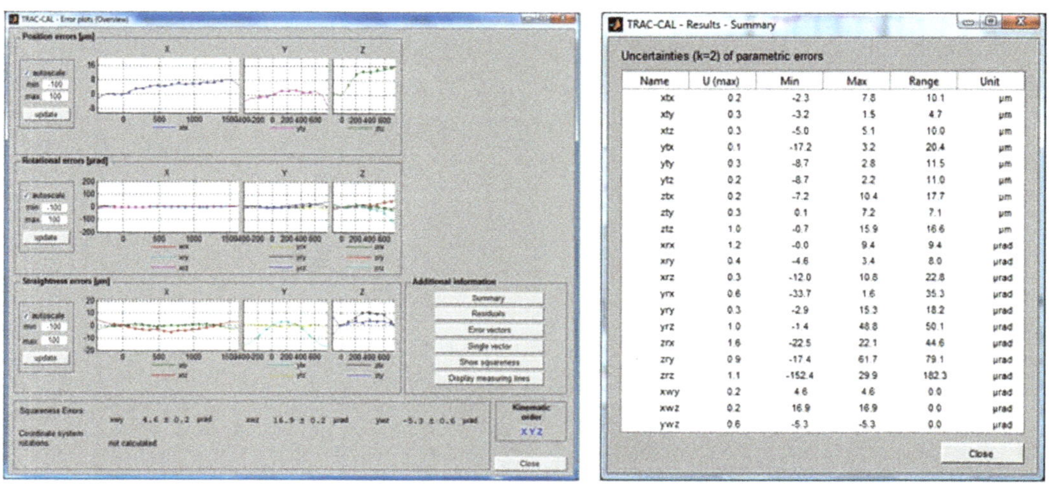

Name	U (max)	Min	Max	Range	Unit
xtx	0.2	-2.3	7.8	10.1	μm
xty	0.3	-3.2	1.5	4.7	μm
xtz	0.3	-5.0	5.1	10.0	μm
ytx	0.1	-17.2	3.2	20.4	μm
yty	0.3	-8.7	2.8	11.5	μm
ytz	0.2	-8.7	2.2	11.0	μm
ztx	0.2	-7.2	10.4	17.7	μm
zty	0.3	0.1	7.2	7.1	μm
ztz	1.0	-0.7	15.9	16.6	μm
xrx	1.2	-0.0	9.4	9.4	μrad
xry	0.4	-4.6	3.4	8.0	μrad
xrz	0.3	-12.0	10.8	22.8	μrad
yrx	0.6	-33.7	1.6	35.3	μrad
yry	0.3	-2.9	15.3	18.2	μrad
yrz	1.0	-1.4	48.8	50.1	μrad
zrx	1.6	-22.5	22.1	44.6	μrad
zry	0.9	-17.4	61.7	79.1	μrad
zrz	1.1	-152.4	29.9	182.3	μrad
xwy	0.2	4.6	4.6	0.0	μrad
xwz	0.2	16.9	16.9	0.0	μrad
ywz	0.6	-5.3	-5.3	0.0	μrad

Fig. 7. Results of geometric errors identification using LaserTracer system

mentioned in this paper. However, there are still fields in which the standard laser interferometer shows its advantage. This situation happens, for example, during compensation of high-accuracy machines. On the other hand, LaserTracer is an invaluable tool for compensation of large volume CMMs and machine tools.

LaserTracer system could also be used for other purposes. One of them is CMMs calibration and checking, which is also simplified comparing to standard methods. The other usage is determination of point reproduction error, which can be then used for uncertainty evaluation using Virtual CMMs models. LT is surely not a common device and its further abilities should be tested in near future as it has a lot of potential.

Acknowledgements

Reported research were realized within confines of project financed by Polish National Centre for Research and Development No: LIDER/06/117/L-3/11/NCBR/2012.

REFERENCES

1. Hocken R., Simpson J., Borchardt B., Lazar J., Stein P. Three dimensional metrology. Annals of the CIRP, 1977, 26M(28): 403-408.
2. Sartori S., Zhang G.X., Geometric Error Measurement and Compensation of Machines, Annals of the CIRP, 1995, 44(2).
3. Evans C.J. Precision Engineering: an evolutionary view. Cranfield Press, Cranfield, UK 1989.
4. Sładek J., Gawlik K. Looking for uncertainty of measurement – Virtual Machines based on the Matrix Method using artificial neural networks. [In:] IV International Congress on Precision Machining, Sandomierz-Kielce 2007.
5. Jakubiec W., Malinowski J. Metrologia wielkości geometrycznych. WNT, Warszawa 2004.
6. Trapet E., Hegelman R., Wäldele F. Coordinate Metrology. PTB-Bericht F-11, Braunschweig, 1992, II.
7. Gawlik J. Kowalski M., Ryniewicz A. Analysis of Measurement of Precision Machines Guides. [In:] 4th International Conference on Ultraprecision in Manufacturing Engineering „Progress in Precision Engineering and Nanotechnology". Braunschweig. Germany, 1997, 1: 319-322.
8. Schwenke H,. Knapp W,. Haitjema H,. Weckenmann A., Schmitt R., Delbressine F. Geometric error measurement and compensation of machines – an update. CIRP Annals Band, 2008, 57(2): 660-675.
9. Sładek J. Accuracy of coordinate measurements. Cracow University of Technology, Krakow 2011.
10. Sładek J., Gąska A., Olszewska M., Kupiec R., Krawczyk M. Virtual Coordinate Measuring Machine built using LaserTracer system and spherical standard. Metrology and Measurement Systems, 2013, 1.

40

EXPERIMENTAL INVESTIGATION OF THE IMPACT OF FLIGHT SPEED ON DRAG FORCE IN THE AUTOGYRO MODEL

Zbigniew Czyż[1], Paweł Magryta[1], Marcin Szlachetka[2]

[1] Department of Thermodynamics, Fluid Mechanics and Aviation Propulsion Systems, Faculty of Mechanical Engineering, Lublin University of Technology, Nadbystrzycka 36, 20-618 Lublin, Poland, e-mail: z.czyz@pollub.pl; p.magryta@pollub.pl

[2] Department of Mechanics and Mechanical Engineering, Pope John Paul II State School of Higher Education in Biala Podlaska, Sidorska 95/97, 21-500 Biała Podlaska, Poland, e-mail: m.szlachetka@dydaktyka.pswbp.pl

ABSTRACT

The paper presents the experimental investigation of the impact of velocity on drag force in the autogyro model. One of the methods which simulate motion of the flying object consists of using a wind tunnel. In this case, test object is stationary and the motion of air is forced by e.g. a special fan. The costs related with renting and the wind tunnel service are still very high. In this paper, the motion of the autogyro with respect to the air, was produced by fixing this model with scale to measure the drag force on the passenger car roof. The position of the object relative to the vehicle was checked on the basis of numerical analysis of the airflow around this vehicle. Based on the investigations, the field of velocity and pressure, and air flow formed around the contour of the vehicle which have been chosen, were determined. In addition, the drag force characteristic was determined as a function of velocity and it was compared with the values from the numerical analysis. This research is a form of verifying opportunities for this type of research on vehicles. The conclusions derived from the analysis of the results will be used in the future to carry out further research.

Keywords: autogyro, aerodynamics, CFD, wind tunnel, airflow.

INTRODUCTION

Aerodynamic experimental investigation allow to obtain aerodynamic characteristics, to verify and investigate the stability, and to verify hazardous flight conditions of aircraft or other research objects. One way to simulate the motion of flying objects is the use of the wind tunnel. In this case, the test object is stationary relative to the ground and the air movement is forced by a special fan. This allows us to perform a lot more research than in real flight conditions. Of course, in addition to the wind tunnel, advanced measuring equipment that measure both the loads acting on the test object and visualization of flow around it are often required. Research and measurement in wind tunnels associated with high costs of rent and service. In addition, the size of the research object is limited by the size of the wind tunnel test section. Often, the object must be scaled in order to ensure the correct proportions in relation to the measuring chamber [1].

The dimensions of the tunnel are the barrier in the case studies, which should be reflected in real scale. If the model test results are to be referenced to the real objects certain dependencies, which are called the similarity laws must be complied. They take into account the geometrical, kinematic and dynamic criteria. Geometric similarity is a consequence of the need to introduce, the so-called, linear scale. Linear scale is defined as the ratio of the linear dimension of the model corresponding to the linear dimension of the real object. Geometric similarity is relatively easy to obtain. But in the case of kinematic similarity, kinematics of individual components including the mapping of all the twists and fluctuations in the position of objects relative to the airflow should also be similar. Dynamic similarity is much harder to keep. This type of similarity is observed when the forces acting on the real object and the model are consistent with the terms of similarities. For this purpose, it is necessary to keep the number of criterion, among which we can find the Mach number, the coefficient of forward velocity, Froude number, Strouhal number or Reynolds and Locke number [2].

The more we reduce the object size, the number of criterial and transform results on a real object are more difficult to keep. Department of Thermodynamics, Fluid Mechanics and Aviation Propulsion Systems at Lublin University of Technology is equipped with a GUNT HM 170 wind tunnel which has the cross-section of the test chamber 292×292 mm respectively. In order to examine EM-11 Orka plane with a span of 13.5 m in this tunnel, it should be reduced at least 67 times. If we do so, the Reynolds number, which is the ratio of the aerodynamic forces of inertia to air viscosity should increase the flow rate by the same times. Even in the case of the minimum velocity for the aircraft, which is 90 km/h, the velocity in wind tunnel should be equal to 6030 km/h. Only in this case, this number does not meet the criterion (with the same working medium). It is also not clear if the fact that the phenomena occurring at the actual flow around the aircraft and the model will be the same. In addition, such a high flow velocity are not possible to achieve using this wind tunnel and we will be not able to meet different criterion numbers. Recognizing the need to study larger objects, it was decided to verify the alternative method, in which the movement of the autogyro (research object) to air was produced by fixing the autogyro with scales to measure the drag force on the car roof. So it is a moving car that forces the air flow around the test object. This method, depending on the used scale makes possible to test individual components provides a measure of the aerodynamic forces and moments.

The results of resistance force obtained from the experiment were compared to the results obtained by numerical calculations. Numerically obtained velocity field of pressure and air flow formed around the contours of the vehicles allowed accurately assess the measurement in conditions that were shown.

RESEARCH OBJECT AND METHODOLOGY

The object of the research was autogyro model, called Gyro - One, which is shown in Figure 1. This is an unmanned RC aircraft, which is propelled using 1000 W electric engine with a maximum take-off weight of about 4 kg. Autogyro is over 1.050 mm in length and 510 mm in height. Main rotor diameter is 1800 mm. More details about the object are added in a different paper [3].

Fig. 1. Research object Gyro - One

Autogyro is mounted on the balance scale equipped with a strain gauge for measuring an additional resistance force (Fig. 2). Autogyro is fixed to the mast (1) mounted on a linear bearing (6). Under the pressure of the air flow it produces the tension on the strain gauge (2) by a rod (7). The signal from the strain gauge is processed by measuring cards and read using a program developed in LabVIEW software.

The whole structure was mounted on a vehicle with a luggage rack. For the experiment, Opel Vectra sedan B was used. Autogyro was po-

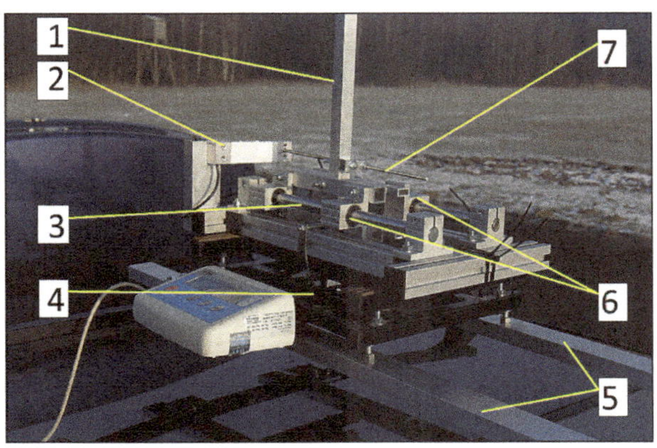

Fig. 2. View of balance scale, 1. mast, 2. strain gauge to measure the power of drag force, 3. rails, 4. platform for scales to measure lift, 5. stringers, 6. linear bearings

sitioned at a height of about 670 mm measured from the surface of the roof. Despite the completed tests including the main rotor rotation, this paper presents only the results of the research object without the main rotor blades.

Tests were carried out in a qualified Technician Improvement Centre of Driving – Moto Park in Ulez. This place is adopted from the old military airport. There is a main starting lane with a length of 1500 m and a width of 30 m. It allowed for carrying out the tests in which the influence of the ground is smaller than on random street. The decision to undertake research on a rented former military airport has also been taken for security reasons.

Fig. 3. View autogyro placed on a vehicle

NUMERICAL METHOD AND BOUNDARY CONDITIONS

Experimental studies have been extended by a simulate flow around the vehicle and autogyro. As the main object of interest is the autogyro model, it was decided that for this object the calculations will be carried out in 3D and in the case of a vehicle a 2D simulation will be made, to find out the field of flow around the vehicle in the longitudinal plane passing through the center of the car, which may help in the analysis of results. For this purpose, the contour of the vehicle (Fig. 4) was traced in Catia v5 and then imported to Design Modeler (Ansys Fluent). After completing the grid, the computing model was loaded into the solver (Ansys Fluent). Table 1 shows the general settings for the numerical analysis of flow around the vehicle geometry.

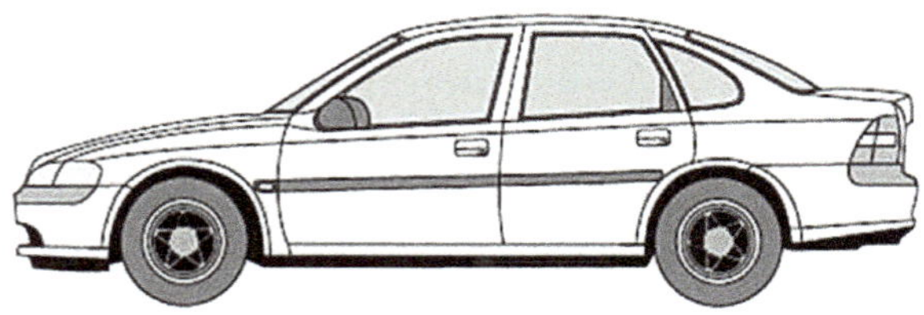

Fig. 4. The two-dimensional draft of a vehicle Opel Vectra B – side view [4]

Autogyro 3D model was done in CATIA V5 software. Figure 5 shows the views of a solid model of the autogyro. The model is without head rotor, blades and an engine. In addition, aircraft interior components, such as accumulators, receiver, cabling were not included in the modeling. It is justified by the type of performed analysis, which is an external flow around the geometry of the tested object. Further simplification involved small, non-significant changes like fasteners –

Table 1. General settings of numerical analysis for the airflow around the vehicle

Basic parameter	Type	Pressure-based	
	Time	Steady	
Models	Viscous model	k-omega (2-eqn) SST	
Materials	Fluid	Air	
	Density	Incompressible-ideal-gas	
	Viscosity	Constant	1.7894e-05
Boundary conditions	Inlet	Velocity	72 [km/h]
		Turbulent intensity [%]	10
		Hydraulic diameter [m]	14
	Outlet	Pressure-outlet	gauge pressure 0 Pa
		Turbulent intensity [%]	10
		Hydraulic diameter [m]	14

Table 2. General settings of numerical analysis for the airflow around the autogyro geometry

Basic parameters	Type	Pressure-based	
	Time	Steady	
Models	Viscous model	k-omega (2-eqn) SST	
Materials	Fluid	Air	
	Density	Incompressible-ideal-gas	
	Viscosity	Constant	1.7894e-05
Boundary conditions	Inlet	Velocity	0–90 [km/h]
		Turbulent intensity [%]	1
		Turbulent length scale [m]	0.28
	Outlet	Pressure-outlet	gauge pressure 0 Pa
		Turbulent intensity [%]	1
		Turbulent length scale [m]	0.28

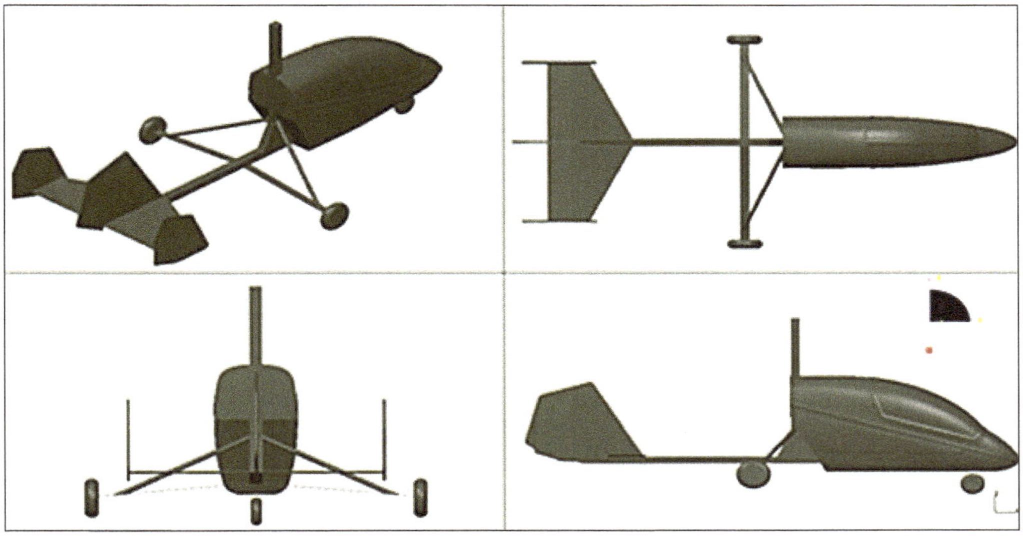

Fig. 5. Geometric model of the autogyro prepared for the CFD analysis

bolts, nuts, and holes. Table 2 shows the general settings for the numerical analysis of flow around the autogyro geometry.

RESULTS AND DISCUSSIONS

Figure 6 shows the drag force as a function of velocity of the moving vehicle. Results recorded from several testes were approximated with a second-degree polynomial line.

The drag force from experimental research as a function of the velocity of vehicle were compared with the results of numerical calculations. It turns out that the drag force of the numerical calculations account for about 50% of the experimental results.

In order to explain such a large differences it was decided to investigate the velocity and the pressure field around the used vehicle. Figure 8 shows the pressure field around the vehicle. We can observe that at the transition between the surface of the glass and the roof area, the pressure is about –0.6713 kPa. Static pressure drop causes an increase in the flow rate as it is shown in Figure 9. Undisturbed air stream velocity is 20 m/s (the boundary condition of numerical analysis), and increases nearly doubled in the largest vacuum.

In order to determine the velocity field around the autogyro, additional figure 10 was shown, in which autogyro model was applied into the velocity field map.

It can be seen that the velocity field, where the autogyro is mounted is not uniform. In the lower part of the object of the velocity is about 29 m/s while at the top it decreases to about 25 m/s. For simplicity, we can assume that the average speed acting on the model is 27 m/s. If we consider this value of the velocity in the calculation of the drag

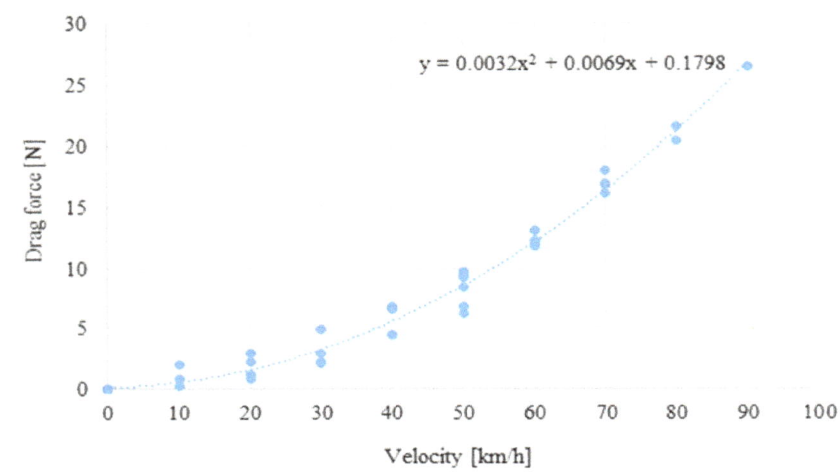

Fig. 6. The drag force as a function of velocity obtained from experimental studies

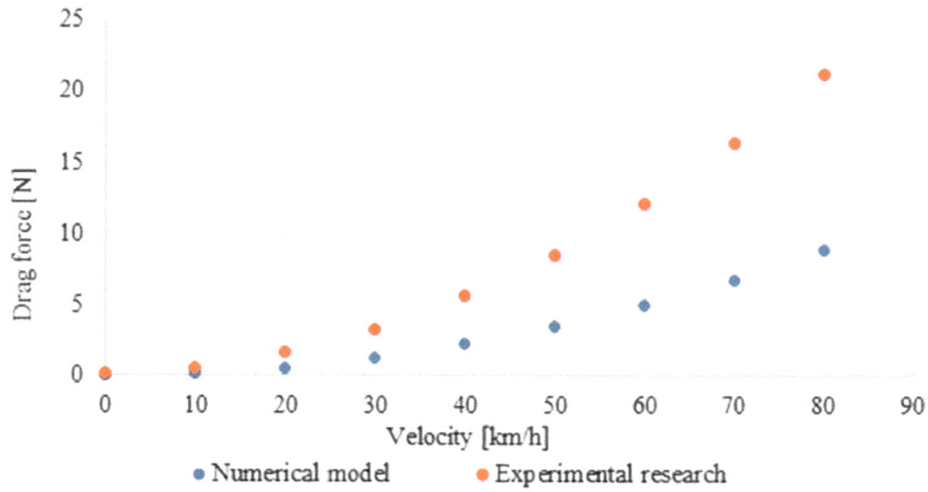

Fig. 7. Comparison of the drag forces as a function of the vehicle velocity for the calculation model and experiment

Fig. 8. Pressure field around the vehicle

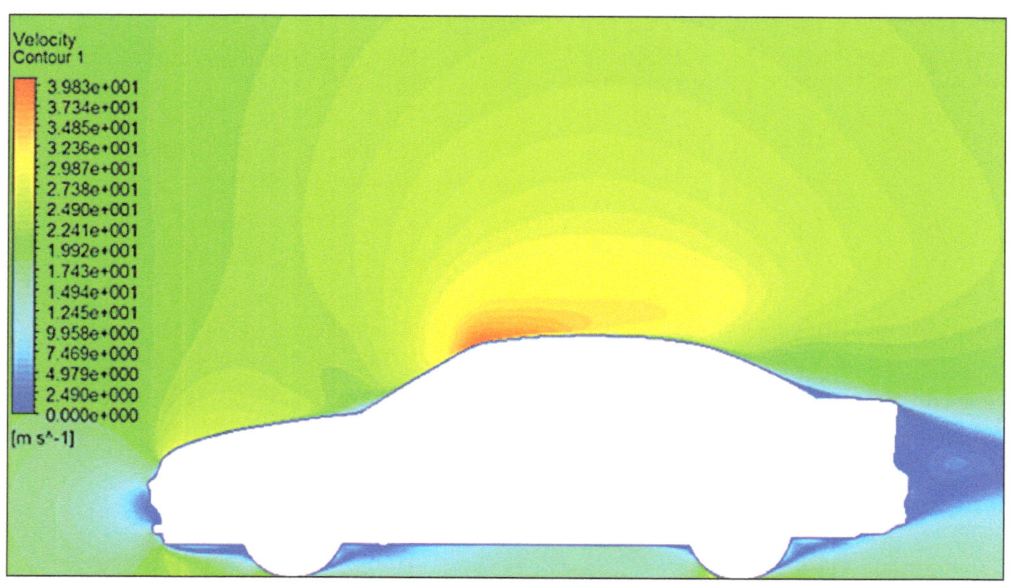

Fig. 9. Velocity field around the vehicle

Fig. 10. Velocity field around the vehicle with the autogyro

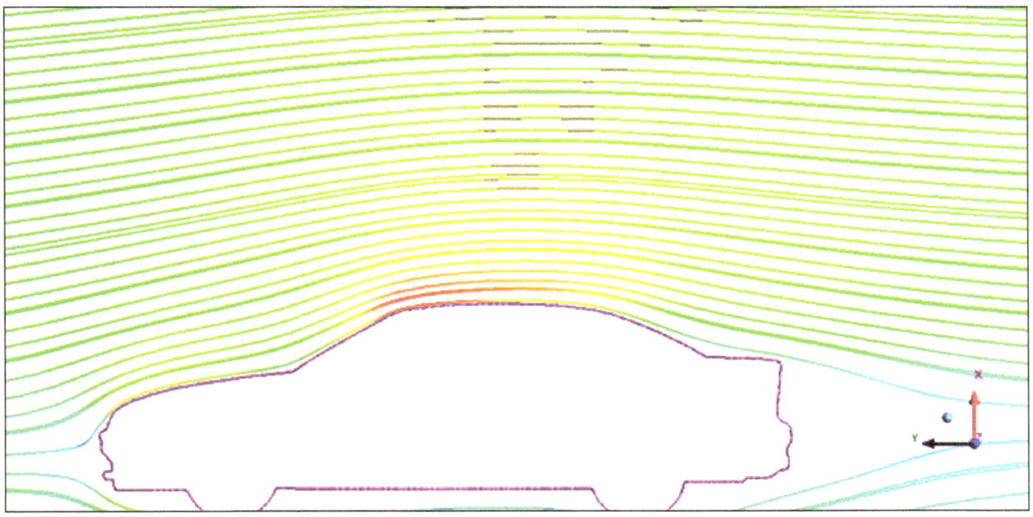

Fig. 11. Air streams around the vehicle

force (1) it turns out, that the component v^2 in numerical computation is equal to 400 m^2/s^2 (for 20 m/s) which corresponds to 54.9% in relation to the actual test (729 m^2/s^2 for the 27 m/s).

$$P_x = 0.5 \cdot C_x \cdot \rho \cdot A \cdot v^2 \quad (1)$$

Except for the velocity and pressure field we should pay attention also to the shape of the air streams which surround the vehicle. It turns out that even the addition of significant changes in velocity and pressure in the cross-sectional shape, the air flow is also affected (there is a deviation of flow) and in a relatively large distance from the roof surface. Deviation of the velocity vector indeed affect the aerodynamic force component.

CONCLUSIONS

Confrontation of experimental methods for measuring aerodynamic forces with the numerical method allows for the verification of the test results. As it is shown in the article, the alternative way of research is less expensive than wind tunnel tests, however, it requires consideration of a number of external factors affecting the results. Computational model almost always differs from real experiments. It has idealized conditions that it makes easy to repeat the steady boundary or initial conditions. In case of real research, we often receive a scatter of results, from which we draw a trend line. Not focusing on the drawbacks of numerical analysis, we can conclude that the disparity of experimental and

CFD simulation results is due to a big impact of air velocity around the vehicle. The geometric model used in the numerical analyzes was slightly different from the object used in the experiment. Some of the elements that produce additional drag force (mast, components mounted on slides of linear bearings, rotor head) were not properly covered.

Research allows to estimate the value of aerodynamic force components. A more detailed analysis with the additional use of CFD allows for the correct interpretation of the results. Further research should take into account all the above-described issues with particular emphasis on the selection of a vehicle and platform on which the test object will be attached.

REFERENCES

1. Krysztofiak G.: Tunelowe badania wpływu usterzenia motylkowego na charakterystyki aerodynamiczne modelu kadłuba wiatrakowca. Modelowanie inżynierskie 42, 2011, 227–232.
2. Krysztofiak G.: Kryteria podobieństwa aerodynamicznego w tunelowych badaniach modeli wirników wiropłatowych statków powietrznych. Modelowanie inżynierskie 42, 2011, 217–226.
3. Czyż Z., Wendeker M., Raczyński R.: Koncepcja hybrydowego statku powietrznego z napędem wielowirnikowym. Logistyka 6, 2014, 2936–2945.
4. http://opelastra24.eu.interiowo.pl/modele/opel_vectra_b_wymiary.html

ROTOR TURBULENCE INFLUENCE ON HELICOPTER FLIGHTS IN HIGH URBAN BUILT-UP AREA

Tomasz Łusiak[1], Andżelika Grudzień[2]

[1] Mechanical Engineering Faculty, Lublin University of Technology, Nadbystrzycka 36, 20-618 Lublin, Poland, e-mail: t.lusiak@pollub.pl

[2] 4th year student in The State School of Higher Education in Chelm, Poland.

ABSTRACT

The paper provides a discussion of the influence of turbulence in the areaa of high urban buildings or in vicinity of fire on safety of helicopter flights. The analysis was conducted using Ansys Fluent software. All the threats considering helicopter flight, landing and hovering in such an environment were shown. As objects of this research helicopter types: W3-A Sokół, W3-A Głuszec and Robinson R-44 were used.

Keywords: helicopter, rotor, turbulence, urban area.

INTRODUCTION

The purpose of this publication is to reveal the threats associated with light helicopter flights in air turbulence induced by building flow round or different types of fires [1, 3]. This work also defines an effect of disturbances due to the resultant of flow interference generators (such as rotor blades, fuselage solids, suspended components and fixed lift generating surfaces) on their loads, dynamics of motion and operational parameters of the helicopter [2, 4]. Dealing with this issue contributed to explanation of previously unrecognized phenomena which have significant impact on helicopter glide ratio or safety during flight in particular conditions as appearance of vortex ring state on lift or tail rotor, flow separation either on rotor blades and fixed lift surfaces or field interference induced by helicopter rotors with obstacles located in the vicinity of flight path (e.g. with ship deck, shape of buildings, nearby which rescue operations are conducted or during take-offs and landings on elevated heliports such as oil rigs and hospitals).

METHODS

Appropriate computational-experimental tools were developed to enable solving aerodynamic and operating problems and thus upgrade helicopters' design and modify nowadays. In model using Navier-Stokes equation solutions, an area of turbulent, disturbed medium flow was determined, introducing only geometrical parameters of environment solids (including helicopter) and models of rotors defined as active step of pressure generators. In order to conduct an accurate analysis of selected aspects of aerodynamic interference the real helicopter model was replaced by a vortex model.

Flow interference around a helicopter is an interaction between disturbances caused by wake turbulence of rotors, fixed lift surfaces, such as stabilizers and auxiliary wing, and presence of fuselage solid. Physical helicopter model used to examine interference issues, for assessment disturbances of field flow using uniform Biot-Savart method, was replaced by vortex meshwork. Another approach concerns mixed methods e.g. flow potential for assessing fuselage flow round. Fuselage, for vortex model is featured as system of vortex panels, and lift generating components as vortex mesh which generates flow field velocity disturbances. Using capabilities of the featured methods, including modal models of rotors and lift surfaces loads, modified in every step of the related vortexes intensity; the presented equation constitutes an interference resolving essence in non-stationary motion of a helicopter. A load waveform in the function of time is also an outcome of the project.

RESULTS

Numerical calculations of the influence aerodynamic interference on helicopter loads were conducted in two stages. For the assumed pressure distribution on rotor (in the first approximation constant distribution was assumed), using FLUENT software, helicopter flow round was determined with consideration for objects in close vicinity. Thus velocity field of flow across the plane of rotor was determined. Subsequently using OBCWN software (developed in Institute of Aviation, Warsaw, and then accordingly modified to meet the requirements of aerodynamic interference analysis) to determine rotor loads. Using this introduced airflow, rotor load characteristics and new pressure distribution o its surface were determined. New pressure distribution was introduced for the recalculation of helicopter flow round in FLUENT software for compensation (Fig. 1-2).

Iteration process was conducted until results' concurrence was obtained. In order to expedite concurrence process, flow velocity field, being an average of previous iteration and current calculations was loaded for subsequent iterations. For rotor blade of high elasticity such exertion was vital due to the significant blade sensitivity for disturbance in the form of blade deflection function. With such an intervention, satisfactory concurrence was obtained in the second calculations step. Numerical calculations of main rotor loads were conducted for a given flow field, calculated with FLUENT software (Fig. 3-4).

Thanks to coupling the results from FLUENT software with OBCWN software, used to determine main rotor blades loads and deformations,

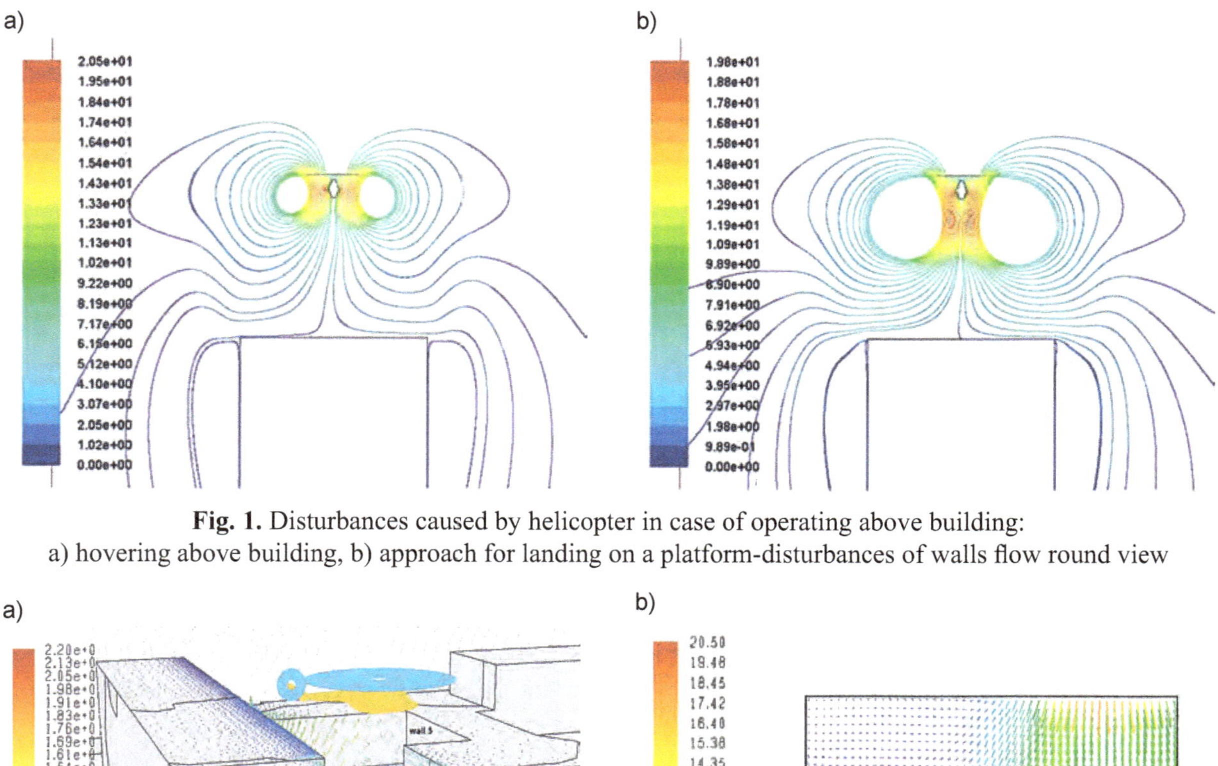

Fig. 1. Disturbances caused by helicopter in case of operating above building:
a) hovering above building, b) approach for landing on a platform-disturbances of walls flow round view

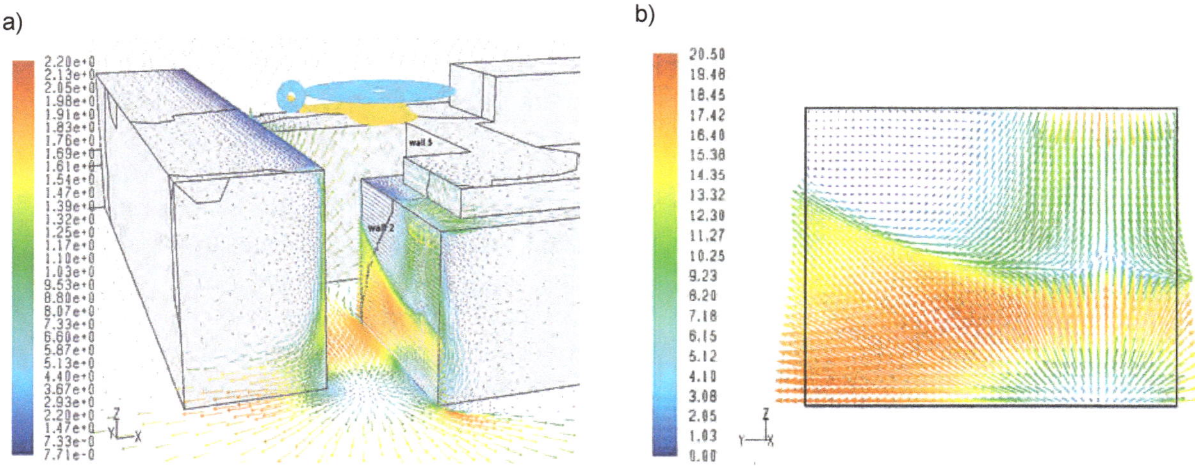

Fig. 2. Another view with walls designations and wall nr 2 flow round view

a) b)

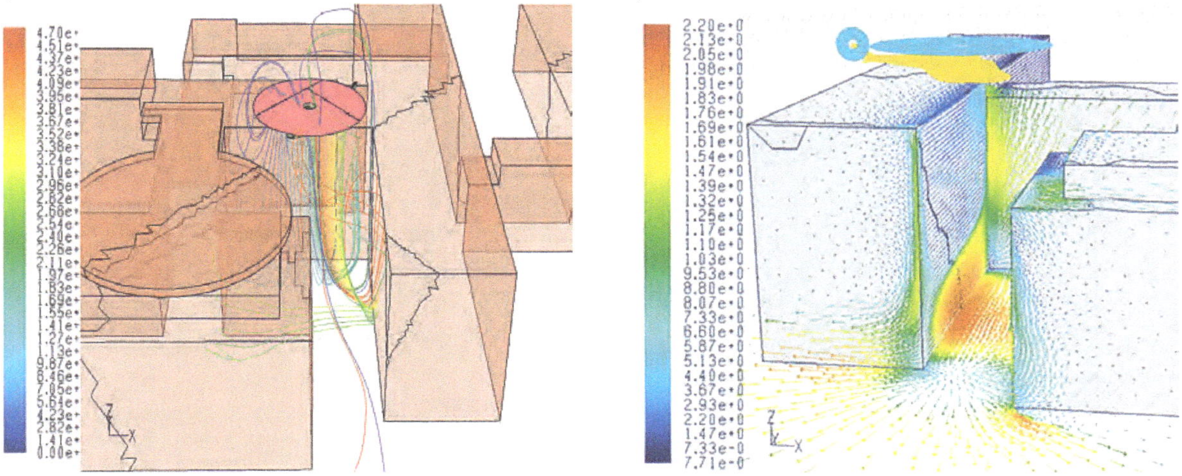

Fig. 3. Disturbances caused by helicopter during roof helipad operations:
a) approach for landing on a platform, b) view of building walls flow round disturbances

a) b)

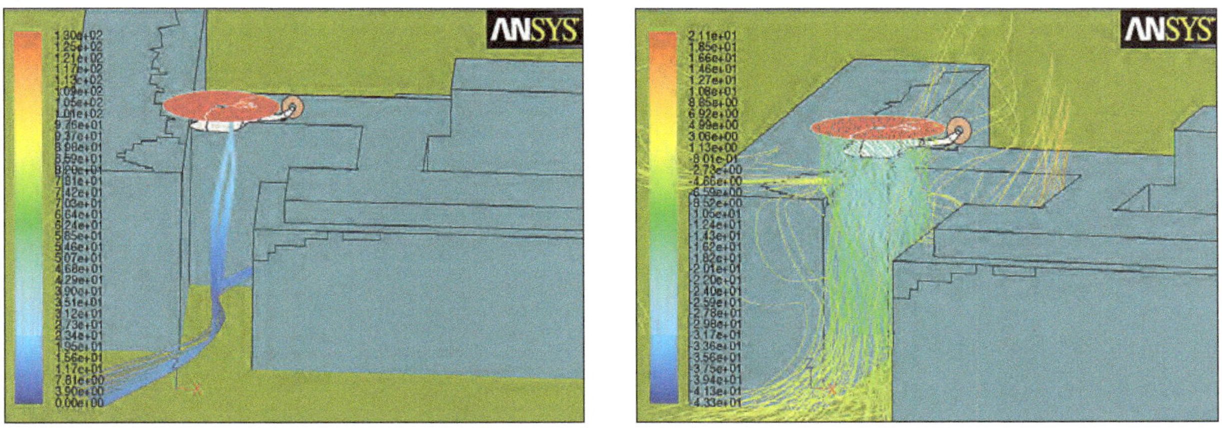

Fig. 4. Disturbances caused by helicopter during roof helipad operations:
a) exhaust streams from engine nozzles, b) rotor stream lines flowing building around

the problem of aerodynamic interference can be analyzed. Thanks to this calculations algorithm, it was possible to analyze rotors with blades of different mass-stiffness and aerodynamic characteristics. For a given moment of time, motion parameters and blades loads, located at various azimuths on rotor disc, were determined. After summing impacts of all the blades, rotor shaft strain, for this particular moment of time, was determined. Aeroelastic blades deformation, aerodynamic load of single blades, net forces of rotor loads and the whole helicopter system, were obtained, using OBCWN software. Empirical studies, using real object – W3-A "Sokół" helicopter, were conducted at PZL Świdnik S.A. factory aerodrome.

SUMMARY

The characteristics of velocity field distribution of main rotor downwash, during hovering at a low altitude, were obtained. The results from researches, mentioned above, were used in comparative analysis with computer simulations results. There were also researches conducted in Nowy Glinnik Air Force Base, using two W3-A "Głuszec" helicopters, during which, horizontal component of downwash characteristics were obtained for interference phenomenon between two helicopters during take-off phase.

The results of the studies on interference phenomenon, may also conduce in helicopter design stage and for aerodynamic characteristics improvement. Thus, by proper selection of system parameters, helicopter's performance may be enhanced and some improvement can be noticed in noise abatement or comfort e.g. lower vibration level. Another area where interference research is vital is the area of pilots operating helicopters. Helicopter flights in high urban build-up agglomeration territory or in the mountains causes nowadays situations, in which helicopter flight in very close vicinity to obstacles of various shapes is necessary. This area will be a further stage of scientific research in the future. Additional influence of wind may cause significant amplification of helicopter aerodynamic interference and may also worsen flight characteristics or lead to dangerous situations.

REFERENCES

1. Bramwell A.R.S.: Helicopter Dynamics. Butterworth-Heinemann, Oxford 2001.
2. Bukowski J.: Mechanika płynów. PWN, Warszawa 1971.
3. Elsner J.W.: Turbulencja przepływów. PWN, Warszawa 1987.
4. Juriew B.N.: Aerodinamiczeskij raszczet wiertoletow. Obarangiz, Moskwa 1956.

Permissions

All chapters in this book were first published in ASTRJ, by Society of Polish Mechanical Engineers and Technicians; hereby published with permission under the Creative Commons Attribution License or equivalent. Every chapter published in this book has been scrutinized by our experts. Their significance has been extensively debated. The topics covered herein carry significant findings which will fuel the growth of the discipline. They may even be implemented as practical applications or may be referred to as a beginning point for another development.

The contributors of this book come from diverse backgrounds, making this book a truly international effort. This book will bring forth new frontiers with its revolutionizing research information and detailed analysis of the nascent developments around the world.

We would like to thank all the contributing authors for lending their expertise to make the book truly unique. They have played a crucial role in the development of this book. Without their invaluable contributions this book wouldn't have been possible. They have made vital efforts to compile up to date information on the varied aspects of this subject to make this book a valuable addition to the collection of many professionals and students.

This book was conceptualized with the vision of imparting up-to-date information and advanced data in this field. To ensure the same, a matchless editorial board was set up. Every individual on the board went through rigorous rounds of assessment to prove their worth. After which they invested a large part of their time researching and compiling the most relevant data for our readers.

The editorial board has been involved in producing this book since its inception. They have spent rigorous hours researching and exploring the diverse topics which have resulted in the successful publishing of this book. They have passed on their knowledge of decades through this book. To expedite this challenging task, the publisher supported the team at every step. A small team of assistant editors was also appointed to further simplify the editing procedure and attain best results for the readers.

Apart from the editorial board, the designing team has also invested a significant amount of their time in understanding the subject and creating the most relevant covers. They scrutinized every image to scout for the most suitable representation of the subject and create an appropriate cover for the book.

The publishing team has been an ardent support to the editorial, designing and production team. Their endless efforts to recruit the best for this project, has resulted in the accomplishment of this book. They are a veteran in the field of academics and their pool of knowledge is as vast as their experience in printing. Their expertise and guidance has proved useful at every step. Their uncompromising quality standards have made this book an exceptional effort. Their encouragement from time to time has been an inspiration for everyone.

The publisher and the editorial board hope that this book will prove to be a valuable piece of knowledge for researchers, students, practitioners and scholars across the globe.

List of Contributors

Radosław Rutkowski
Faculty of Maritime Technology and Transport, West Pomeranian University of Technology, 41 Piastów Str., 71-065 Szczecin, Poland

Ewa Siemionek
Department of Automotive Vehicles, Faculty of Mechanical Engineering, Lublin University of Technology, 36 Nadbystrzycka Str., 20-618 Lublin, Poland

Matúš Košinár
Department of Automation and Production Systems, Faculty of Mechanical Engineering, University of Žilina, SK-010 26 Žilina, Slovak Republic

Ivan Kuric
Department of Automation and Production Systems, Faculty of Mechanical Engineering, University of Žilina, SK-010 26 Žilina, Slovak Republic

Mustafa Arda
Trakya University, 22030 Balkan Yerleşkesi, Edirne, Turkey

Metin Aydogdu
Trakya University, 22030 Balkan Yerleşkesi, Edirne, Turkey

Gabriel Nowacki
Military University of Technology, Cybernetics Faculty, Sylwester Kaliski Street 2, 00-908 Warsaw 49, Poland

Anna Niedzicka
Motor Transport Institute, Management and Telematics Center, Jagiellońska 80, 03-301 Warsaw, Poland

Cezary Krysiuk
Motor Transport Institute, Section of Planning and Budget, Jagiellońska 80, 03-301 Warsaw, Poland

Remigiusz Romuald Iwańkowicz
Faculty of Maritime Technology and Transport, West Pomeranian University of Technology, Al. Piastów 41, 71-065 Szczecin, Poland

Mariusz Sobolak
Department of Mechanical Engineering, Rzeszow University of Technology, al. Powstańców Warszawy 12, 35-959 Rzeszów, Poland

Piotr Strojny
Department of Mechanical Engineering, Rzeszow University of Technology, al. Powstańców Warszawy 12, 35-959 Rzeszów, Poland

Andrzej Puchalski
University of Technology and Humanities in Radom, Malczewskiego 29, 26-600 Radom, Poland

Leszek Semotiuk
Department of Production Engineering, Mechanical Engineering Faculty, Lublin University of Technology, 36 Nadbystrzycka Str., 20-816 Lublin, Poland

Jerzy Józwik
Department of Production Engineering, Mechanical Engineering Faculty, Lublin University of Technology, 36 Nadbystrzycka Str., 20-816 Lublin, Poland

Kuric Ivan
Department of Automation and Production Systems, Faculty of Mechanical Engineering, University of Žilina, 1 Univerzitná Str., 010-26 Žilina, Slovakia

Arkadiusz Macheta
Department of Hematooncology and Bone Marrow Transplantation, Medical University of Lublin, Al. Racławickie 1, 20-059 Lublin, Poland

Katarzyna Radko
Department of Hematooncology and Bone Marrow Transplantation, Medical University of Lublin, Al. Racławickie 1, 20-059 Lublin, Poland

Agnieszka Szymczyk
Students Scientific Association at Department of Hematooncology and Bone Marrow Transplantation Medical University of Lublin, Al. Racławickie 1, 20-059 Lublin, Poland

Monika Podhorecka
Department of Hematooncology and Bone Marrow Transplantation, Medical University of Lublin, Al. Racławickie 1, 20-059 Lublin, Poland

Piotr Ignaciuk
Institute for Transport, Internal Combustion Engines and Ecology, Lublin University of Technology, Nadbystrzycka 36, 20-618 Lublin, Poland

Leszek Gil
Department of Transport and Computer Science, University of Economics and Innovation in Lublin, Projektowa 4, 20-209 Lublin, Poland,

Anna Rudawska
Department of Production Engineering, Faculty of Mechanical Engineering, Technical University of Lublin, 36 Nadbystrzycka Str., 20-618 Lublin, Poland

Agnieszka Seń
Graduate student of Faculty of Mechanical Engineering,, Technical University of Lublin, Poland

Leszek Semotiuk
Department of Production Engineering, Faculty of Mechanical Engineering, Technical University of Lublin, 36 Nadbystrzycka Str., 20-618 Lublin, Poland

Filip Górski
Chair of Management and Production Engineering, Poznań University of Technology, Piotrowo 3, 60-965 Poznań, Poland

Radosław Wichniarek
Chair of Management and Production Engineering, Poznań University of Technology, Piotrowo 3, 60-965 Poznań, Poland

Wiesław Kuczko
Chair of Management and Production Engineering, Poznań University of Technology, Piotrowo 3, 60-965 Poznań, Poland

Przemysław Zawadzki
Chair of Management and Production Engineering, Poznań University of Technology, Piotrowo 3, 60-965 Poznań, Poland

Paweł Buń
Chair of Management and Production Engineering, Poznań University of Technology, Piotrowo 3, 60-965 Poznań, Poland

Paweł Lonkwic
Lublin Factory of Personal Lifts LIFT SERVICE S.A., Roztocze 6, 20-722 Lublin, Poland

Kamil Szydło
Lublin Factory of Personal Lifts LIFT SERVICE S.A., Roztocze 6, 20-722 Lublin, Poland

Kasper Górny
Institute of Machines and Motor Vehicles, Poznan University of Technology, Piotrowo 3, 60-965 Poznań, Poland

Przemysław Tyczewski
Institute of Machines and Motor Vehicles, Poznan University of Technology, Piotrowo 3, 60-965 Poznań, Poland

Wiesław Zwierzycki
Institute of Machines and Motor Vehicles, Poznan University of Technology, Piotrowo 3, 60-965 Poznań, Poland

Jacek Caban
Institute of Transport, Combustion Engines and Ecology, Faculty of Mechanical Engineering, Lublin University of Technology, 36 Nadbystrzycka Str., 20-618 Lublin, Poland

Agata Gniecka
Maria Curie-Skłodowska University, Plac Marii Skłodowskiej-Curie 5. 20-031 Lublin, Poland

Lukáš Holeša
Department of Road and Urban Transport, Faculty of Operation and Economics of Transport and Communications, University of Žilina, Univerzitná 1, SK-010 26 Žilina, Slovakia

Maciej Gruza
Laboratory for Coordinate Metrology (M-10), Faculty of Mechanical Engineering, Cracow University of Technology, Jana Pawła II 37, 31-864 Kraków, Poland

Wiktor Harmatys
Laboratory for Coordinate Metrology (M-10), Faculty of Mechanical Engineering, Cracow University of Technology, Jana Pawła II 37, 31-864 Kraków, Poland

Piotr Gąska
Laboratory for Coordinate Metrology (M-10), Faculty of Mechanical Engineering, Cracow University of Technology, Jana Pawła II 37, 31-864 Kraków, Poland

Adam Gąska
Laboratory for Coordinate Metrology (M-10), Faculty of Mechanical Engineering, Cracow University of Technology, Jana Pawła II 37, 31-864 Kraków, Poland

Jerzy Sładek
Laboratory for Coordinate Metrology (M-10), Faculty of Mechanical Engineering, Cracow University of Technology, Jana Pawła II 37, 31-864 Kraków, Poland

Tadeusz Markowski
Departament of Mechanical Engineering, The Faculty of Mechanical Engineering and Aeronautics, Rzeszow University of Technology, al. Powstańców Warszawy 8, 35-959 Rzeszów, Poland

Jacek Mucha
Departament of Mechanical Engineering, The Faculty of Mechanical Engineering and Aeronautics, Rzeszow University of Technology, al. Powstańców Warszawy 8, 35-959 Rzeszów, Poland

Waldemar Witkowski
Departament of Mechanical Engineering, The Faculty of Mechanical Engineering and Aeronautics, Rzeszow University of Technology, al. Powstańców Warszawy 8, 35-959 Rzeszów, Poland

G.N. Ezeh
Electrical and Electronic Engineering, Federal University of Technology Owerri, P.M.B. 1526, Owerri, Imo State, Nigeria

N.S. Chukwuneke
Electrical and Electronic Engineering, Federal University of Technology Owerri, P.M.B. 1526, Owerri, Imo State, Nigeria

N.C. Ogujiofor
Electrical and Electronic Engineering, Federal University of Technology Owerri, P.M.B. 1526, Owerri, Imo State, Nigeria

U.H. Diala
Electrical and Electronic Engineering, Federal University of Technology Owerri, P.M.B. 1526, Owerri, Imo State, Nigeria

Valeriy Martynyuk
Depatrment of Radio Engineering and Communications, Faculty of Programming, Computer and Telecommunication Systems, Khmelnytskyi National University, Khmelnytskyi, Ukraine

Mykola Fedula
Depatrment of Radio Engineering and Communications, Faculty of Programming, Computer and Telecommunication Systems, Khmelnytskyi National University, Khmelnytskyi, Ukraine

Oleksii Balov
Depatrment of Radio Engineering and Communications, Faculty of Programming, Computer and Telecommunication Systems, Khmelnytskyi National University, Khmelnytskyi, Ukraine

Tomasz Garbacz
Lublin University of Technology, 36 Nadbystrzycka St., 20-618 Lublin, Poland

Ludmila Dulebova
Technical University in Košice, 74 Mäsiarska St., 040 01 Košice, Slovakia

Volodymyr Krasinskyi
Lviv Polytechnic National University, 12 St. Bandery St., Lviv, Ukraine

Lenka Markovicova
University of Zilina, 1 Univerzitna St., 010 26 Zilina, Slovakia

Zdzisław Kamiński
Department of Thermodynamics, Fluid Mechanics and Aviation Propulsion Systems, Mechanical Engineering Faculty, Lublin University of Technology, 36 Nadbystrzycka Str., 20-618 Lublin, Poland

Zbigniew Czyż
Department of Thermodynamics, Fluid Mechanics and Aviation Propulsion Systems, Mechanical Engineering Faculty, Lublin University of Technology, 36 Nadbystrzycka Str., 20-618 Lublin, Poland

Mirosław Wendeker
Department of Thermodynamics, Fluid Mechanics and Aviation Propulsion Systems, Mechanical Engineering Faculty, Lublin University of Technology, 36 Nadbystrzycka Str., 20-618 Lublin, Poland

Leszek Gardyński
Faculty of Mechanical Engineering, Lublin University of Technology, 36 Nadbystrzycka Str., 20-618 Lublin, Poland

Paweł Powroźnik
Institute of the Foundations of Electrotechniqu and Electrotechnology, Faculty of Electric and Information Technologies, Lublin University of Technology, Nadbystrzycka 38A, 20-618 Lublin, Poland

Mirosław Malec
Fundamentals Technology Faculty, Lublin University of Technology, 38 Nadbystrzycka Str., 20-618 Lublin, Poland

Tomasz Cieplak
Management Faculty, Lublin University of Technology, 38 Nadbystrzycka Str., 20-618 Lublin, Poland

Sławomir Walczuk
Graduate student of Fundamentals Technology Faculty, Lublin University of Technology, Poland

Tomasz Garbacz
Lublin University of Technology, 36 Nadbystrzycka St., 20-618 Lublin, Poland

Ludmila Dulebova
Technical University in Košice, 74 Mäsiarska St., 040 01 Košice, Slovakia

Volodymyr Krasinskyi
Lviv Polytechnic National University, 12 St. Bandery St., Lviv, Ukraine

Katarzyna Grzegorczyk
Institute of Aviation, Al. Krakowska 110/114, 02-256 Warsaw, Poland

Ewa Wajszczak
The State School of Higher Education in Chełm, Pocztowa 54, 22-100 Chełm, Poland

Dominik Galas
The State School of Higher Education in Chełm, Pocztowa 54, 22-100 Chełm, Poland

Stefan Fijałkowski
Lublin University of Technology, Nadbystrzycka 36, 20-608 Lublin, Poland

Ewa Siemionek
Mechanical Engineering Faculty, Lublin University of Technology, Nadbystrzycka 36, 20-618 Lublin, Poland

Mieczysław Dziubiński
Mechanical Engineering Faculty, Lublin University of Technology, Nadbystrzycka 36, 20-618 Lublin, Poland

Michał Charlak
Fundamentals Technology Faculty, Lublin University of Technology, 38 Nadbystrzycka Str., 20-618 Lublin, Poland

Marek A. Jakubowski
Management Faculty, Lublin University of Technology, 38 Nadbystrzycka Str., 20-618 Lublin, Poland

Antoni Świć
The Institute of Information Technology Systems, Lublin University of Technology, ul. Nadbystrzycka 36, 20-618 Lublin, Poland

Oleg Draczew
Togliatti State Technical University, Russia

Wiktor Taranenko
Luck State Technical University, Ukraine

Sadi Majdalawi
Luck State Technical University, Ukraine

Ján Král' Jr.
Faculty of Mechanical Engineering, Centrum of Informatics, Technical University of Košice, Slovakia

Ján Král'
Faculty of Mechanical Engineering, Department of Technologies and Materials, Technical University of Košice, Slovakia

Leszek Gardyński
Department of Material Engineering, Lublin University of Technology, 36 Nadbystrzycka Str., 20-618 Lublin, Poland

Jolanta Kałdonek
Department of Material Engineering, Lublin University of Technology, 36 Nadbystrzycka Str., 20-618 Lublin, Poland

Tomasz Muszyński
The State School of Higher Education in Chełm, Pocztowa 54, 22-100 Chełm, Poland

Piotr Strzelczyk
Faculty of Mechanical Engineering and Aeronautics, Rzeszow University of Technology, Al. Powstańców Warszawy 8, 35-959 Rzeszów, Poland

Jan Gilewski
Elektroprojekt S.A., Lublin, Poland

Jerzy Montusiewicz
Institute of Computer Science, Lublin University of Technology, Nadbysrtrzycka 36b, 20-618 Lublin, Poland

Arkadiusz Gola
Department of Enterprise Organization, Faculty of Management, Lublin University of Technology, 38 Nadbystrzycka Str., 20-618 Lublin, Poland

Weronika Konczal
Management of Production Students' Scientific Association, Lublin University of Technology, 38 Nadbystrzycka Str., 20-618 Lublin, Poland

Piotr Penkała
Institute of Technological Systems of Information, Faculty of Mechanical Engineering, Nadbystrzycka 36, 20-781 Lublin, Poland

Beata Płowaś
Institute of Technical Sciences and Aviation, The State School of Higher Education in Chełm, Pocztowa 54, 22-100 Chełm, Poland

Maciej Gruza
Laboratory of Coordinate Metrology, Mechanical Department, Cracow University of Technology, Al. Jana Pawła II 37, 31-834 Kraków, Poland

Michał Karpiuk
Laboratory of Coordinate Metrology, Mechanical Department, Cracow University of Technology, Al. Jana Pawła II 37, 31-834 Kraków, Poland

Zbigniew Czyż
Department of Thermodynamics, Fluid Mechanics and Aviation Propulsion Systems, Faculty of Mechanical Engineering, Lublin University of Technology, Nadbystrzycka 36, 20-618 Lublin, Poland

Paweł Magryta
Department of Thermodynamics, Fluid Mechanics and Aviation Propulsion Systems, Faculty of Mechanical Engineering, Lublin University of Technology, Nadbystrzycka 36, 20-618 Lublin, Poland

Marcin Szlachetka
Department of Mechanics and Mechanical Engineering, Pope John Paul II State School of Higher Education in Biala Podlaska, Sidorska 95/97, 21-500 Biała Podlaska, Poland

Tomasz Łusiak
Mechanical Engineering Faculty, Lublin University of Technology, Nadbystrzycka 36, 20-618 Lublin, Poland

Andżelika Grudzień
4th year student in The State School of Higher Education in Chelm, Poland